Progress in Surface and Membrane Science

VOLUME 14

Academic Press Rapid Manuscript Reproduction

Progress in Surface and Membrane Science

VOLUME 14

EDITED BY

D. A. CADENHEAD

DEPARTMENT OF CHEMISTRY
STATE UNIVERSITY OF NEW YORK
BUFFALO, NEW YORK

J. F. DANIELLI

WORCESTER POLYTECHNIC INSTITUTE
WORCESTER, MASSACHUSETTS

1981
ACADEMIC PRESS
A Subsidiary of Harcourt Brace Jovanovich, Publishers
New York London Toronto Sydney San Francisco

ACADEMIC PRESS, INC.
111 Fifth Avenue, New York, New York 10003

United Kingdom Edition published by
ACADEMIC PRESS, INC. (LONDON) LTD.
24/28 Oval Road, London NW1 7DX

ISSN No. 0096-4298

ISBN: 0-12-571814-4

PRINTED IN THE UNITED STATES OF AMERICA

81 82 83 84 9 8 7 6 5 4 3 2 1

CONTENTS

COUPLING OF IONIC AND NONELECTROLYTE FLUXES IN ION SELECTIVE MEMBRANES

O. K. Stefanova and M. M. Shultz

ELECTROCATALYTIC PROPERTIES OF METALLOPORPHINS AT THE INTERFACE

M. R. Tarasevich and K. A. Radyushkina

ADSORPTION FROM BINARY GAS AND LIQUID PHASES

K. V. Chmutov and O. G. Larionov

PHASE TRANSITIONS IN TWO-DIMENSIONAL AMPHIPHILIC SYSTEMS

J. Francois Baret

THE WETTABILITY OF SOLIDS BY LIQUID METALS

Ju. V. Naidich

CONTRIBUTORS

Numbers in parentheses indicate the pages on which the authors' contributions begin.

J. FRANCOIS BARET (291), *Departement de Physique des Liquides, Université de Provence, Marseille, France*

MAŁGORZATA BORÓWKO (1), *Department of Theoretical Chemistry, Institute of Chemistry, M. Curie-Skłodowska University, 20031 Lublin, Poland*

K. V. CHMUTOV (237), *Institute of Physical Chemistry, Academy of Sciences of the USSR, Moscow B-71, Leninsky Pr. 31, USSR*

N. V. CHURAEV (69), *Institute of Physical Chemistry, Academy of Sciences of the USSR, Moscow B-71, Leninski Pr. 31, USSR*

B. V. DERJAGUIN (69), *Institute of Physical Chemistry, Academy of Sciences of the USSR, Moscow B-71, Leninski Pr. 31, USSR*

MIECYSŁAW JARONIEC (1), *Department of Theoretical Chemistry, Institute of Chemistry, M. Curie-Skłodowska University, Nowotki 12, 20031 Lublin, Poland*

O. G. LARIONOV (237), *Institute of Physical Chemistry, Academy of Sciences of the USSR, Moscow B-71, Leninski Pr. 31, USSR*

JU. V. NAIDICH (353), *Institute of Material Problems, Academy of Sciences, Ukranian SSR, Kiev, USSR*

ANDRZEJ PATRYKIEJEW (1), *Department of Theoretical Chemistry, Institute of Chemistry, M. Curie-Skłodowska University, Nowotki 12, 20031 Lublin, Poland*

K. A. RADYUSHKINA (175), *Institute of Electrochemistry, Academy of Sciences of the USSR, Moscow B-71, Leninski Pr. 31, USSR*

M. M. SHULTZ (131), *Institute of Silicate Chemistry, Academy of Sciences of the USSR, Leningrad, USSR*

O. K. STEFANOVA (131), *Department of Physical Chemistry, Leningrad State University, Leningrad, USSR*

M. R. TARASEVICH (175), *Institute of Electrochemistry, Academy of Sciences of the USSR, Moscow B-71, Leninski Pr. 31, USSR*

CONTENTS OF RECENT VOLUMES

VOLUME 12

VOLUME 13

STATISTICAL THERMODYNAMICS OF MONOLAYER ADSORPTION FROM GAS AND LIQUID MIXTURES ON HOMOGENEOUS AND HETEROGENEOUS SOLID SURFACES

Mieczysław Jaroniec
Andrzej Patrykiejew
Małgorzata Borówko
Department of Theoretical Chemistry
Institute of Chemistry
M. Curie-Skłodowska University
Lublin, Poland

PROGRESS IN SURFACE AND MEMBRANE
SCIENCE, VOL. 14

ISBN 0-12-571814-4

I. GENERAL CONSIDERATIONS

A. Introduction

Adsorption from multicomponent mixtures on solids underlies a number of extremely important processes of utilitarian significance. It is enough to mention here only such important processes in industry as heterogeneous catalysis, separation of mixtures, and purification of water and air. These processes can be controlled on an industrial scale, on the condition that their mechanism is known and can be mathematically described. Because of the fact that experimental studies of adsorption from multicomponent mixtures is difficult and laborious, theoretical methods, which make possible the prediction of adsorption from multicomponent mixtures by means of parameters characterizing simpler systems, are of great value, particularly in engineering applications. Adsorption from gas mixtures can be predicted by means of parameters characterizing adsorption of single gases, as well as adsorption from multicomponent liquid mixtures on the basis of data characterizing adsorption from binary solutions, or on the basis of parameters describing adsorption of gas mixtures.

A great number of experimental and theoretical papers have been published on adsorption of single gases and adsorption from binary liquid mixtures on solids. This problem has also been dealt with in numerous monographs. Most of the papers concerning this problem have been based on the assumption of energetic homogeneity of the adsorbent surface. However, in the last few years a considerable number of papers have appeared, in which adsorption of single gases on heterogeneous surfaces of solids is discussed. These papers have contributed to the formulation of a general theory of adsorption from multicomponent gas mixtures on heterogeneous surfaces, and have shown a direction of study on the adsorption from multicomponent liquid mixtures on surfaces of this type.

The purpose of this chapter is to present the present state of study on the monolayer adsorption from multicomponent gas and liquid mixtures on the surfaces of solids. The most promising theories are statistical ones, which need not involve any particular model of adsorption process, but only require a knowledge of the adsorbate-adsorbent and adsorbate-adsorbate interaction potentials, although at present, they are not of much use in engineering applications. However, a complete statistical description of actual multicomponent adsorption systems is at present, in the state of generalized formulas. For these reasons, let us confine ourselves to the discussion of selected adsorption models, which have already been generalized, or may be extended to adsorption from multicomponent mixtures on heterogeneous surfaces, and which are frequently used for the physicochemical interpretation of experimental data. In the problem of adsorption on homogeneous surfaces, the cases that have not been broadly discussed in literature will be described in detail; the so-called "mixed adsorption model" as well as the model of "associated adsorbate" are involved here. A special consideration will be given to the problems connected with adsorption of multicomponent gas and liquid mixtures on heterogeneous surfaces, as there is a lack of papers dealing with them. As far as adsorption from liquid mixtures is concerned, only those problems will be discussed that are directly associated with adsorption of gas mixtures. Other problems connected with the model of localized and mobile adsorption will be treated schematically. Mathematical discussion of the selected adsorption models will be carried out on the basis of the statistical thermodynamics.

B. Concept of Surface Heterogeneity in Physical Adsorption

At the end of the 1940s and the beginning of the 1950s, when extensive investigations on the effect of heterogeneity of surfaces were started (Young and Crowell, 1962), there appeared papers in which attention was drawn to the great difficulties connected with the evaluation of the effects of surface heterogeneity in adsorption phenomena (Roginsky and Todes, 1946; Honig and Rosenbloom, 1955), and a considerable role of lateral interactions in adsorption was pointed out (Halsey, 1948). This problem caused controversial discussions still in later years (Hsieh, 1964; Adamson *et al.*, 1966; Smutek, 1976). It is obvious at present that surface heterogeneity and molecular interactions in the adsorption layer play a very important role in adsorption; although they act in opposite directions, they do not compensate one another, because they depend differently on the degree of surface coverage (Hill, 1949a; Halsey, 1952; Ross and Olivier, 1964; House and Jaycock, 1974). For small surface coverings, effects of surface heterogeneity dominate, whereas with the increase in surface coverage, the role of the lateral interactions grows, and surface heterogeneity will undergo obliteration due to the occupation of the most energetic surface areas by adsorbate molecules. From the studies carried out so far, it follows that the effects of heterogeneity of the adsorbent surfaces are greater than previously supposed. Recently, Ross (1971) wrote: "Any attempt to face realistically the problem posed by interpreting the data of gas-solid adsorption must take into account the energetic heterogeneity of the solid substrate."

In a large majority of the papers known so far, the adsorbent structure is considered as completely stiff, not changing during contact with the adsorption phase. Adsorption potential, considered as an unimolecular function of the variables x, y, and z, serves then as the classification criterion of adsorbent surfaces (Pierotti, 1971). On this basis, two principal surface types are distinguished: homogeneous and heterogeneous. For the former, the adsorption potential is a function only of the distance z from the surface in a vertical direction (uniform surfaces), or at a definite z, it is the function in relation to the vector of the parallel shift to the adsorbent surface (nonuniform surfaces). For the case when adsorption potential is more complex, dependent on the variables x, y, and z, the adsorbent surface is considered as being heterogeneous. In the theory of adsorption based on two-dimensional models of adsorption, the adsorption energy ε is often used, which is equal to minimum potential value taken with a countersign. By this terminology, the adsorption energy has a positive value. Differential distribution of adsorption energy $\chi(\varepsilon)$ has been accepted as a quantitative characteristic of adsorbent surface, where $\chi(\varepsilon)d\varepsilon$ is the fraction of the surface with adsorption energies between ε and $\varepsilon + d\varepsilon$. Of course, the function $\chi(\varepsilon)$ characterizes quantitatively only the global heterogeneity of the adsorbent surface, giving no information at all as regard the distribution topography of this heterogeneity. It is adequate to take into account energetic heterogeneity of the adsorbent surface in the description of localized monolayer adsorption without lateral interactions in the adsorption phase. For the case of mobile or partially mobile adsorption, even when lateral interactions are neglected, additional information about the topography of adsorption sites on the adsorbent surface is necessary (Tompkins, 1950); this problem will be discussed more thoroughly in Section III.A. The assumption of a definite model of heterogeneous surface, which is synonymous with the assumption of a definite distribution of adsorption sites on the adsorbent surface, becomes indispensable in the description of adsorption with lateral interactions on heterogeneous surfaces (Jaroniec and Borówko, 1978).

The present studies on the adsorption of gases on surfaces of solids, has allowed us to distinguish three models of heterogeneous surface (Steele, 1963; Pierotti and Thomas, 1973; Rudziński *et al.*, 1977):

(1) Patchwise model (PM) - the adsorption sites of equal adsorption energies are assumed to be grouped together into patches; these patches are so large such that the case when two molecules adsorbed on different patches are still at interaction distances can be neglected,

(2) Random model (RM) - the adsorption sites of equal adsorption energies are assumed to be distributed fully at random over a heterogeneous surface; and

(3) Medial model (MM) - a part of the adsorption sites of equal adsorption energies are grouped together into patches, whereas the rest of adsorption sites of equal adsorption energies are distributed fully at random between the existing surface patches.

In the first case, there is a total spatial correlation of sites of equal adsorption energies, whereas in the second case the spatial correlation between adsorption sites of equal adsorption energies does not exist. In the light of Pierotti's terminology (Pierotti, 1971) the surfaces of ideal crystals belong to nonuniform surfaces, because their adsorption sites of equal energies are distributed regularly. In the theoretical description of adsorption on nonuniform surfaces, the periodical character of potential field of adsorbent is usually utilized (Steele, 1973a,b; Doll and Steele, 1974; Steele, 1976). This problem can be approached differently. In the dependence on the dimensions of the crystal lattice and adsorbate molecules, several kinds of adsorption sites accessible for adsorbate molecules can be distinguished on the nonuniform surfaces. Moreover, the surface properties of crystalline solids are frequently dependent upon surface defects. Therefore, adsorption systems with nonuniform surfaces should be well characterized by a discrete adsorption energy distribution, or by continuous distributions showing sharp maxima. For a theoretical description of such systems, formalism used in the adsorption theory on heterogeneous surfaces (Jackson and Davis, 1974; House and Jaycock, 1977a; Ross, 1971) can be applied. Because of the general character of this chapter, a simpler division of real adsorbents will be adopted: homogeneous surfaces (they will be identified with uniform surfaces) and heterogeneous surfaces (including nonuniform and heterogeneous surfaces according to Pierotti's classification).

In the studies of adsorption on heterogeneous surfaces, one of the two following approaches is usually applied: either the heterogeneity of adsorbent surfaces is determined from experimental data, or a definite model of heterogeneous surface is *a priori* assumed, and for this model equations of adsorption isotherms and isosteric adsorption heats are obtained, which are then confronted with experimental data. As far as the determination of heterogeneity of the adsorbent surface is concerned, the first approach seems more justified. The methods for the determination of the surface heterogeneity known so far can be classified with regard to the type of the experimental data used. Here, we can distinguish methods based on:

(1) temperature-dependence of adsorption data;
(2) pressure-dependence of adsorption data;
(3) calorimetric data; and
(4) thermal desorption data.

The methods of the first two groups are most popular and they have been shortly characterized by Rudziński and Jaroniec (1974); however, the methods based on calorimetric data have been described in the monograph of Young and Crowell (1962). As regard the method of the fourth group, the literature is very scanty (Kisluk *et al.*, 1975). It is beyond doubt that the modern methods for investigations of adsorbent surfaces (for example, low-energy electron diffraction, scanning electron microscope, etc.) make it possible to determine surface heterogeneity more accurately; this applies particularly to the study of the topography of adsorption sites on the adsorbent surface (Pierotti, 1971; Steele, 1974; Dash, 1975).

II. LOCALIZED ADSORPTION OF GASES ON SOLIDS

A. Adsorption of Single Gases on Solids

"The Langmuir equation occupies a central position in the theoretical discussion of monolayer adsorption because of its simplicity of form and directness of derivation" - such was the opinion of Stebbins and Halsey (1964). The assumptions leading to the Langmuir adsorption isotherm are commonly known; they define the localized adsorption model without attractive interactions on uniform solid surfaces. The taking into account of the lateral interactions in this model results in the Fowler - Guggenheim isotherm (1939). Both adsorption isotherms were derived on the basis of statistical thermodynamics; the literature concerning this problem can be found almost in every monograph on adsorption (e.g., Clark, 1970). There are other known approaches to localized adsorption with lateral interactions besides that of Fowler and Guggenheim (1939). One of the more popular models is that of "associated adsorbate," proposed by Kiselev (1958). Within this model, lateral interactions are represented by the so-called associates parallel to the adsorbent surface; this model as well as its application in the theory of adsorption from mixtures will be given closer consideration in Section VI.

Another different approach to monolayer adsorption has been presented by Jovanovic (1969). He derived an equation for a monolayer adsorption isotherm on homogeneous surfaces, which may be presented in the form

$$\theta = 1 - \exp(-a^1 p) \qquad (1)$$

Equation (1) for small values of adsorbate pressures p gives the Langmuir equation

$$\theta = a^1 p(1+a^1 p)^{-1} \qquad (2)$$

In Eqs. (1) and (2), θ is the relative coverage of a homogeneous surface and a^1 is the Langmuir constant. The Jovanovic approach, is kinetic in character, and considers the collisions of gas-phase molecules with adsorbed molecules. Rudziński and Jaroniec (1974, 1975) expressed the opinion that the Jovanovic adsorption isotherm is probably the result of the effect of mechanical contact between the adsorbed and bulk phases. Attempts toward a reduction of the kinetic derivation to a statistico-mechanical derivation have been made by Jaroniec (1976a), Rudziński and Wojciechowski (1977), and Cerofolini (1978). In their considerations, it was shown that Eq. (1) represents another approximation for monolayer-localized adsorption without lateral attractive interactions on homogeneous surfaces. The attempts to derive Eq. (1), have not represented a purely statistical development. Let us quote here the opinion of Rudziński and Wojciechowski (1977): "Our development of the Jovanovic adsorption isotherm, does not represent a purely statistical development, since we introduce the kinetic argument of Jovanovic. Our semistatistical derivation shows only a general way of how, in the case of other adsorption systems, to take into account the effects considered by Jovanovic." The present state of investigations into monolayer-localized adsorption on homogeneous surfaces can be summarized as follows (Cerofolini, 1976a; Rudziński and Wojciechowski, 1977; Dondi *et al.*, 1977):

- submonolayer localized adsorption
  - mechanical contact neglected
    - without lateral interactions: Langmuir (1918)
    - with lateral interactions: Fowler-Guggenheim (1939), Kiselev (1958)
  - mechanical contact included
    - without lateral interactions: Jovanovic (1969)
    - with lateral interactions: Fowler-Guggenheim-Jovanovic (Jaroniec, Sokołowski, and Cerofolini, 1976)

We shall now discuss the more important problems concerning adsorption of single gases on heterogeneous surfaces. The majority of the papers devoted to localized adsorption of single gases on heterogeneous surfaces have been based on the well-known integral equation (Ross and Olivier, 1964):

$$\Theta(p) = \int_{\Delta} \theta(p,\varepsilon)\chi(\varepsilon)\,d\varepsilon \tag{3}$$

where Θ is the relative coverage of a heterogeneous surface, $\chi(\varepsilon)$ the distribution function of adsorption energy ε, Δ the region of possible variations of adsorption energy ε. The function $\theta(p,\varepsilon)$, appearing in Eq. (3) describes adsorption on sites possessing adsorption energy ε; it is called the local adsorption isotherm. However, the function $\Theta(p)$ describes the adsorption process on the whole surface; it is called the overall adsorption isotherm. The equation for the local adsorption isotherm is assumed *a priori*; it can be any adsorption isotherm derived on assuming homogeneity of the adsorbent surface. From Eq. (3) one may obtain:

(i) a numerical solution with regard to the function $\chi(\varepsilon)$, when the overall adsorption is taken in experimental form;

(ii) an analytical solution with regard to the function $\chi(\varepsilon)$ (or $\Theta(p)$), when the other function $\Theta(p)$ (or $\chi(\varepsilon)$) is assumed *a priori* in analytical form.

As regard the first point, several numerical methods for determining the distribution function $\chi(\varepsilon)$ from experimental adsorption isotherms are known:

(a) Ross and Olivier method (1964) based on Gaussian distribution (Jaycock and Waldsax, 1971; House and Jaycock, 1974);

(b) Van Dongen method (1973) based on the distribution of an exponential higher degree polynomial;

(c) Adamson and Ling iterative method (1961), which was used by Dormant and Adamson (1972), House and Jaycock (1974), Waksmundzki *et al.* (1975);

(d) Hobson method (1965) based on the differentiation of experimental data (Rudzinski *et al.*, 1974; Waksmundzki *et al.*, 1975; Hsu *et al.*, 1975);

(e) methods based on the development of the functions $\theta(p,\varepsilon)$, $\Theta(p)$ and $\chi(\varepsilon)$ in orthonormal series (Cerofolini, 1974; Jaroniec and Rudzinski, 1975a; Rudzinski *et al.*, 1975; Jaroniec and Rudzinski, 1976);

(f) Ross and Morrison method (1973) uses a program called CAEDMON; and

(g) House and Jaycock method (1978) uses a program called HILDA (Sidebottom *et al.*, 1976; House and Jaycock, 1977a,b).

The two last methods seem to be the most useful from the practical point of view, because they were numerically tested in detail. It has been also shown (House and Jaycock, 1977b) that the program HILDA is found to yield more details about the heterogeneity distribution than CAEDMON.

Analytical solutions of Eq. (3) are possible for the Langmuir and Jovanovic local adsorption isotherms. Application of the Langmuir equation for local adsorption isotherm in the integral equation (3) leads to the Stieltjes transform (Sips, 1948, 1950; Misra, 1969, 1970) or the Hilbert transform (Prasad and Doraiswamy, 1977), whereas for the Jovanovic equation the Laplace transform is obtained (Rudziński and Jaroniec, 1974). Using the method of the Stieltjes transform (Langmuir local isotherm), the energy distribution functions corresponding to the most popular empirical adsorption isotherms have been found. In this way, physical justification for Freundlich (Sips, 1948), generalized Freundlich (Sips, 1950; Misra, 1970), Dubinin - Radushkevich (Misra, 1969; Jaroniec, 1975a), and Tóth adsorption isotherms (Tóth *et al.*, 1974) have been found. The theoretical studies of Hobson (1974), Cerofolini (1972, 1975a,b, 1976a,b) and Jaroniec (1975a) pointed to the special role of the equation of Freundlich and Dubinin - Radushkevich in the theory of adsorption on nonporous adsorbents. Cerofolini (1976a) suggests that these adsorption isotherms are due to an equilibrium structure of the adsorbing surfaces, this structure being that of the equilibrium at their formation temperature. The general expression for an adsorption isotherm on heterogeneous surfaces, permitting the Freundlich and Dubinin - Radushkevich behavior, is the "exponential adsorption isotherm" proposed by Jaroniec *et al.* (1975) for which a statistico-mechanical derivation has been found by Jaroniec (1975a). The Dubinin - Radushkevich equation (1947) undoubtedly occupies a central role in the theory of physical adsorption on porous and microporous adsorbents; the studies brought forth here show that this equation plays a less significant role in the theory of adsorption on nonporous, energetic heterogeneous surfaces (as discussed by Dubinin, 1972).

The method of the Laplace transform (Jovanovic local behavior) generates adsorption isotherm equations of small advantage and is more complicated than the method of Stieltjes transform (Misra, 1973). One of more interesting results is the Freundlich isotherm generated by the exponential distribution of adsorption energy (Misra, 1973).

The application of the Fowler - Guggenheim equation in the character of local adsorption isotherm in Eq. (3) does not lead to analytical expressions for overall adsorption isotherm and, moreover, it limits the considerations only to patchwise heterogeneous surfaces (Jackson and Davis, 1974; House and Jaycock, 1977a). The most advanced studies on localized adsorption with lateral interactions on heterogeneous surfaces, having adsorption sites distributed patchwise and at random, were those of Hill (1949a). Hill also introduced analytical expressions for thermodynamic functions characterizing adsorption of single gases on heterogeneous surfaces with patchwise and random distribution of adsorption sites.

Concluding this section, we shall mention the method of condensation approximation, which played an important role in the theory of adsorption on heterogeneous surfaces. In this method the local adsorption isotherm $\theta(p,\varepsilon)$ is replaced by a condensation isotherm, which supposes that at a pressure p all sites are filled if their energy exceeds a critical energy $\varepsilon_c(p)$, and are completely free otherwise. The mathematical development of this method is due to the papers of Harris (1968, 1969a,b), Cerofolini (1971, 1972, 1975c-e, 1976a), Van Dongen and Broekhoff (1969), and Jaroniec *et al.* (1976). A detailed discussion of condensation approximation method is given in the review of Cerofolini (1974).

B. *Adsorption of Gas Mixtures on Homogeneous Surfaces*

Adsorption of gas mixtures on homogeneous solid surfaces has been frequently discussed. A number of general adsorption monographs exist that contain sections on this subject. Among monographs on physical adsorption, those of Young and Crowell (1962, particularly good for old references), Ponec *et al.* (1974), Szabó and Kall (1976), and especially Dubinin and Serpinsky (1972) should be noted. A summary describing adsorption from gas mixtures on homogeneous surfaces was presented in the review by Dubinin *et al.*

(1964), Van Ness (1969), Bülow and Schirmer (1972), Sircar and Myers (1973), Nodzeński (1976), and Jaroniec (1978a).

A literature survey of the subject of adsorption from gaseous mixtures indicates that the majority of the works published so far can be classified into two main groups. In the first group those papers are included, which take advantage of the formalism of classical thermodynamics of adsorbed solutions; the second group of papers, however, deal with the statistical thermodynamics of two-dimensional adsorption models. The methods of estimating adsorption equilibria of gas mixtures, based on thermodynamics of adsorbed solutions, have been excellently described in the review of Sircar and Myers (1973). The fundamental papers of this group have been published by Myers and Prausnitz (1965), Hoory and Prausnitz (1967), and Bering and Serpinsky (1952, 1961, 1969, 1972a,b). Recently, Bering *et al.* (1977a,b) applied the theory describing the osmotic equilibrium of the volume solution to the adsorption of binary gas mixtures on homogeneous surfaces. Of the methods based on classical thermodynamics, that of Myers and Prausnitz (1965), which has often been used to predict adsorption equilibria of gas mixtures by means of the pure-gas adsorption parameters (Friederich and Mullins, 1972; Sloan and Mullins, 1975; Perfetti and Wightman, 1975, 1976), has become popular. This method has also been extended to nonideal adsorbed solutions on homogeneous and heterogeneous surfaces by Hoory and Prausnitz (1967).

Of the second group of papers, only those will be summarized here which deal with localized adsorption from gas mixtures on homogeneous surfaces. The adsorption isotherm and other thermodynamic functions, characterizing localized monolayer adsorption of n-component gas mixtures on homogeneous surfaces, may be obtained from the complete partition function

$$Q^1_{(\boldsymbol{n})} = \frac{B!}{\prod_{i=1}^{n} N_i! \left(B - \sum_{i=1}^{n} N_i\right)!} \prod_{i=1}^{n} \left(q_i^1\right)^{N_i} \tag{4}$$

where

$$q_i^1 = \hat{q}_i^1(T)\exp(\varepsilon_i/kT) \qquad \text{for } i=1,\, 2,\, \ldots,\, n \tag{6}$$

and B is the total number of sites having the adsorption energy ε_i with regard to the ith component, $\hat{q}_i^1$ the partition function of isolated localized adsorbed molecule of the ith component, N_i the number of sites occupied by adsorbed molecules of the ith component, the vector $\boldsymbol{n} = (1,2,\ldots,n)$ refers to n-component gas mixtures. The partition function (4) generates the well-known equation of adsorption isotherm

$$\theta_{i(\boldsymbol{n})} = \frac{a_i^1 p_i}{1 + \boldsymbol{a}^1\boldsymbol{p}} \qquad \text{for } i=1,\, 2,\, \ldots,\, n \tag{6}$$

where

$$\theta_{i(\boldsymbol{n})} = N_i/B, \qquad \boldsymbol{a}^1\boldsymbol{p} = \sum_{i=1}^{n} a_i^1 p_i$$

$$\boldsymbol{a}^1 = \left(a_1^1,\, a_2^1,\, \ldots,\, a_n^1\right) \tag{7}$$

$$\boldsymbol{p} = (p_1,\, p_2,\, \ldots,\, p_n)$$

and $\theta_{i(\boldsymbol{n})}$ is the relative surface coverage of the ith component adsorbed from n-component gas mixture. The sum of all $\theta_{i(\boldsymbol{n})}$, i.e.,

$$\theta_{(\boldsymbol{n})} = \sum_{i=1}^{n} \theta_{i(\boldsymbol{n})} \tag{8}$$

defines a relative coverage for all components of the mixture adsorbed on a homogeneous surface. The vector notation used here is clearly explained in the papers of Jaroniec and Rudziński (1975b, 1977). The Langmuir constant a_i^1 is connected with the function

q_i^1 and the standard chemical potential of the *i*th component, μ_i^{0G} as follows (Clark, 1970):

$$a_i^1 = q_i^1 \exp\left(\mu_i^{0G}/kT\right) = \hat{q}_i^1(T)\exp\left(\mu_i^{0G}/kT\right)\exp\left(\varepsilon_i/kT\right)$$

$$= \tilde{q}_i^1(T)\exp(\varepsilon_i/kT) \quad \text{for } i=1, 2, \ldots, n \tag{9}$$

The other thermodynamic functions for localized adsorption of *n*-component gas mixtures on homogeneous solid surfaces may be calculated by using Eq. (4) and the well-known relationships between partition and thermodynamic functions.

Another equation for monolayer adsorption of *n*-component gas mixtures has been obtained on the basis of the Jovanovic adsorption model by Jaroniec (1975b). The application of this equation to the analysis of experimental data gives favorable results (Jaroniec and Tóth, 1976a). This equation is a simple generalization of Eq. (1):

$$\theta_{(\boldsymbol{n})} = 1 - \exp(-\boldsymbol{a}^1\boldsymbol{p}) \tag{10}$$

for small values of the components of the vector $\boldsymbol{p}$, we have Eq. (6). No papers dealing with localized adsorption of *n*-component gas mixtures with lateral interactions are known as yet. An exceptional paper in this respect is that of Ivanov and Martinov (1973), in which the authors have derived the analog of Fowler - Guggenheim equation (1939) for two-component gas mixtures on homogeneous surfaces.

C. Adsorption of Gas Mixtures on Heterogeneous Surfaces

Localized binary monolayers without lateral interactions on heterogeneous surfaces have been considered by Roginsky and Todes (1945a,b), Tompkins and Young (1951), Glueckauf (1953), and Bering and Serpinsky (1953). The most advanced theoretical results are due to Roginsky and Todes (1945a) and Glueckauf (1953). Roginsky and Todes (1945a,b) extended the approximate control band method (Roginsky and Yanovsky, 1952) to the adsorption of binary gas mixtures. However, Glueckauf (1953) derived equations for adsorption isotherms of binary gas mixtures by assuming exponential distribution for the characterization of the surface heterogeneity. A common feature of the approaches to adsorption from gas mixtures on heterogeneous surfaces mentioned here is the assumption of an additional functional relationship between the adsorption energies of both gases. Then, the adsorption isotherm for binary gas mixtures on heterogeneous surfaces can be presented in the form of a single integral, which is quite analogous to that used in the theory of pure gas adsorption.

Detailed analysis of the adsorption process from multicomponent gas mixtures on heterogeneous solid surfaces leads to the equation of the overall adsorption isotherm, which has the form of a multiple integral (Jaroniec, 1975c, 1976b, 1977a, 1978a, Jaroniec and Rudziński, 1975b,c, 1977; Jaroniec and Tóth, 1976a, 1977; Jaroniec *et al.*, 1978):

$$\Theta_{(\boldsymbol{n})}(\boldsymbol{p}) = \int_{\Delta_{(\boldsymbol{n})}} \theta_{(\boldsymbol{n})}\chi_{(\boldsymbol{n})}(\boldsymbol{\varepsilon})\,d\boldsymbol{\varepsilon} \tag{11}$$

where

$$\Theta_{(\boldsymbol{n})} = \sum_{i=1}^{n} \Theta_{i(\boldsymbol{n})} \tag{12}$$

is the relative coverage of a heterogeneous surface for n-component gas mixtures; $\Theta_{i(\boldsymbol{n})}$ refers to the coverage of the ith component on a heterogeneous surface; $\boldsymbol{\varepsilon} = (\varepsilon_1, \varepsilon_2, \ldots, \varepsilon_n)$ is an n-dimensional vector, the component ε_i of which characterizes the interactions of the ith component of the gas mixture with the solid surface; $\chi_{(\boldsymbol{n})}(\boldsymbol{\varepsilon})$ is the n-dimensional energy distribution function characterizing the global heterogeneity of the adsorbent surface; $\Delta_{(\boldsymbol{n})}$

is the n-dimensional integration region. In such a treatment, each adsorption site is characterized unambiguously by the vector $\boldsymbol{\varepsilon}$. So far, Gaussian (Jaroniec, 1976b; Jaroniec and Rudziński, 1975b,c, 1977; Jaroniec and Tóth, 1976a, 1977) and log-normal (Jaroniec and Borówko, 1977) n-dimensional distributions were used for the function $\chi_{(\boldsymbol{n})}(\boldsymbol{\varepsilon})$; however, Eqs. (6) and (10) were applied for the local adsorption isotherm $\theta_{(\boldsymbol{n})}(\boldsymbol{p},\boldsymbol{\varepsilon})$ (references for Eq. (6): Jaroniec, 1977a, 1978a; Jaroniec and Rudziński, 1975b,c, 1977; references for Eq. (10): Jaroniec, 1975c; Jaroniec and Tóth, 1976a, 1977; Jaroniec, *et al.*, 1978). These problems have been discussed in detail in a review of Jaroniec and Rudziński (1977).

It is difficult to obtain analytical expressions for $\Theta_{(\boldsymbol{n})}(\boldsymbol{p})$ by means of Eq. (11); analytical solutions are usually not possible for the local adsorption isotherms (6) and (10) as well as for more real energy distribution functions. Only in the case of Eq. (10) and a special form of the function $\chi_{(\boldsymbol{n})}(\boldsymbol{\varepsilon})$ such solutions are possible (Jaroniec, 1975c). Accordingly, using the Jovanovic equation (10), and function $\chi_{(\boldsymbol{n})}(\boldsymbol{\varepsilon})$ of the following type

$$\chi_{(\boldsymbol{n})}(\boldsymbol{\varepsilon}) = \prod_{i=1}^{n} \chi_i(\varepsilon_i) \tag{13}$$

in Eq. (11), we obtain the general analytical expression for $\Theta_{(\boldsymbol{n})}(\boldsymbol{p})$ (Jaroniec, 1975c):

$$\Theta_{(\boldsymbol{n})}(\boldsymbol{p}) = 1 - \prod_{i=1}^{n} [1 - \Theta_i(p_i)] \tag{14}$$

The symbols Θ_i and χ_i, refer to the adsorption of single gases. It is of interest to note that the equation of the adsorption isotherm for binary gas mixtures obtained by Roginsky (1948), apart from the use of another approximation, is a peculiar case of Eq. (14).

Equation (11) and other thermodynamic functions, characterizing monolayer-localized adsorption without lateral interactions on heterogeneous surfaces, may be obtained from the following canonical partition function:

$$Q^{1}_{(\boldsymbol{n};\boldsymbol{r})} = \prod_{\boldsymbol{r}} \frac{B_{\boldsymbol{r}}! \prod_{i=1}^{n} \left(q^{1}_{i,r_i}\right)^{N_{i,\boldsymbol{r}}}}{\prod_{i=1}^{n} (N_{i,\boldsymbol{r}})! \left[B_{\boldsymbol{r}} - \sum_{i=1}^{n} N_{i,\boldsymbol{r}}\right]!} \tag{15}$$

where $\boldsymbol{r} = (r_1, r_2, \ldots, r_n)$ is an n-dimensional vector defining the $\boldsymbol{r}$th type of adsorption sites. The physical meaning of the symbols appearing in Eq. (15) is analogous to that from Eq. (4); the subscript $\boldsymbol{r}$ refers to the adsorption process of gas mixture on the $\boldsymbol{r}$th type of adsorption sites. The thermodynamic functions for localized adsorption without interactions of gas mixtures on heterogeneous surfaces have been summarized by Jaroniec (1977a,b). The assumption of a functional relation between adsorption energies, i.e.,

$$\varepsilon_i = \psi_i(\varepsilon_1) \quad \text{for } i = 2, 3, \ldots, n \tag{16}$$

in Eq. (11) leads to the single integral (Jaroniec, 1977c; Jaroniec and Rudziński, 1977):

$$\theta_{(\boldsymbol{n})}(\boldsymbol{p}) = \int_{\Delta_1} \theta^{*}_{(\boldsymbol{n})}(\boldsymbol{p},\varepsilon_1)\chi_1(\varepsilon_1)d\varepsilon_1 \tag{17}$$

which is analogous to that used by Roginsky and Todes (1945a) and Glueckauf (1953). The functions $\theta^{*}_{(\boldsymbol{n})}(\boldsymbol{p},\varepsilon_1)$ and $\chi_1(\varepsilon_1)$, appearing in Eq. (17) are defined as follows:

$$\theta^{*}_{(\boldsymbol{n})}(\boldsymbol{p},\varepsilon_1,\psi_2(\varepsilon_1),\psi_3(\varepsilon_1),\ldots,\psi_n(\varepsilon_1)) \tag{18}$$

and

$$\chi_1(\varepsilon_1) = \int_{\Delta^*_{(\boldsymbol{n})}} \chi_{(\boldsymbol{n})}(\boldsymbol{\varepsilon})\, d\varepsilon_2 d\varepsilon_3 \ldots d\varepsilon_n \tag{19}$$

The symbol $\Delta^*_{(\boldsymbol{n})} = \Delta_2 \times \Delta_3 \times \ldots \times \Delta_n$ denotes the $(n-1)$-dimensional integration region. Equation (17) is considerably simpler than Eq. (11) and gives a series of analytical solutions for $\Theta_{(\boldsymbol{n})}(\boldsymbol{p})$. Taking advantage of this equation, the best-known equations of single-gas adsorption isotherm for the case of adsorption of n-component mixtures have been generalized; the Freundlich equation is involved here (Tiemkin, 1975; Snagovsky, 1975; Jaroniec *et al.*, 1978a), Dubinin-Radushkevich (Jaroniec *et al.*, 1978a) and Tóth (Jaroniec, 1977d; Jaroniec *et al.*, 1976b; 1978b). In the theoretical considerations, linear relationship between adsorption energies ε_i and ε_1 have frequently been assumed. Results of particular interest were obtained on assuming that

$$\varepsilon_i - \varepsilon_1 = \eta_i \quad \text{for } i = 2, 3, \ldots, n \tag{20}$$

is a constant, characteristic for the whole surface. Then, Eq. (17) is formally identical with the integral equation (3) describing adsorption of single gases on heterogeneous surfaces (Jaroniec, 1977c). For the case of validity of Eq. (20), the scalar product in Eqs. (6) and (10) may be rewritten in the following form:

$$\boldsymbol{a}^1\boldsymbol{p} = a_1^1\left(p_1 + \sum_{i=2}^{n} \xi_{i1}p_i\right) = a_1^1 z \tag{21}$$

where

$$\xi_{i1} = a_i^1/a_1^1 = \left(\tilde{q}_i^1/\tilde{q}_1^1\right) \exp(\eta_i/kT) \tag{22}$$

is constant for the whole surface. Thus, the substitution of the term $\boldsymbol{a}^1\boldsymbol{p}$ by $a_1^1 z$ in Eqs. (6) and (10) gives equations that are formally identical with those characterizing adsorption of single gases. The procedure for obtaining the equations of mixed-gas adsorption isotherms is very simple; it consists in substituting pressure p by variable z in isotherm equations of single-gas adsorption isotherms (Jaroniec, 1977c; Jaroniec *et al.*, 1978a,b). As it appears from the literature cited here, equations of mixed-gas adsorption isotherms obtained by means of Eq. (17) and the relationship (20) are frequently used for the interpretation of experimental data of adsorption from gas mixtures. The studies have shown that these equations can be used to characterize multicomponent adsorption systems and to predict adsorption from gas mixtures by means of parameters describing adsorption of single gases. The interpretation of experimental data from gas mixtures is considerably facilitated by the fact that they are very often measured with a fixed parameter (Jaroniec *et al.*, 1978a). Then, it is possible to obtain a much less complicated expression describing adsorption equilibrium (Jaroniec, 1975d, 1976c, 1977e; Jaroniec and Rudziński, 1976; Jaroniec *et al.*, 1978a). The generalized integral equation (11) also creates new possibilities for the investigation of mixed gas adsorption kinetics on heterogeneous solid surfaces. This problem has recently been discussed by Jaroniec (1978b-d) in terms of Eq. (11) without interactions. Although the thermodynamics of localized adsorption of n-component gas mixtures on heterogeneous surfaces without lateral interactions can be relatively easily formulated in terms of the generalized integral equation (11), the consideration of lateral interactions makes theoretical investigations considerably difficult. For this reason, this problem has not yet been studied. Of course, it is possible to use the Fowler - Guggenheim equation, describing the adsorption of two-component mixtures on homogeneous surfaces (derived by Ivanov and Martinov, 1973) in the integral equation (11) on assuming a patchwise model of a heterogeneous surface. This problem has not been investigated, because of the fact that the equation derived by Ivanov and Martinov (1973) has not yet been experimentally tested with respect to adsorption of gas mixtures on homogeneous surfaces.

III. MOBILE ADSORPTION OF GASES ON SOLIDS

A. Adsorption of Single Gases on Solids

Although the success of the theory of the Langmuir localized adsorption is beyond doubt, it should be stressed here that the probability of mobile adsorption is much greater at temperatures normally used in laboratories; even at temperatures close to liquid air, partially mobile adsorption is to be expected more frequently than a completely localized one (Hill, 1946; Young and Crowell, 1962). This fact is the reason for the great popularity of the model of mobile adsorption. In this model, it is assumed that molecules adsorbed in the monolayer are not affected by field periodicity on the adsorbent surface, which enables them to move in the directions parallel to the surface. Volmer (1925) first derived the isotherm equation for mobile adsorption treating the adsorption phase as a two-dimensional perfect gas. The Volmer model was widely discussed on the basis of statistical mechanics (see Young and Crowell, 1962; Clark, 1970). Thermodynamic aspects of this model, were also the subject of detailed considerations in Hill's work (1949b).

The introduction of the isotherm equation describing mobile adsorption with lateral interactions was a great success (Hill, 1946; 1952). Hill (1946), using methods of statistical mechanics, showed that the application of the analog of van der Waals equation to two-dimensional adsorption layer gave an equation of adsorption isotherm which is known in the literature as the Hill - de Boer isotherm (Ross and Olivier, 1964). The Hill - de Boer equation has been used by many authors as the local isotherm in integral equation (3) for studying the adsorption of gases on patchwise heterogeneous surfaces. Particular achievements in this respect have to be attributed to Ross and co-workers, these were discussed in detail in the monograph by Ross and Olivier (1964), as well as in the excellent review article by Ross (1971). Studies on the heterogeneity of adsorbent surfaces by using Eq. (3) with the Hill - de Boer local isotherm, have been initiated by Ross, and have been successfully continued by House and Jaycock (1977a,b; 1978).

Concluding this section, we cannot omit Tompkin's work (1950), which deals with mobile and localized monolayer adsorption without lateral interactions of gases on heterogeneous surfaces with patchwise and random distribution of adsorption sites. This problem has been discussed in terms of statistical thermodynamics and has been fully presented in the monograph of Young and Crowell (1962), and Clark (1970). According to the treatment of Tompkins (1950), the canonical partition function for monolayer mobile adsorption without lateral interactions of gases on patchwise heterogeneous surfaces is given by the following expression:

$$Q_r^{\mathrm{m},p} = \prod_{r=1}^{r_0} \frac{A_{\mathrm{f},r}^{N_r}}{N_r!} q_r^{\mathrm{m}}(T)^{N_r} \tag{23}$$

where

$$A_{\mathrm{f},r} = A_r - \beta N_r$$

and N_r is the number of molecules adsorbed on the rth surface patch, $A_{\mathrm{f},r}$ the free area available to molecules that are adsorbed on the rth surface patch, A_r the surface area of the rth patch, β the area occupied by one adsorbed molecule, $q_r^{\mathrm{m}}(T)$ the partition function of the mobile adsorbed molecule on the rth surface patch, and r_0 the number of types of adsorption sites. In the case of mobile adsorption of the surface having adsorption sites randomly distributed, the canonical partition function proposed by Tompkins (1950) is

$$Q_r^{\mathrm{m},r} = \prod_{r=1}^{r_0} \frac{\left(A - \sum_{r=1}^{r_0} \beta N_r\right)^{N_r}}{N_r!} q_r^{\mathrm{m}}(T)^{N_r} \tag{24}$$

where A is the adsorbent surface area. Equation (23) corresponds to the adsorption model in which, the adsorbed molecules are mobile within the patches in which they occur; however, the acceptance of function (24) is synonymous with the assumption of complete mobility of adsorbed molecules, i.e., the adsorbed molecules can move freely over the whole surface. In Section I.B, we discussed that for the case of localized adsorption without lateral interactions, the integral equation (3) was valid for various models of a heterogeneous surface; as far as mobile adsorption is concerned, the topography of adsorption sites on the adsorbent surface should be taken into considerations even if lateral interactions are neglected.

B. Adsorption of Gas Mixtures on Solids

The first work dealing with the mobile adsorption of gas mixtures on homogeneous surfaces of solids was that of Kemball *et al.* (1948), in which these authors extended Volmer's equation (1925) to two-component mobile monolayers. The studies of mobile monolayers were further continued by Schay (1956), Schay *et al.* (1957), and Hansen and Stage (1956).

In the case of mobile adsorption without lateral interactions of n-component gas mixtures on homogeneous surfaces, the Volmer equation may be written in the following form:

$$a_i^m p_i = \frac{\theta_{i(\boldsymbol{n})}}{1 - \theta_{(\boldsymbol{n})}} \exp\left(\frac{\sum_{j=1}^{n} \kappa_{ij}\theta_{i(\boldsymbol{n})}}{1 - \theta_{(\boldsymbol{n})}}\right) \quad \text{for } i = 1,2,\ldots,n \tag{25}$$

where

$$a_i^m = \beta_i q_i^m(T)\exp\left(\mu_i^{0G}/kT\right) \tag{26}$$

Equation (25) has been obtained from the following canonical partition function:

$$Q^{\mathrm{m}}_{(\boldsymbol{n})} = \frac{\left(A - \sum_{i=1}^{n} \beta_i N_i\right)^{\sum_{i=1}^{n} N_i}}{\prod_{i=1}^{n} N_i!} \prod_{i=1}^{n} \left(q_i^{\mathrm{m}}\right)^{N_i} \tag{27}$$

and by using the notation:

$$\theta_{i(\boldsymbol{n})} = N_i \beta_i / A \quad \text{and} \quad \kappa_{ij} = \beta_i / \beta_j \tag{28}$$

The problem of mobile adsorption with interactions of gas mixtures was dealt with by Drenan and Hill (1950). Based on the theory of three-dimensional van der Waals fluids, these authors formulated the canonical partition function for the mobile two-component monolayer:

$$\begin{aligned} Q^{\mathrm{m},in}_{(1,2)} &= \frac{(A - N_1\beta_1 - N_2\beta_{21})^{N_1}\left(q_1^{\mathrm{m}}\right)^{N_1}}{N_1!} \\ &\times \frac{(A - N_1\beta_{12} - N_2\beta_2)^{N_2}\left(q_2^{\mathrm{m}}\right)^{N_2}}{N_2!} \\ &\times \exp\left[\left(\alpha_1 N_1^2 + 2\alpha_{12} N_1 N_2 + \alpha_2 N_2^2\right)/AkT\right] \end{aligned} \tag{29}$$

where α and β are the two-dimensional analogs of the three-dimensional van der Waals constants. The use of the approximation:

$$\beta_{21} = \beta_2 \quad \text{and} \quad \beta_{12} = \beta_1$$

in Eq. (29) gives the following adsorption isotherm:

$$a_1^{\mathrm{m}} p_1 = \frac{\theta_{1(1,2)}}{1 - \theta_{(1,2)}} \exp\left[\frac{\sum_{j=1}^{2} \kappa_{1j}\theta_{1(1,2)}}{1 - \theta_{(1,2)}} - \frac{2\alpha_1\theta_{1(1,2)}}{\beta_1 kT} - \frac{2\alpha_{12}\theta_{2(1,2)}}{\beta_2 kT}\right] \tag{31}$$

Equation (31) is a simple extension of the Hill - de Boer adsorption isotherm.

Thermodynamics of the mobile adsorption without lateral interactions of n-component gas mixtures on heterogeneous solid surfaces has not yet been discussed. This problem may be considered in terms of the treatment of Tompkins (1950). Generalization of Eqs. (23) and (24) to n-component gas mixtures give the following expressions for the canonical partition function:

$$Q^{\mathrm{m,p}}_{(\boldsymbol{n},\boldsymbol{r})} = \prod_{\boldsymbol{r}} \frac{\left(A_{\boldsymbol{r}} - \sum_{i=1}^{n} \beta_i N_{i,\boldsymbol{r}}\right)^{\sum_{i=1}^{n} N_{i,\boldsymbol{r}}}}{\prod_{i=1}^{n} N_{i,\boldsymbol{r}}!} \times \prod_{i=1}^{n} \left(q^{\mathrm{m}}_{i,\boldsymbol{r}}\right)^{N_{i,\boldsymbol{r}}} \quad \text{(for PM)} \tag{32}$$

and

$$Q^{\mathrm{m,r}}_{(\boldsymbol{n},\boldsymbol{r})} = \prod_{\boldsymbol{r}} \frac{\left(A - \sum_{\boldsymbol{r}} \sum_{i=1}^{n} \beta_i N_{i,\boldsymbol{r}}\right)^{\sum_{i=1}^{n} N_{i,\boldsymbol{r}}}}{\sum_{i=1}^{n} N_{i,\boldsymbol{r}}!} \times \prod_{i=1}^{n} \left(q^{\mathrm{m}}_{i,\boldsymbol{r}}\right)^{N_{i,\boldsymbol{r}}} \quad \text{(for RM)} \tag{33}$$

Applying the well-known thermodynamic relationships to Eqs. (32) and (33), equations of thermodynamic functions such as energy, entropy, heat of adsorption, and adsorption isotherms may be obtained.

IV. PARTIALLY MOBILE ADSORPTION OF GASES ON SOLIDS

A. Adsorption of Single Gases on Solids

In the majority of papers concerning the monolayer adsorption on solids, two extreme idealized models are discussed; they are the completely localized and completely mobile adsorption models (cf. Sections II and III). The studies of Hill (1946) show, however, that the partially mobile adsorption on solid surfaces seems to be more close to reality. Also, the theoretical considerations of Stebbins and Halsey (1964) and Tsien and Halsey (1967) pointed to the occurrence of localized adsorption having surpassed mobile with increase surface coverage. The support for these theoretical studies are numerous experimental adsorption measurements (Thomy and Duval, 1969; 1970a,b; Matecki *et al.*, 1974a,b, 1977; Duval and Tomy, 1975; Regnier *et al.*, 1975; Regnier *et al.*, 1977; Menaucourt *et al.*, 1977). It follows from the above cited papers that during the formation of the monolayer a change from nonlocalized to localized adsorption occurs.

Although, the partial localization may not be too important for equilibrium properties of adsorption systems satisfying the conditions $\phi/kT < 0.5$ and $\phi/kT > 5$ (Steele and Ross, 1961; Steele, 1961), it can play an important role in surface transport (Lee and O'Connell, 1972). Therefore, further studies devoted to partially mobile adsorption are to be expected in the near future. They may explain the phase transitions occurring in adsorbed monolayers and may be useful in describing surface transport phenomena. The partially mobile adsorption model is a general one, which in boundary cases becomes the model for completely mobile or completely localized adsorption.

Attempts have been made to describe partially mobile adsorption; literature concerning this problem is rather scanty. One of the first approaches to "mixed" adsorption was the McAlpin and Pierotti treatment (1964, 1965), which was based on the significant structures theory (SST) of liquid state proposed by Eyring *et al.* (1958). For the clear and short presentation of the significant structures theory, we cite three sentences of Pierotti (1971), which excellently characterizes this theory: "The SST is based upon the intuitive recognition of certain significant structures which must be considered when constructing the partition function for a system. Each significant structure contributes to the complete partition function of the two-dimensional phase in proportion to the availability of that structure under certain specified conditions. The availability of each degree of freedom will be dependent upon the mole fraction of vacancies in the system." The SST leads to the following canonical partition function:

$$Q = Q^{SL} Q^{GL}$$

$$= [q^{SL}(T)]^{N_m \theta^2} [q^{GL}(T)]^{N_m \theta(1-\theta)} \tag{34}$$

where N_m is the number of molecules in the surface phase at $\theta = 1$; $q^{SL}(T)$ is the molecular partition function for the two-dimensional solidlike structure, which may be due to the presence of nearest-neighbor vacancies, or in which a molecule is restrained by its neighbors for occupying a given equilibrium site; $q^{GL}(T)$ is the molecular partition function for the two-dimensional gaslike structure, in which a molecule is free to translate over a portion of the surface. The partition function for the two-dimensional solidlike structure can be described by the Lennard-Jones and Devonshire partition function (1937), whereas the gaslike structure can be treated as an ideal two-dimensional gas (Pierotti, 1971). Equation (34) leads to relatively simple equation of adsorption isotherm, which gives good agreement with experimental data [the adsorption of argon and krypton on graphite (McAlpin and Pierotti, 1964)].

One year after publication of the significant structures theory for physical adsorption of gases on solids (McAlpin and Pierotti, 1964), Holland (1965a,b) published a simple model for monolayer adsorption without lateral interactions. In this model, the states of an adsorbed molecule fall into two classes: the localized states, having adsorption energies greater than ϕ, the energy for translational motion parallel to the surface, and the unlocalized mobile states with adsorption energies less than ϕ. The canonical partition function for Holland's adsorption model may be written as follows:

$$Q^{\mathrm{ml}} = Q^{\mathrm{l}} Q^{\mathrm{m}} \tag{35}$$

where

$$Q^{\mathrm{l}} = \frac{B!}{L!(B-L)!} (q^{\mathrm{l}})^{L} \tag{36}$$

$$Q^{\mathrm{m}} = \frac{(A-\beta N)^{M}}{M!} (q^{\mathrm{m}})^{M} \tag{37}$$

and

$$N = M + L \tag{38}$$

In Eqs. (36)-(38), N denotes the total number of adsorbed molecules, whereas L and M denote the number of localized and mobile molecules, respectively. Equation (35) generates the following conditions for the adsorption equilibrium:

$$\frac{\theta^{l}}{(1-\nu\theta^{l})a^{l}} \exp\left(\frac{\theta^{m}}{1-\theta}\right) = \frac{\theta^{m}}{(1-\theta)a^{m}} \exp\left(\frac{\theta^{m}}{1-\theta}\right) = p \tag{39}$$

where

$$\theta^{l} = L\beta/A; \ \theta^{m} = M\beta/A; \ \theta = \theta^{m} + \theta^{l} \tag{40}$$

and

$$\nu = \beta_s/\beta \tag{41}$$

The constants a^l and a^m are defined in Eqs. (9) and (26), and β_s is the area of an adsorption site. The dependences θ, θ^m, and θ^l versus adsorbate pressure p may be calculated by means of Eq. (39).

A slightly different approach to partially mobile adsorption than that by Holland has been recently proposed by Lee and O'Connell (1972, 1974). Their treatment differs from that of Holland (1965a) in four respects: (a) they consider three different states of adsorbed molecules; mobile molecules not over sites, mobile molecules over sites, and localized molecules on sites; (b) they assume that the ratio θ^m/θ^l is independent over the surface coverage; (c) they used the scaled particle theory with the virial theorem to express the configurational area A_f; and (d) they take into account the lateral interactions in the adsorbed phase.

A simplified version of Lee and O'Connell model, i.e., the model with assumptions (b), (d), and modified assumption (a), was recently discussed by Patrykiejew and Jaroniec (1978). In their treatment, two different states of adsorbed molecules are distinguished, i.e., mobile and localized molecules. This point of the model of Patrykiejew and Jaroniec is identical with that by Holland. Thus, the canonical partition function is similar to that used by Holland (1965a);

$$Q^{ml,in} = \frac{B^*!\,(q^1)^{Nf^1}}{(Nf^1)\ (B^*-Nf^1)!}\ \frac{[(A-\beta N)q^m]^{Nf^m}}{(Nf^m)!}$$

$$\times\ \exp\left[-\frac{N}{2kT}\left(\phi_1 f^m + \phi_2 f^1\right)\right] \tag{42}$$

where

$$f^1 + f^m = 1;\ f^1,\ f^m = \text{const} \tag{43}$$

and

$$B^* = \begin{cases} B & (44a) \\ B - Nf^m & (44b) \end{cases}$$

The symbols f^m and f^1 denote fractions of mobile and localized molecules in the adsorption phase, respectively; ϕ_1 and ϕ_2 characterize the lateral interactions, they have been identified in the paper of Lee and O'Connell (1972).

The first term of the function (42) characterizes the localized adsorption, the second, mobile adsorption, and the third term is due to the mutual interactions between adsorbed molecules. The number of sites available in localized adsorption may be defined in two different ways (cf. Eqs. (44a) and (44b): condition (44a) denotes that the localized adsorption can occur on all adsorption sites, even if they are occupied by mobile molecules, whereas Eq. (44b) limits the localized adsorption; it does not occur on the surface covered by mobile molecules. The canonical partition function (42) with the conditions $\beta_s = \beta$ and (44a) generate the following equation of adsorption isotherm:

$$p = \frac{1}{a^{ml}}\,\frac{\theta}{1-\theta}\left(\frac{1-\theta}{1-f^1\theta}\right)^{f^1}\exp\left[\frac{f^m\theta}{1-\theta} - \alpha^{ml}\theta\right] \tag{45}$$

where

$$a^{ml} = (a^{l})^{-1}(a^{m}/a^{l})^{f^{m}}(f^{m})^{-f^{m}}(f^{l})^{-f^{l}} \tag{46}$$

and

$$\alpha^{ml} = \alpha^{m}f^{m}(2-f^{m}) + \alpha^{l}(f^{l})^{2} \tag{47}$$

However, assuming the conditions $\beta_s = \beta$ and (44b), we obtain

$$p = \frac{1}{a^{ml}}\,\frac{\theta}{1-\theta}\left(\frac{1-f^{m}\theta}{1-\theta}\right)^{f^{m}} \exp\left[\frac{f^{m}\theta}{1-\theta} - \alpha^{ml}\theta\right] \tag{48}$$

The adsorption isotherms (45) and (48) for boundary cases of f^{m} becomes the Fowler - Guggenheim and Hill - de Boer equations. Figure 1 presents the theoretical adsorption isotherms (45) (cf. Fig. 1a) and (48) (cf. Fig. 1b) calculated for $\alpha^{ml} = 0$ at different values of f^{m}. The former equation (45) predicts very regular changes of the surface coverage at a constant adsorbate pressure with increasing parameter f^{m}, whereas the other (48) is strongly affected by the parameter f^{m}. Such behavior of the adsorption isotherms results directly from the physical meaning of conditions (44a) and (44b). Figure 2a-c show the dependences θ^{l} versus θ, θ^{m} versus θ, and θ versus p/a^{ml}, respectively; they have been calculated for the adsorption isotherm (45) by assuming $f^{m} = 0.2$, 0.5, 0.8 and $\alpha^{ml} = 5$. The assumption of f^{m} = const leads to the linear behavior of θ^{m} versus θ and θ^{l} versus θ (cf. Fig. 2a,b). The influence of the parameter f^{m} on the critical properties of the adsorption isotherm (45) is presented in Fig. 2c. However, Fig. 2d shows the contributions of θ^{l} and θ^{m} to the total adsorption θ.

Next, Patrykiejew and Jaroniec (1978) extended the Lee and O'Connell adsorption model, taking into account the dependence of the degree of mobility of adsorbed molecules upon surface coverage. They assumed that

$$\theta^{m}/\theta^{l} = F(\theta, T) \tag{49}$$

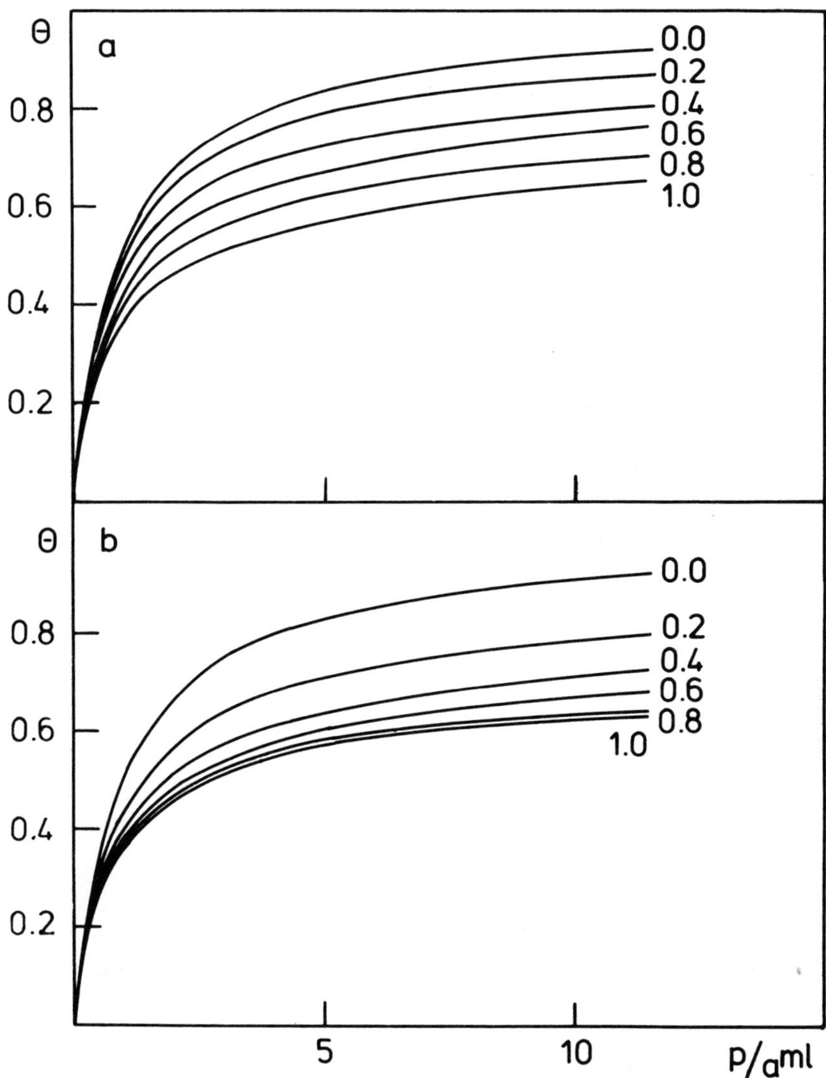

Fig. 1. Theoretical adsorption isotherms calculated according to Eqs. (45) (cf. Fig. 1a) and (48) (cf. Fig. 1b) for $\alpha^{ml} = 0$ *and different values of* f^m.

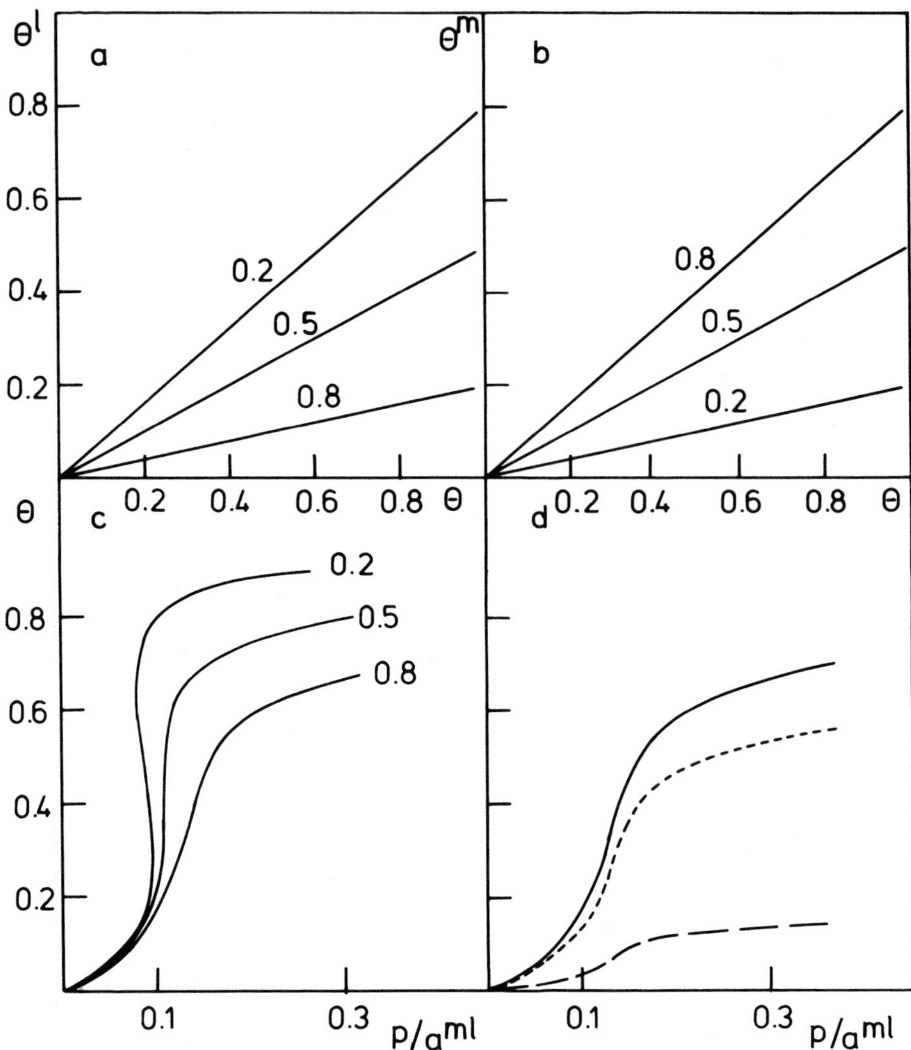

Fig. 2. Model calculations for Lee - O'Connell adsorption isotherm (45); a,b - contributions of localized (a) and mobile adsorption (b) in the total adsorption; c - theoretical adsorption isotherms (45) calculated for $\alpha^{ml} = 5$ and different values of f^m, d - theoretical adsorption isotherm (45) with $\alpha^{ml} = 5$, $f^m = 0.8$ (solid line) and partial isotherms for localized (-----) and mobile (— — —) adsorption.

Then,

$$\theta^{m} = \theta \frac{F(\theta, T)}{1 + F(\theta, T)} \quad \text{and} \quad \theta^{1} = \theta \frac{1}{1 + F(\theta, T)} \tag{50}$$

Using definitions (50) in the canonical partition function (35), they derived a general equation for the adsorption isotherm, which for $F(\theta, T) = F^*(T)$ becomes Eq. (45). The assumption,

$$F(\theta, T) = (1 - \theta)/\theta \tag{51}$$

leads to an adsorption isotherm, which is similar to that obtained for the significant structures theory (McAlpin and Pierotti, 1964). A more realistic definition of $F(\theta, T)$ is given by

$$F(\theta, T) = K^*(1 - \theta)/(1 + K^*\theta) \tag{52}$$

where

$$K^* = a^{1}/a^{m} \tag{53}$$

Then, the dependences θ^{m}/θ versus θ and θ^{1}/θ versus θ have linear courses:

$$\theta^{m}/\theta = K^*(1 - \theta)/(1 + K^*) \tag{54}$$

$$\theta^{1}/\theta = (1 + K^*\theta)/(1 + K^*) \tag{55}$$

The function $F(\theta, T)$ assumes the value K^* at $\theta = 0$, and with increasing θ, it decreases to zero. Figure 3a and b present the dependences θ^{1} versus θ and θ^{m} versus θ, respectively. The curves θ^{1} versus θ increase monotonically for different values of K^* (cf. Fig. 3a), whereas, the curves θ^{m} versus θ reach a maximum value at the point $\theta = 0.5$, and have zero values at the points $\theta = 0$ and $\theta = 1$. The course of the isotherms θ^{m} versus p/a^{ml} obtained by assuming Eq. (52) is more realistic than in the case of the constancy of f^{m} (cf. Figs. 2d and 4); Equation (52) guarantees an in-

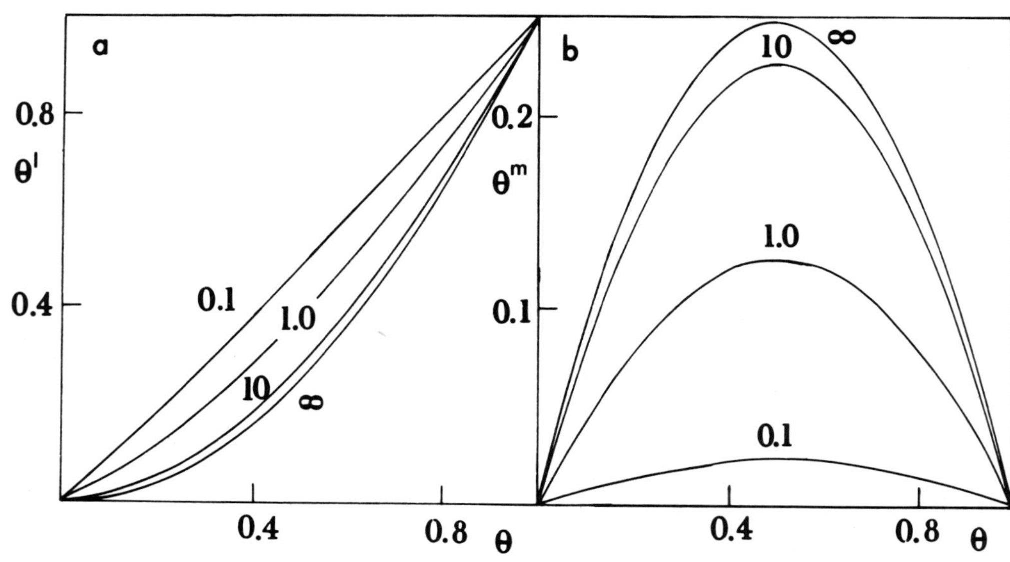

Fig. 3. Contributions of localized (cf. Fig. 3a) and mobile (cf. Fig. 3b) adsorption in the total adsorption for Patrykiejew and Jaroniec model (1978) calculated according to Eq. (50) for different values of K^*.

crease of mobile adsorption θ^m at small pressures and a decrease at higher adsorbate pressures.

Although, the experimental examination of the partially mobile adsorption model gives satisfactory results (Lee and O'Connell, 1972; 1974; Patrykiejew and Jaroniec, 1978), further studies in this direction are to be expected. This model creates new possibilities for the investigation of adsorption from gas mixtures on solid surfaces.

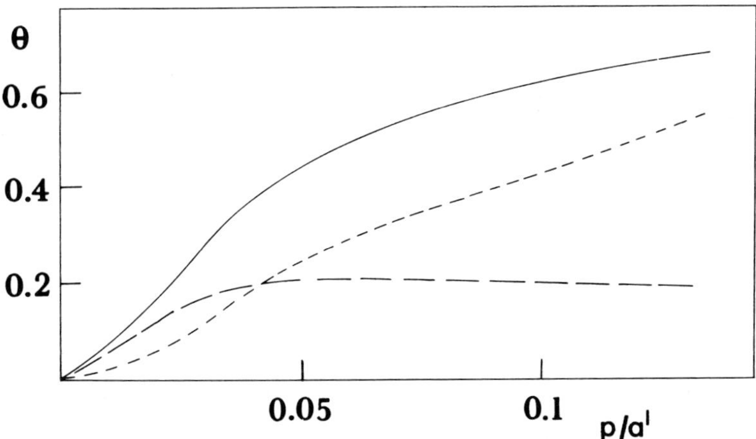

Fig. 4. Theoretical adsorption isotherm calculated according to Patrykiejew and Jaroniec model (1978) for $K^* = 2.31$, $\alpha = 4.65$, and partial isotherms for localized (-----) and mobile (— — —) adsorption.

B. Adsorption of Gas Mixtures on Homogeneous Surfaces

The possibilities of the extension of the Holland's adsorption model (1965a) to adsorption from gas mixtures, without lateral interactions, on homogeneous solid surfaces have been recently discussed by Łajtar *et al.* (1978). They start from the following canonical partition function:

$$Q^{\mathrm{ml}}_{(\boldsymbol{n})} = \frac{\left(B - \sum_{i=1}^{n} M_i\right)! \left(A - \sum_{i=1}^{n} \beta_i N_i\right)^{\sum_{i=1}^{n} M_i}}{\left(B - \sum_{i=1}^{n} N_i\right)!} \times \prod_{i=1}^{n} \frac{\left(q_i^{\mathrm{m}}\right)^{M_i} \left(q_i^{\mathrm{l}}\right)^{L_i}}{M_i!\,L_i!} \tag{56}$$

where

$$N_i = M_i + L_i \qquad \text{for } i=1,\ 2,\ \ldots,\ n \tag{57}$$

The function (56) leads to a system of equations, which are analogous to Eq. (39), obtained for adsorption of the ith component. This system of equations, may be solved with respect to θ_i^{m} and θ_i^{l}, giving the total adsorption θ_i for the ith component. Analytical solutions are possible for two-component gas mixtures.

The statistical mechanical model for partially mobile adsorption of single adsorbates on homogeneous surfaces, developed by Lee and O'Connell (1972), has been extended to mixture adsorption equilibria (Lee and O'Connell, 1974). This model gives good agreement with experimental adsorption data when simple mixing rules are used. The simplified version of the Lee and O'Connell adsorption model (1972), discussed by Patrykiejew and Jaroniec (1978) for the case of single adsorption, may be extended to the adsorption of gas mixtures. The canonical partition function for Patrykiejew and Jaroniec model of mixed-gas adsorption may be obtained from Eq. (56) by assuming

$$L_i = f_i^{\mathrm{l}} N_i \;; \qquad M_i = f_i^{\mathrm{m}} N_i \tag{58}$$

where $f_i^{\mathrm{l}} + f_i^{\mathrm{m}} = 1$ for $i = 1, 2, \ldots, n$ and f_i^{m} is the fraction of mobile molecules of the ith component. Assuming that $\beta_s = \beta_1 = \beta_2 = \ldots = \beta_n$, we obtain from Eqs. (56) and (58) the following equation for the adsorption equilibrium:

$$p_i a_i^{\mathrm{ml}} = \frac{\theta_i \left(1 - \sum_{j=1}^{n} f_j^{\mathrm{m}} \theta_j\right)^{f_i^{\mathrm{m}}}}{\left(1 - \sum_{j=1}^{n} \theta_j\right)^{1 + f_i^{\mathrm{m}}}} \exp\left[\frac{\sum_{j=1}^{n} f_j^{\mathrm{m}} \theta_j}{1 - \sum_{j=1}^{n} \theta_j}\right] \tag{59}$$

The selectivity coefficient s obtained from Eq. (59) is equal to

$$s = \frac{\theta_1 p_2}{\theta_2 p_1} = \frac{a_1^{\mathrm{ml}}}{a_2^{\mathrm{ml}}} \left(\frac{1 - f_1^{\mathrm{m}}\theta_1 - f_2^{\mathrm{m}}\theta_2}{1 - \theta_1 - \theta_2}\right)^{f_2^{\mathrm{m}} - f_1^{\mathrm{m}}} \quad \text{for } n = 2 \tag{60}$$

If the properties of both components are similar, then $f_1^{\mathrm{m}} = f_2^{\mathrm{m}} = f^{\mathrm{m}}$ and Eq. (60) becomes

$$s = a_1^{\mathrm{ml}} / a_1^{\mathrm{ml}} \tag{61}$$

For the case of differing chemical properties for both components, the factors f_1^{m} and f_2^{m} are different; in the boundary case, when $f_1^{\mathrm{m}} = 1$ and $f_2^{\mathrm{m}} = 0$, i.e., when molecules of 1-st component are mobile and molecules of 2-nd component are localized, we obtain

$$s = \frac{a_1^{\mathrm{m}}}{a_2^{\mathrm{l}}} \frac{1 - \theta_1 - \theta_2}{1 - \theta_1} \tag{62}$$

Figure 5 presents the dependence of the selectivity coefficient s upon the surface coverage θ_1 by assuming constancy of f_2^m and θ_2. For the case of identical parameters f_1^m and f_2^m, the dependence s versus θ_1 at constant θ_2 is linear; such a result is observed for completely mobile and completely localized adsorption of gas mixtures. If $f_1^m \neq f_2^m$, then the curves s versus θ_1 are not linear. Thus, the factors which are expected to influence nonlinear behavior of the function s versus θ_1 may be summarized as

(i) adsorbate interaction energy (cf. Eq. (31));
(ii) adsorbent heterogeneity (cf. Eq. (11));
(iii) partially mobile adsorption (cf. Eq. (60)) for $f_1^m \neq f_2^m$.

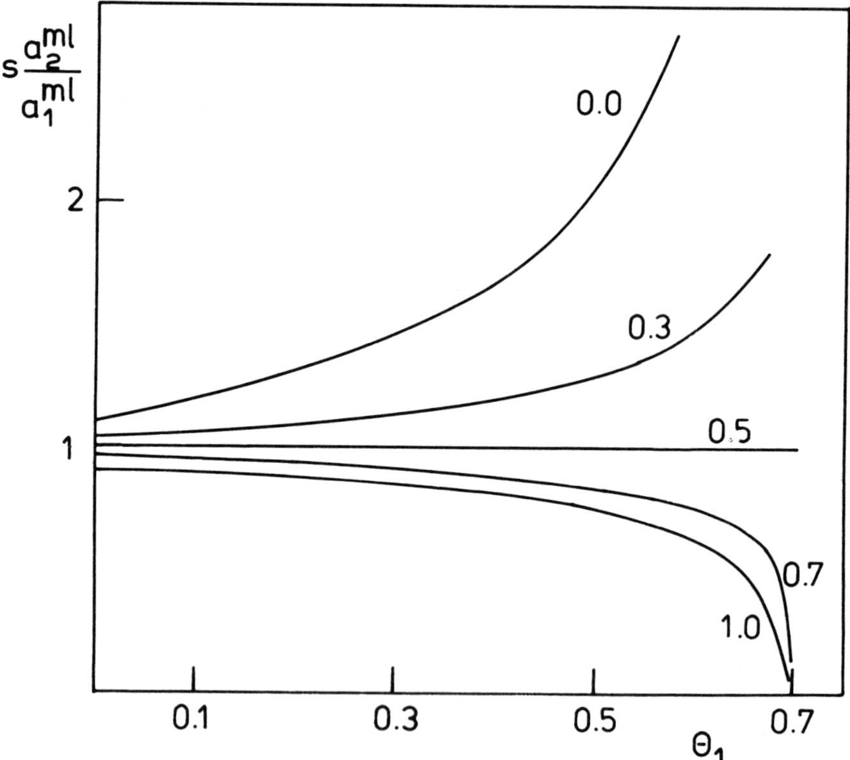

Fig. 5. Dependence of the selectivity coefficient upon the partial adsorption θ_1 for $\theta_2 = 0.3$, $f_2^m = 0.5$, and different values of f^m.

Further modification of the adsorption isotherm (59), is possible by assuming that f_i^m is a function of $\theta_{(\boldsymbol{n})}$, i.e.,

$$\theta_i^m/\theta_i^1 = F_i(\theta_{(\boldsymbol{n})}, T) \quad \text{for } i=1, 2, \ldots, n \tag{63}$$

Equation (63) means that we assume the partially mobile adsorption for all components.

The particular form of the adsorption isotherm, derived from the canonical partition function (56), depends on what assumptions about the functions F_i have been made. If F_i = const for all surface coverages, then the adsorption isotherm (59) is obtained. The most simple equation for F_i, taking into account the dependence of F_i upon $\theta_{(\boldsymbol{n})}$, may be deduced from Eq. (51); it is

$$F_i = \frac{1 - \sum_{j=1}^{n} \theta_j}{\sum_{j=1}^{n} \theta_j} \quad \text{for } i=1, 2, \ldots, n \tag{64}$$

The physical meaning of Eq. (64) is the following: The dependence θ_i^m/θ_i^1 versus $\theta_{(\boldsymbol{n})}$ is identical for all components of the mixture and is dependent upon the total surface coverage. Assumption of other equations for F_i leads to very complicated expressions for mixed-gas adsorption isotherm.

For the adsorption of binary gas mixtures, the components of which have different chemical properties, we can use the "boundary" model in which the adsorption of first component is localized, whereas adsorption of the other component is mobile. For such a model, the equations of partial adsorption isotherm are considerably simpler.

C. *Partially Mobile Adsorption on Heterogeneous Surfaces*

Extension of the partially mobile adsorption model to heterogeneous surfaces is difficult (cf. Section VI.A). Useful equations may be obtained for the Patrykiejew and Jaroniec model (1978), in which the constancy of f_i^m for $i = 1, 2, \ldots, n$ is assumed and the lateral interactions are neglected. Then, the total fraction of mobile molecules, $\hat{f}_i^m$ is defined as follows:

$$\hat{f}_i^m = \frac{\int_{\Delta(\boldsymbol{n})} f_i^m \theta_{i(\boldsymbol{n})}(\boldsymbol{p}, \boldsymbol{\varepsilon}) \chi_{(\boldsymbol{n})}(\boldsymbol{\varepsilon})\, d\boldsymbol{\varepsilon}}{\int_{\Delta(\boldsymbol{n})} \theta_{i(\boldsymbol{n})}(\boldsymbol{p}, \boldsymbol{\varepsilon}) \chi_{(\boldsymbol{n})}(\boldsymbol{\varepsilon})\, d\boldsymbol{\varepsilon}} \quad \text{for } i=1, 2, \ldots, n \tag{65}$$

An analogous equation may be written for the total fraction of localized molecules, $\hat{f}_i^1$. Although the parameters f_i^m and f_i^1, defining the fractions of mobile and localized molecules of the *i*th component on a given surface patch, are independent upon the surface coverages θ_i ($i = 1, 2, \ldots, n$), for the case of heterogeneous solid surfaces, they are dependent upon the partial pressures of the components, i.e.,

$$\hat{f}_i^m = \hat{f}_i^m(\boldsymbol{p}) \tag{66}$$

This means that for heterogeneous solid surfaces $\hat{f}_i^m$ for $i = 1, 2, \ldots, n$ is not constant, and is dependent upon the surface coverage.

V. ADSORPTION FROM LIQUID MIXTURES ON SOLIDS

A. *Homogeneous Surfaces*

The monograph of Kipling (1965) gives a particularly clear presentation of the theoretical framework of adsorption from liquid mixtures on homogeneous solid surfaces. A number of excellent reviews devoted to this problem have been published (Schay, 1969, 1976; Everett, 1973; Brown and Everett, 1975; Rusanov, 1971). The majority of papers published on adsorption from solutions concern the adsorption thermodynamics of binary liquid mixtures on homogeneous surfaces. The fundamental papers on this field have been summarized in the reviews of Everett (1973) and Brown and Everett (1975); they were published by Kiselev and Pavlova (1962, 1965), Everett (1964, 1965), Schay and Nagy (1961, 1972), Larionov *et al.* (1967), Larionov and Myers (1971), Sircar and Myers (1970), and Tóth (1970, 1974). The thermodynamic relationships have also been used for the correlation of the behavior of liquid/solid and gas/solid systems. The purpose of this correlation is to establish the basis upon which adsorption from liquid mixtures can be related to adsorption from a single component and from gas mixtures; studies in this direction have been initiated by Myers and Sircar (1972, 1973), and Bering *et al.* (1973).

In contrast to the extensive literature from binary liquid mixtures, the multicomponent liquid adsorption systems have been hardly investigated (cf. Brown and Everett, 1975). The adsorption from dilute multicomponent solutions were studied by Martynov and Maevskaya (1969, 1972), Radke and Prausnitz (1972), and Jossens *et al.* (1978). The classical thermodynamics of adsorption of a given component present in a very low concentration from a multicomponent solvent has been widely discussed by Ościk (1961a,b, 1965, 1972). However, Gryazev (1958), Adler *et al.* (1971), and Rakhlevskaya *et al.* (1971) studied the effects of the competition between several similar adsorptives contained in a multicomponent liquid mixture.

As concerns the adsorption from multicomponent liquid mixtures in the whole concentration region, the most advanced papers have been presented by Minka and Myers (1973) and Nassonov (1971, 1972). The equation for the mole fraction of a given component in the adsorbed phase, derived by Minka and Myers (1973) may be obtained in terms of the statistical mechanics. Let us assume, following Borówko *et al.* (1979), that

(a) only one molecule can be adsorbed on one adsorption site;

(b) molecular sizes of all components are assumed to be identical;

(c) the total number of adsorbed molecules is still constant; and

(d) the adsorbent surface is homogeneous.

Then, the canonical partition function may be analogous to that used for adsorption from multicomponent gas mixtures (cf. Eq. (4)); that is,

$$Q_{(\boldsymbol{n})} = \frac{B!}{\left(\prod_{i=1}^{n-1} N_i!\right)\left(B - \sum_{i=1}^{n-1} N_i\right)!} \left[\prod_{i=1}^{n-1} \left(q_i^1\right)^{N_i}\right] \left(q_n^1\right)^{B - \sum_{i=1}^{n-1} N_i} \tag{67}$$

where

$$\sum_{i=1}^{n} N_i = B = \text{const} \tag{68}$$

and N_i is the number of adsorbed molecules of the ith component from n-component liquid mixture. The canonical partition function (67) leads to the following equation:

$$y_{i(\boldsymbol{n})} = \frac{c_{in} x_{in}}{1 + \sum_{j=1}^{n-1} c_{jn} x_{jn}} \qquad \text{for } i=1, 2, \ldots, n-1 \tag{69}$$

where $y_{i(\boldsymbol{n})}$ is the mole fraction of the ith component in the adsorbed phase, c_{in} is the constant analogous to the constant $a_{in}^1 = a_i^1/a_n^1$, and $x_{in} = x_i/x_n$; x_i is the mole fraction of the ith component in the bulk phase. Equation (69) describes the adsorption from solution when both the adsorbed and bulk phases are ideal.

For the case of nonideal bulk phase and ideal adsorbed phase, Eq. (69) becomes (Minka and Myers, 1973):

$$y_{i(\mathbf{n})} = \frac{c_{in} x_{in} \gamma_{in}}{1 + \sum_{j=1}^{n-1} c_{jn} x_{jn} \gamma_{jn}} \quad \text{for } i=1, 2, \ldots, n-1 \tag{70}$$

where $\gamma_{in} = \gamma_i / \gamma_n$ and γ_i is the activity coefficient of the *i*th component. The main thermodynamic functions characterizing the adsorption from solutions may be obtained by means of the canonical partition function (67).

B. Heterogeneous Surfaces

Although the adsorbent heterogeneity in adsorption from solutions plays a smaller role than in adsorption from gaseous phases, its influence on adsorption equilibrium cannot be neglected. This problem, very interesting and important from both theoretical and practical reasons, was little studied.

Schuchovitzky (1938) was the first to investigate qualitatively the heterogeneity effects in adsorption from solutions. He considered the adsorbent surfaces with a discrete distribution of adsorption energy. A similar procedure was used later by Šiškova and Erdös (1960). However, Hansen and May (1957), Delmas and Patterson (1960), and Coltharp and Hackerman (1973) saw in the surface heterogeneity of the adsorbent a source of imperfection of the adsorbed phases. They suggested that the change of the sign of excess adsorption isotherms should be associated with the surface heterogeneity. Their suggestions were supported by the studies of Coltharp and Hackerman (1973).

The most advanced papers deal with theory of adsorption from binary ideal liquid mixtures on heterogeneous solid surfaces by assuming the ideality of the adsorbed phase (Rudziński *et al.*, 1973; Ościk *et al.*, 1974-1977). Their studies are based on integral representation for adsorption isotherm from binary solutions, which is formally identical with Eq. (3):

$$Y_{1(1,2)}(x_{12}) = \int_{\Omega_1} y_{1(1,2)}(x_{12}, \varepsilon_{12}) X_1(\varepsilon_{12}) d\varepsilon_{12} \tag{71}$$

where $x_{12} = x_1/x_2$, $\varepsilon_{12} = \varepsilon_1 - \varepsilon_2$, $X_1(\varepsilon_{12})$ is the distribution function of ε_{12} normalized to unity, $Y_{1(1,2)}$ is the mole fraction of 1-st component in the adsorbed phase for a heterogeneous surface, and Ω_1 is the integration region for ε_{12}.

Expressing the excess adsorption isotherms by the following equations:

$$N^{\sigma s}_{1(1,2)} = N^{\sigma}_{1(1,2)}/N^{s} = Y_{1(1,2)} - x_1 \tag{72}$$

$$n^{\sigma s}_{1(1,2)} = n^{\sigma}_{1(1,2)}/n^{s} = y_{1(1,2)} - x_1 \tag{73}$$

we obtain an alternative equation to Eq. (71):

$$N^{\sigma s}_{1(1,2)} = \int_{\Omega_1} n^{\sigma s}_{1(1,2)} X_1(\varepsilon_{12}) d\varepsilon_{12} \tag{74}$$

The symbols $n^{\sigma}_{1(1,2)}$ and $N^{\sigma}_{1(1,2)}$ denote the excess adsorption isotherms for 1-st component from binary liquid mixtures on homogeneous and heterogeneous surfaces, respectively; n^{s} and N^{s} are the total number of moles in the adsorbed phase for a homogeneous surface patch and for the whole surface. Equation (71) was solved for the local adsorption isotherm

$$y_{1(1,2)} = \frac{c_{12}x_{12}}{1 + c_{12}x_{12}}; \qquad c_{12} \approx \exp(\varepsilon_{12}/kT) \tag{75}$$

which is a special case of Eq. (69), and by using the different distribution functions (Ościk *et al.*, 1975-1977). Ościk *et al.* (1975) and Jaroniec (1976e), show that Eq. (71) with the local adsorption isotherm (75) relates to Eq. (3) with the Langmuir equation (2).

Thus, the analytical expressions for the mole fraction $Y_{1(1,2)}$ may be obtained from the equations of adsorption isotherms for single gases by replacing the adsorbate pressure by the variable x_{12} (Ościk *et al.*, 1975, 1976). In this way, the Freundlich and Dubinin - Radushkevich equations, known in adsorption from gaseous phase, have been generalized.

The thermodynamics of adsorption from multicomponent liquid mixtures on heterogeneous solid surfaces have been recently discussed by Borówko *et al.* (1979). This treatment is based on the thermodynamics of monolayer localized adsorption from multicomponent gas mixtures (Jaroniec, 1977a). The generalization of the canonical partition function (67) to adsorption on heterogeneous solid surfaces gives

$$Q_{(\boldsymbol{n};\boldsymbol{r})} = \prod_{\boldsymbol{r}} \frac{B_{\boldsymbol{r}}!}{\left(\prod_{i=1}^{n-1} N_{i,\boldsymbol{r}}!\right)\left(B_{\boldsymbol{r}} - \sum_{i=1}^{n-1} N_{i,\boldsymbol{r}}\right)!} \times \left[\prod_{i=1}^{n-1}\left(q^{1}_{i,r_i}\right)^{N_{i,\boldsymbol{r}}}\right]\left(q^{1}_{n,r_n}\right)^{B_{\boldsymbol{r}} - \sum_{i=1}^{n-1} N_{i,\boldsymbol{r}}} \tag{76}$$

Equation (76) generates the following adsorption isotherm

$$Y_{i(\boldsymbol{n})} = \int_{\Delta_{(\boldsymbol{n})}} y_{i(\boldsymbol{n})}(\boldsymbol{x},\boldsymbol{\varepsilon})\,\chi_{(\boldsymbol{n})}(\boldsymbol{\varepsilon})\,d\boldsymbol{\varepsilon} \quad \text{for } i=1,\ 2,\ \ldots,\ n-1 \tag{77}$$

where

$$\sum_{j=1}^{n} Y_{j(\boldsymbol{n})} = 1 \; ; \qquad \sum_{j=1}^{n} y_{j(\boldsymbol{n})} = 1 \; ; \qquad \sum_{j=1}^{n} x_j = 1$$

and $\mathbf{x} = (x_1, x_2, \ldots, x_n)$; this equation is formally identical with Eq. (3). The adsorption from solution has a competitive character, i.e., differences between adsorption energies play an important role in this process. It follows from Eq. (69) that the adsorption isotherm $y_{i(\boldsymbol{n})}$ is a function of $(n-1)$-dimensional vectors: $\mathbf{x}^* = (x_{1n}, x_{2n}, \ldots, x_{n-1,n})$ and $\boldsymbol{\varepsilon}^* = (\varepsilon_{1n}, \varepsilon_{2n}, \ldots, \varepsilon_{n-1,n})$. Introducing a new distribution function:

$$X(\boldsymbol{\varepsilon}^*) = \int_{\Delta(n)} \chi_{(\boldsymbol{n})}(\boldsymbol{\varepsilon}^*, \varepsilon_n)\, d\varepsilon_n \tag{78}$$

into Eq. (77), we obtain

$$Y_{i(\boldsymbol{n})} = \int_{\Omega_{(\boldsymbol{n}^*)}} \frac{c_{in} x_{in}}{1 + \mathbf{c}^* \mathbf{x}^*} X_{(\boldsymbol{n}^*)}(\boldsymbol{\varepsilon}^*)\, d\boldsymbol{\varepsilon}^* \tag{79}$$

where $X_{(\boldsymbol{n}^*)}(\boldsymbol{\varepsilon}^*)$ is the $(n-1)$-dimensional distribution function normalized to unity,

$$\mathbf{c}^* = (c_{1n}, c_{2n}, \ldots, c_{n-1,n})$$

$$\mathbf{x}^* = (x_{1n}, x_{2n}, \ldots, x_{n-1,n}) \tag{80}$$

are $(n-1)$-dimensional vectors, and $\Omega_{(\boldsymbol{n}^*)} = \Omega_{1n} \times \Omega_{2n} \times \ldots \times \Omega_{n-1,n}$ is $(n-1)$-dimensional integration region.

Concluding the above, it can be stated that both excess and individual adsorption isotherms for n-component ideal and nonideal liquid mixtures on heterogeneous surfaces may be represented by means of the integral (77). In this treatment, the heterogeneity of the adsorbent surface is characterized by an n-dimensional distribution function $\chi_{(\boldsymbol{n})}(\boldsymbol{\varepsilon})$. For the case of adsorption without lateral interactions in the adsorbed phase, when Eq. (69) may be used to describe the local adsorption on a homogeneous surface patch, an $(n-1)$-dimensional distribution function of differences of adsorption energies (cf. Eq. (78)) may be introduced to characterize the adsorbent heterogeneity. Then the adsorption from n-component liquid mixtures is described in $(n-1)$-dimensional space (cf. Eq. (79)). For example, the adsorption isotherm for binary liquid mixtures is represented by a single integral (cf. Eq. (71)). From formal analogy between the Langmuir equation for single gases (2) and the isotherm (75) for binary liquid mixtures, it shows that a similar analogy exists between Eq. (11) for n-component gas mixtures and Eq. (79) for $(n-1)$-component liquid mixtures.

The adsorption from multicomponent liquid mixtures on solids is one of the very little investigated problems of physical adsorption, therefore further studies in this direction can be expected.

VI. APPROXIMATE MODELS FOR ADSORPTION OF MIXTURES WITH LATERAL INTERACTIONS

A. Bragg - Williams Approximation

In contrast to the extensive literature on adsorption from gas and liquid mixtures without lateral interactions on solids, particularly on homogeneous solid surfaces, there has been comparatively little work done on the simultaneous adsorption of two or more components with lateral interactions. Moreover, these studies concern homogeneous solid surfaces. Assumption of heterogeneity of the adsorbent surface considerably complicates the theoretical considerations.

The thermodynamics of monolayer adsorption of mixtures of heterogeneous solid surfaces, taking into account lateral interactions, may be easily formulated in terms of the Bragg - Williams approximation (BWA) (1934). The BWA is a crude approximation; therefore it may be used to describe adsorption systems with small effects of adsorbent heterogeneity and lateral interactions. For the purpose of the formulation of adsorption from mixtures on heterogeneous surfaces in terms of the BWA, we introduce the following notation.

Let the symbols Q^{m} and Q^{l} denote the canonical partition functions for mobile and localized adsorption without lateral interactions, respectively; Q^{in} is a term of the canonical partition function describing the effects of lateral interactions in the adsorbed phase; the subscript $(\boldsymbol{n})$ at these symbols refers to adsorption from n-component mixtures; the subscript $(\boldsymbol{n})\boldsymbol{r}$ at these symbols refers to adsorption from n-component mixtures on $\boldsymbol{r}$th type of adsorption sites; $Q^{m,p}$ and $Q^{m,r}$ refer to mobile adsorption, without lateral interactions, on the surface showing patchwise and random distribution of adsorption sites (cf. Section III.A). The canonical partition function Q^{in} describes the average effects of lateral interactions on a whole surface (RM surfaces), or on a given surface patch (PM surfaces) (cf. Section I.B). According to this notation, the equations of canonical partition functions for adsorption from mixtures on different models of adsorbent surfaces are

(a) localized adsorption

$$Q^{l,in}_{(\boldsymbol{n})} = Q^{l}_{(\boldsymbol{n})}(\boldsymbol{L},B)\,Q^{in}_{(\boldsymbol{n})}(\boldsymbol{L},B) \qquad \text{for HM}$$

$$Q^{l,p,in}_{(\boldsymbol{n};\boldsymbol{r})} = \prod_{\boldsymbol{r}} Q^{l}_{(\boldsymbol{n})\boldsymbol{r}}(\boldsymbol{L}_{\boldsymbol{r}},B_{\boldsymbol{r}})\,Q^{in}_{(\boldsymbol{n})\boldsymbol{r}}(\boldsymbol{L}_{\boldsymbol{r}},B_{\boldsymbol{r}}) \qquad \text{for PM}$$

$$Q^{l,r,in}_{(\boldsymbol{n};\boldsymbol{r})} = Q^{in}_{(\boldsymbol{n})}(\boldsymbol{L},B)\prod_{\boldsymbol{r}} Q^{l}_{(\boldsymbol{n})\boldsymbol{r}}(\boldsymbol{L}_{\boldsymbol{r}},B_{\boldsymbol{r}}) \qquad \text{for RM}$$

(b) mobile adsorption

$$Q^{m,in}_{(\boldsymbol{n})} = Q^{m}_{(\boldsymbol{n})}(\boldsymbol{M},A)\,Q^{in}_{(\boldsymbol{n})}(\boldsymbol{M},A) \qquad \text{for HM}$$

$$Q^{m,p,in}_{(\boldsymbol{n};\boldsymbol{r})} = \prod_{\boldsymbol{r}} Q^{m,p}_{(\boldsymbol{n})\boldsymbol{r}}(\boldsymbol{M}_{\boldsymbol{r}},A_{\boldsymbol{r}})\,Q^{in}_{(\boldsymbol{n})\boldsymbol{r}}(\boldsymbol{M}_{\boldsymbol{r}},A_{\boldsymbol{r}}) \qquad \text{for PM}$$

$$Q^{m,r,in}_{(\boldsymbol{n};\boldsymbol{r})} = Q^{in}_{(\boldsymbol{n})}(\boldsymbol{M},A) \prod_{\boldsymbol{r}} Q^{m,r}_{(\boldsymbol{n})\boldsymbol{r}}(\boldsymbol{M}_{\boldsymbol{r}},A) \qquad \text{for RM}$$

(c) partially mobile adsorption for $\beta_s = \beta_1 = \beta_2 = \dots = \beta_n$

$$Q^{ml,in}_{(\boldsymbol{n})} = Q^{m}_{(\boldsymbol{n})}(\boldsymbol{M},B)\,Q^{l}_{(\boldsymbol{n})}(\boldsymbol{L},B)\,Q^{in}_{(\boldsymbol{n})}(\boldsymbol{M},\boldsymbol{L},B) \;;$$

$$\bigwedge_i M_i + L_i = N_i \qquad \text{for HM}$$

$$Q^{ml,p,in}_{(\boldsymbol{n};\boldsymbol{r})} = \prod_{\boldsymbol{r}} Q^{m,p}_{(\boldsymbol{n})\boldsymbol{r}}(\boldsymbol{M}_{\boldsymbol{r}},B_{\boldsymbol{r}})\,Q^{l}_{(\boldsymbol{n})\boldsymbol{r}}(\boldsymbol{L}_{\boldsymbol{r}},B_{\boldsymbol{r}})\,Q^{in}_{(\boldsymbol{n})\boldsymbol{r}}(\boldsymbol{M}_{\boldsymbol{r}},\boldsymbol{L}_{\boldsymbol{r}},B_{\boldsymbol{r}}) \;;$$

$$\bigwedge_i \bigwedge_{\boldsymbol{r}} M_{i,\boldsymbol{r}} + L_{i,\boldsymbol{r}} = N_{i,\boldsymbol{r}} \qquad \text{for PM}$$

$$Q^{ml,r,in}_{(\boldsymbol{n};\boldsymbol{r})} = Q^{in}_{(\boldsymbol{n})}(\boldsymbol{M},\boldsymbol{L},B) \prod_{\boldsymbol{r}} Q^{m,r}_{(\boldsymbol{n})\boldsymbol{r}}(\boldsymbol{M}_{\boldsymbol{r}},B)\,Q^{l}_{(\boldsymbol{n})\boldsymbol{r}}(\boldsymbol{L}_{\boldsymbol{r}},B_{\boldsymbol{r}}) \qquad \text{for RM}$$

where n-dimensional vectors $\boldsymbol{L}$, $\boldsymbol{L}_{\boldsymbol{r}}$, and $\boldsymbol{M}$, $\boldsymbol{M}_{\boldsymbol{r}}$ are defined as follows:

$$\boldsymbol{L} = (L_1, L_2, \dots, L_n)$$

$$\boldsymbol{L}_{\boldsymbol{r}} = (L_{1,\boldsymbol{r}}, L_{2,\boldsymbol{r}}, \dots, L_{n,\boldsymbol{r}})$$

$$\boldsymbol{M} = (M_1, M_2, \dots, M_n)$$

$$\boldsymbol{M}_{\boldsymbol{r}} = (M_{1,\boldsymbol{r}}, M_{2,\boldsymbol{r}}, \dots, M_{n,\boldsymbol{r}})$$

The thermodynamical functions may be derived by using these general expressions for the canonical partition functions. The analytical expressions for Q^{m}, Q^{l}, $Q^{m,p}$, $Q^{m,r}$, and Q^{in} may be constructed in an analogous way to that discussed in Sections II-V. An interesting conclusion results from this discussion; that the application of the BWA for the surface showing patchwise distribution of adsorption sites (PM) is more exact from a theoretical point of view than for RM. For the case of PM, the function Q^{in} describes an average effect of the lateral interactions for a given homogenous surface patch, whereas for the case of RM this function characterizes the average effect of lateral interactions for the whole heterogeneous surface.

B. Quasichemical Approximation

A relatively simple and somewhat more accurate treatment of the effect of lateral interactions on the properties of a lattice gas is embodied in the quasichemical approximation. The known monographs devoted to the physical adsorption contain sections on quasichemical approximations (Young and Crowell, 1962; Steele, 1974; Clark, 1970).

In this section the literature concerning a special case of the quasichemical adsorption model, i.e., the model of associated adsorbate (AA), will be summarized. This model is characterized by its great universality and comparatively simple mathematical formalism. In this model, the adsorbent-adsorbate and adsorbate-adsorbate interactions and formations of multilayers are represented by suitable quasichemical reactions. Therefore, the Langmuir equation is derived by means of the equilibrium constant characterizing the quasichemical reaction of a molecule with a free site on the adsorbent surface.

The BET equation is obtained by assuming the formation of linear chains of associates normal to the surface. However, the adsorbate-adsorbate interactions can be represented by suitable quasichemical reactions giving associates parallel to the surface.

The AA model was first applied by Kiselev (1958) and later by Berezin and Kiselev (1972a) to describe the localized adsorption with adsorbate-adsorbate interactions on homogeneous surfaces. Their treatment showed the good agreement of the theoretical results with the experimental ones (Berezin and Kiselev, 1972b). Critical discussions of this model have been made by Karpiński and Garbacz (1974) and later by Garbacz (1975, 1977).

Recently, Jaroniec (1976f, 1978e) and Jaroniec *et al.* (1977) proposed a generalization of the AA model to the localized adsorption of gas mixtures on homogeneous surfaces. This problem has also been discussed by Dent (1977). However, Bernache-Assollant and Thomas (1975, 1976) considered adsorption of gas and gas mixtures by zeolites in terms of the quasichemical model. Their treatment concerns both localized and mobile adsorption. The detailed discussion of mobile adsorption in terms of the AA model has been presented by Berezin and Kiselev (1974) and Garbacz (1975).

Particularly interesting results were obtained for the adsorption of single gases on heterogeneous surfaces (Jaroniec and Borówko, 1978, Borówko *et al.*, 1977). Assuming a discrete distribution of adsorption energy, a general equation describing the adsorption equilibrium has been derived (Jaroniec and Borowko, 1978). This equation was then used to calculate the adsorption on surfaces having patchwise, random, and regular distribution of adsorption sites. Such numerical calculations can give information on the sequence of the filling of adsorption sites and the contribution of adsorption on the particular kinds of sites in relation to the overall adsorption for different models of heterogeneous surfaces.

For further studies on the AA model, detailed interpretation on the basis of statistical thermodynamics would be useful. For this purpose, papers concerning three-dimensional associated fluids should be very useful (Heidemann and Prausnitz, 1976).

LIST OF SYMBOLS

A: total area of an adsorbent surface

A_f: free area available to molecules which are adsorbed

a: Henry's adsorption constant

B: total number of adsorption sites

c: constant characterizing adsorption equilibrium of a binary liquid mixture

F: ratio of mobile and localized adsorbed molecules

f^l, f^m: fractions of localized and mobile molecules on a homogeneous surface

$\hat{f}^l$, $\hat{f}^m$: fractions of localized and mobile molecules on a heterogeneous surface

K^*: ratio of the constants a^l and a^m

k: Boltzmann constant

L: number of localized adsorbed molecules

M: number of mobile adsorbed molecules

N: total number of adsorbed molecules

N_m: capacity of the adsorbed phase

N^s: total number of moles in surface phase for liquid adsorption on heterogeneous surfaces

N^σ: excess adsorption isotherm for a heterogeneous surface

n^s: total number of moles in surface phase for liquid adsorption on homogeneous surfaces

n^σ: excess adsorption isotherm for a homogeneous surface

$N^{\sigma s}$: N^σ/N^s

$n^{\sigma s}$: n^σ/n^s

p: adsorbate pressure

Q: canonical partition function

$Q_{(\boldsymbol{n})}$: canonical partition function for n-component mixtures and a homogeneous surface

$Q_{(\boldsymbol{n};\boldsymbol{r})}$: canonical partition function for n-component mixtures and a heterogeneous surface *q*: partition function $\hat{q}$ multiplied by the exponential factor of adsorption energy

$\hat{q}$: partition function of isolated adsorbed molecule

$\tilde{q}$: $\hat{q}$ multiplied by the exponential factor of the standard chemical potential of a gas

r_o: number types of adsorption sites on a surface

s: selectivity coefficient for adsorption of two gases

T: absolute temperature

X: distribution function of differences of adsorption energies

x: mole fraction of a given component in the bulk phase

Y: mole fraction of a given component in the surface phase for liquid adsorption on a heterogeneous surface

y: mole fraction of a given component in the surface phase for liquid adsorption on a homogeneous surface

z: variable introduced in Eq. (21)

Greek Symbols

α: two-dimensional analog of the van der Waals constant

β: area occupied by one adsorbed molecule

β_s: area of an adsorption site

γ: activity coefficient

Δ: integration region for adsorption energy

η: difference of two adsorption energies, Eq. (20)

Θ: relative coverage of a heterogeneous surface

θ: relative coverage of a homogeneous surface

θ^*: function introduced in Eq. (18)

κ: ratio of two constants β_i and β_j

μ^{oG}: standard chemical potential of a gas

ν: β_s/β

ξ: constant defined by Eq. (22)

Φ: function characterizing lateral interactions in the surface phase

ϕ: height of energy barrier

χ: energy distribution function

ψ: function introduced in Eq. (16)

Ω: integration region for differences of adsorption energies

Vectors

$\boldsymbol{a} = (a_1, a_2, \ldots, a_n)$

$\boldsymbol{c}^* = (c_{1n}, c_{2n}, \ldots, c_{n-1,n})$

$\boldsymbol{n} = (1, 2, \ldots, n)$

$\boldsymbol{n}^* = (1, 2, \ldots, n-1)$

$\boldsymbol{p} = (p_1, p_2, \ldots, p_n)$

$\boldsymbol{r} = (r_1, r_2, \ldots, r_n)$

$\boldsymbol{x}^* = x_{1n}, x_{2n}, \ldots, x_{n-1,n})$

$\boldsymbol{\varepsilon} = (\varepsilon_1, \varepsilon_2, \ldots, \varepsilon_n)$

$\boldsymbol{\varepsilon}^* = (\varepsilon_{1n}, \varepsilon_{2n}, \ldots, \varepsilon_{n-1,n})$

Superscripts

in: lateral interactions

l: localized adsorption

m: mobile adsorption

ml: partially mobile adsorption

p: patchwise surfaces

r: random surfaces

GL: gaslike structure

SL: solidlike structure

Subscripts

i: ith component

j: jth component

ij: the ratio of two quantities in the case of x, a, c, γ, and κ or to the difference of two adsorption energies in the case of ε

$i(\boldsymbol{n})$: adsorption of the ith component from n-component mixture

i,r_i: the quantities associated with adsorption energy ,

$i,\boldsymbol{r}$: adsorption of the ith component on rth surface patch

$(\boldsymbol{n})$: n-component mixtures

$(\boldsymbol{n}^*)$: $(n-1)$-component mixtures

$(\boldsymbol{n})\boldsymbol{r}$: adsorption of n-component mixtures on $\boldsymbol{r}$th surface patch

$\boldsymbol{r}$: $\boldsymbol{r}$th surface patch

Abbreviations

AA: model of associated adsorbate

BWA: Bragg - Williams approximation

HM: model of a homogeneous surface

PM: model of a patchwise surface

RM: model of a random surface

MM: model of a surface having medial distribution of adsorption sites on the surface

SST: significant structures theory

REFERENCES

Adamson, A. W., and Ling, J. (1961). *Adv. Chem. 33,* 51.

Adamson, A. W., Ling, L., Dormant, L., and Orem, M. (1966). *J. Colloid Interface Sci. 21,* 445.

Adler, Yu. P., Gryazev, N. N., Rakhlevskaya, M. N., and Rumyantseva, G. A. (1971). *Dokl. Akad. Nauk SSSR 200,* 1123.

Berezin, J. G., and Kiselev, A. V. (1972a). *J. Colloid Interface Sci. 38,* 227.

Berezin, J. G., and Kiselev, A. V. (1972b). *J. Colloid Interface Sci. 38,* 335.

Berezin, J. G., and Kiselev, A. V. (1974). *J. Colloid Interface Sci. 46,* 203.

Bering, B. P., and Serpinsky, V. V. (1952). *Izv. Akad. Nauk SSSR,* 997.

Bering, B. P., and Serpinsky, V. V. (1953). *Izv. Akad. Nauk SSSR, Otd. Khim.,* 37.

Bering, B. P., and Serpinsky, V. V. (1961). *Izv. Akad. Nauk SSSR, Otd. Khim.,* 1947.

Bering, B. P., and Serpinsky, V. V. (1969). *Izv. Akad. Nauk SSSR, Otd. Khim.,* 1222.

Bering, B. P., and Serpinsky, V. V. (1972a). *Izv. Akad. Nauk SSSR, Otd. Khim.,* 169.

Bering, B. P., and Serpinsky, V. V. (1972b). *Izv. Akad. Nauk SSSR, Otd. Khim.,* 171.

Bering, B. P., Serpinsky, V. V., and Surinova, S. J. (1973). *Izv. Akad. Nauk SSSR, Ser. Khim., 22,* 3.

Bering, B. P., Serpinsky, V. V., and Jakubov, T. S. (1977a). *Izv. Akad. Nauk SSSR, Otd. Khim.,* 727.

Bering, B. P., Serpinsky, V. V., and Jakubov, T. S. (1977b). *Izv. Akad. Nauk SSSR, Otd. Khim.,* 991.

Bernache-Assolant, D., and Thomas, G. (1975). *J. Chim. Phys. 72,* 1241.

Bernache-Assolant, D., and Thomas, G. (1976). *J. Chim. Phys. 73,* 967, 975.

Borówko, M., Jaroniec, M., and Rudzinski, W. (1977). *Thin Solid Films 46,* 239.

Borówko, M., Jaroniec, M., and Rudziński, W. (1979). *Colloid Polymer Sci.,* in press.

Bragg, W. L., and Williams, E. J. (1934). *Proc. Roy. Soc. A154,* 699.

Brown, C. E., and Everett, D. H. (1975). *In* "Colloid Science" (D. H. Everett, ed.) Vol. 2, pp. 52-100. Specialist Periodical Reports, The Chemical Society, London.

Bülow, M., and Schirmer, W. (1972). *Z. Chemie 12,* 161.

Cerofolini, G. F. (1971). *Surface Sci. 24,* 391.

Cerofolini, G. F. (1972). *J. Low Temperature Phys. 6,* 473.

Cerofolini, G. F. (1974). *Thin Solid Films 23,* 129.

Cerofolini, G. F. (1975a). *Vuoto (Italy) 8,* 178.

Cerofolini, G. F. (1975b). *Surface Sci. 51,* 333.

Cerofolini, G. F. (1975c). *Thin Solid Films 26,* 53.

Cerofolini, G. F. (1975d). *Surface Sci. 47,* 469.

Cerofolini, G. F. (1975e). *Surface Sci. 52,* 195.

Cerofolini, G. F. (1976a). *J. Low Temperature Phys. 23,* 687.

Cerofolini, G. F. (1976b). *Surface Sci. 61,* 678.

Cerofolini, G. F. (1978). *Z. Phys. Chem. 259,* 314.

Clark, A. (1970). "The Theory of Adsorption and Catalysis." Academic Press, New York and London.

Coltharp, M. T., and Hackerman, N. (1973). *J. Colloid Interface Sci. 43,* 176, 185.

Dash, J. G. (1975). "Films on Solid Surfaces." Academic Press, New York and London.

Delmas, G., and Patterson, D. (1960). *J. Phys. Chem. 64,* 1827.

Dent, R. W. (1977). *Textile Res. J. 47,* 145, 188.

Doll, J. J., and Steele, W. A. (1974). *Surface Sci. 44,* 449.

Dondi, F., Gonnord, M. F., and Guiochon, G. (1977). *J. Colloid Interface Sci. 62,* 316.

Dormant, L. M., and Adamson, A. W. (1972). *J. Colloid Interface Sci. 38,* 285.

Drenan, J. W., and Hill, T. L. (1950). *J. Phys. Chem. & Colloid Chem. 54,* 1132.

Dubinin, M. M. (1972). "Adsorption-Desorption Phenomena" (F. Ricca, ed.), pp. 3-19. Academic Press, London and New York.

Dubinin, M. M., and Radushkevich, L. V. (1947). *Dokl. Akad. Nauk SSSR 55,* 331.

Dubinin, M. M., and Serpinsky, V. V. (Editors), (1972). "Physical Adsorption from Multicomponent Phases.: Nauka, Moscow.

Dubinin, M. M., Bering, B. P., and Serpinsky, V. V. (1964). *In* "Progress in Surface and Membrane Science," Vol. 2, pp. 1-45. Academic Press, New York and London.

Duval, X., and Thomy, A. (1975). *Carbon 13,* 242.

Everett, D. H. (1964). *Trans. Faraday Soc. 60,* 1803.

Everett, D. H. (1965). *Trans. Faraday Soc. 61,* 2478.

Everett, D. H. (1973). *In* "Colloid Science" (D. H. Everett, ed.), Vol. 1, Chap. 2. Specialist Periodical Reports, The Chemical Society, London.

Eyring, H., Ree, T., and Hirai, N. (1958). *Proc. Natl. Acad. Sci. US 44,* 683.

Fowler, R. H., and Guggenheim, E. A. (1939). "Statistical Thermodynamics." Cambridge, New York.

Friederich, R. O., and Mullins, J. C. (1972). *Ind. Eng. Chem. Fundamentals 11,* 439.

Garbacz, J. (1975). *J. Colloid Interface Sci. 51,* 352.

Garbacz, J. (1977). *Ann. Soc. Chim. Polonorum 51,* 1421, 2375.

Garbacz, J., and Karpiński, K. (1975). *Ann. Soc. Chim. Polonorum 49,* 597.

Glueckauf, E. (1953). *Trans. Faraday Soc. 49,* 1066.

Gryazev, N. N. (1958). *Dokl. Akad. Nauk SSSR 118,* 317.

Halsey, G. D. (1948). *J. Chem. Phys. 16,* 931.

Halsey, G. D. (1952). *Adv. Catal. 4,* 259.

Hansen, R. S., and May, U. H. (1957). *J. Phys. Chem. 61*, 573.

Hansen, R. S., and Stage, D. V. (1956). *Iowa State Coll. J. Sci. 31*, 33.

Harris, L. B. (1968). *Surface Sci. 10*, 129.

Harris, L. B. (1969a). *Surface Sci. 13*, 377.

Harris, L. B. (1969b). *Surface Sci. 15*, 187.

Heidemann, R. A., and Prausnitz, J. M. (1976). *Proc. Natl. Acad. Sci. USA 73*, 1773.

Hill, T. L. (1946). *J. Chem. Phys. 14*, 441.

Hill, T. L. (1949a). *J. Chem. Phys. 17*, 762.

Hill, T. L. (1949b). *J. Chem. Phys. 17*, 520.

Hill, T. L. (1952). *Adv. Catal. 4*, 211.

Hobson, J. P. (1965). *Can. J. Phys. 43*, 1934.

Hobson, J. P. (1974). *CRC Crit. Rev. Solid State Sci. 4*, 221.

Holland, B. W. (1965a). *Trans. Faraday Soc. 61*, 546.

Holland, B. W. (1965b). *Trans. Faraday Soc. 61*, 555.

Honig, J. M., and Rosenbloom, P. C. (1955). *Can. J. Chem. 33*, 193.

Hoory, S. E., and Prausnitz, J. M. (1967). *Chem. Eng. Sci. 22*, 1025.

House, W. A., and Jaycock, J. M. (1974). *J. Colloid Interface Sci. 47*, 50.

House, W. A., and Jaycock, J. M. (1977a). *J. Colloid Interface Sci. 59*, 252.

House, W. A., and Jaycock, J. M. (1977b). *Trans. Faraday Soc. 73*, 942.

House, W. A., and Jaycock, J. M. (1978). *Colloid Polymer Sci. 256*, 52.

Hsieh, P. Y. (1964). *J. Phys. Chem. 68*, 1068.

Hsu, C. C., Rudziński, W., and Wojciechowski, B. W. (1975). *Chromatographia 8*, 633.

Ivanov, J. B., and Martinov, A. (1973). *In* "Physical Adsorption from Multicomponent Phases" (Dubinin, M. M., and Serpinsky, V. V., ed.), pp. 191-201, Nauka, Moscow.

Jackson, D. J., and Davis, B. W. (1974). *J. Colloid Interface Sci. 47*, 499.

Jaroniec, M. (1975a). *Surface Sci. 50*, 553.

Jaroniec, M. (1975b). *Chemickie Zvesti 29*, 512.

Jaroniec, M. (1975c). *J. Colloid Interface Sci. 53*, 422.

Jaroniec, M. (1975d). *J. Colloid Interface Sci. 52*, 41.

Jaroniec, M. (1976a). *Colloid Polymer Sci. 254*, 601.

Jaroniec, M. (1976b). *Phys. Lett. 56A*, 53.

Jaroniec, M. (1976c). *Z. Phys. Chem. 257*, 449.

Jaroniec, M. (1976d). *Vuoto (Italy) 9*, 57.

Jaroniec, M. (1976e). *J. Low Temperature Phys. 24*, 253.

Jaroniec, M. (1976f). *Vuoto (Italy 9*, 36.

Jaroniec, M. (1977a). *Trans. Faraday Soc. II 73*, 933.

Jaroniec, M. (1977b). *J. Colloid Interface Sci. 59*, 371.

Jaroniec, M. (1977c). *Colloid Polymer Sci. 255*, 176.

Jaroniec, M. (1977d). *Colloid Polymer Sci. 255*, 32.

Jaroniec, M. (1977e). *J. Colloid Interface Sci. 59*, 230.

Jaroniec, M. (1978a). *Thin Solid Films 50*, 163.

Jaroniec, M. (1978b). *Vacuum 28*, 17.

Jaroniec, M. (1978c). *Rect. Kinet. Catal. Letters 8*, 425.

Jaroniec, M. (1978d). *Trans. Faraday Soc. II 74*, 1292.

Jaroniec, M. (1978e). *Colloid Polymer Sci. 256*, 1089.

Jaroniec, M., and Borówko, M. (1977). *Surface Sci. 66*, 652.

Jaroniec, M., and Borówko, M. (1978). *J. Colloid Interface Sci. 63*, 362.

Jaroniec, M., and Rudziński, W. (1975a). *Colloid Polymer Sci. 253*, 683.

Jaroniec, M., and Rudziński, W. (1975b). *Surface Sci. 52*, 641.

Jaroniec, M., and Rudziński, W. (1975c). *Phys. Lett. 53A*, 59.

Jaroniec, M., and Rudziński, W. (1976). *Acta Chim. Hung. 88*, 351.

Jaroniec, M., and Rudziński, W. (1977). *J. Catal.* (Hokkaido Univ.) 25, 197.

Jaroniec, M., and Tóth, J. (1976a). *Acta Chim. Hung. 91*, 153.

Jaroniec, M., and Tóth, J. (1976b). *Colloid Polymer Sci. 254*, 643.

Jaroniec, M., and Tóth, J. (1977). *Acta Chim. Hung. 94,* 35.

Jaroniec, M., Borówko, M., and Rudziński, W. (1977). *AIChE J. 23,* 605.

Jaroniec, M., Narkiewicz, J., Borówko, M., and Rudziński, W. (1978a). *J. Colloid Interface Sci. 65,* 9.

Jaroniec, M., Narkiewicz, J., Borówko, M., and Rudziński, W. (1978b). *Polish J. Chem. 52,* 197.

Jaroniec, M., Rudziński, W., Sokołowski, S., and Smarzewski, R. (1975). *Colloid Polymer Sci. 253,* 164.

Jaroniec, M., Sokołowski, S., and Cerofolini, G. F. (1976). *Thin Solid Films 31,* 321.

Jaroniec, M., Sokołowski, S., and Rudziński, W. (1978). *Polish J. Chem. 52,* 395.

Jaycock, J. M., and Waldsax, J. C. R. (1971). *J. Colloid Interface Sci. 37,* 462.

Jossens, S., Prausnitz, J. M., Fritz, W., Schlünder, E. U., and Myers, A. L. (1978). *Chem. Eng. Sci. 33,* 1097.

Jovanovic, D. S. (1969). *Colloid Polymer Sci. 235,* 1203.

Karpiński, K., and Garbacz, J. (1974). *Ann. Soc. Chim. Polonorum 48,* 803.

Kemball, C., Rideal, E. K., and Guggenheim, E. A. (1948). *Trans. Faraday Soc. 44,* 948.

Kipling, J. J. (1965). "Adsorption from Solutions of Nonelectrolytes." Academic Press, New York.

Kiselev, A. V. (1958). *Kolloid Zh. SSSR 20,* 338.

Kiselev, A. V., and Pavlova, L. E. (1962). *Izv. Akad. Nauk SSSR, Otd. Khim. Nauk* 2121.

Kiselev, A. V., and Pavlova, L. E. (1965). *Izv. Akad. Nauk SSSR, Otd. Khim. Nauk* 18.

Kisluk, M. U., Sklarov, T. M., and Dangan, T. M. (1975). *Izv. Akad. Nauk SSSR, Ser. Khim.* 2161.

Langmuir, J. (1918). *J. Am. Chem. Soc. 40,* 1361.

Larionov, O. G., and Myers, A. L. (1971). *Chem. Eng. Sci. 26,* 1025.

Larionov, O. G., Chmutov, K. V., and Yudilevich, M. D. (1967). *Zh. Fiz. Khim. SSSR 41,* 1011, 2616.

Lee, C. S., and O'Connell, J. P. (1972). *J. Colloid Interface Sci. 41,* 415.

Lee, C. S., and O'Connell, J. P. (1974). *Ind. Eng. Chem. Fundamentals 13,* 165.

Lee, C. S., and O'Connell, J. P. (1975). *J. Phys. Chem. 79,* 885.

Lennard-Jones, J. E., and Devonshire, A. F. (1937). *Proc. Roy. Soc. A163,* 53.

Łajtar, L., Jaroniec, M., and Sokołowski, S. (1978). *Thin Solid Films 50,* L21.

McAlpin, J. J., and Pierotti, R. A. (1964). *J. Chem. Phys. 41,* 68.

McAlpin, J. J., and Pierotti, R. A. (1965). *J. Chem. Phys. 42,* 1842.

Martynov, Yu. M., and Maevskaya, B. M. (1969). *Zh. Fiz. Khim. 43,* 1255.

Martynov, Yu. M., and Maevskaya, B. M. (1972). *Zh. Fiz. Khim. 46,* 1265.

Matecki, M., Thomy, A., and Duval, X. (1974a). *J. Chim. Phys. 71,* 1484.

Matecki, M., Thomy, A., and Duval, X. (1974b). *C. R. Acad. Sci. Paris C278,* 647.

Matecki, M., Thomy, A., and Duval, X. (1977). *Surface Sci. 69,* 596.

Menaucourt, J., Thomy, A., and Duval, X. (1977). *J. Physique C4-38,* 195.

Minka, C., and Myers, A. L. (1973). *AIChE J. 19,* 453.

Misra, D. N. (1969). *Surface Sci. 18,* 367.

Misra, D. N. (1970). *J. Chem. Phys. 52,* 5499.

Misra, D. N. (1973). *J. Colloid Interface Sci. 43,* 85.

Myers, A. L., and Prausnitz, J. M. (1965). *AIChE J. 11,* 121.

Myers, A. L., and Sircar, S. (1972). *J. Phys. Chem. 76,* 3412, 3415.

Myers, A. L., and Sircar, S. (1973). *AIChE J. 19,* 159.

Nasonov, P. M. (1971). *Zh. Fiz. Khim. SSSR 45,* 1593.

Nasonov, P. M. (1972). *Zh. Fiz. Khim. SSSR 46*, 387.

Nodzeński, A. (1976). *Wiadomosci Chem. 30*, 387.

Ościk, J. (1961a). *Przemysł Chem. 40*, 279.

Ościk, J. (1961b). *Bull. Acad. Polon. Sci. 9*, 29 and 33.

Ościk, J. (1965). *Przemysł Chem. 44*, 129.

Ościk, J. (1972). *In* "Physical Adsorption from Multicomponent Phases" (M. M. Dubinin, ed.), p. 138. Nauka, Moscow (in Russian).

Ościk, J., Rudzinski, W., and Dąbrowski, A. (1974). *Ann. Soc. Chim. Polonorum 48*, 1991.

Ościk, J., Dąbrowski, A., Sokołowski, S., and Jaroniec, M. (1975). *J. Catal. (Hokkaido Univ.) 23*, 91.

Ościk, J., Dąbrowski, A., Jaroniec, M., and Rudziński, W. (1976). *J. Colloid Interface Sci. 56*, 403.

Ościk, J., Dąbrowski, A., and Rudziński, W. (1977). *Colloid Polymer Sci. 255*, 50.

Patrykiejew, A., and Jaroniec, M. (1978). *Surface Sci. 77*, 365.

Perfetti, G. A., and Wightman, J. P. (1975). *Carbon 13*, 473.

Perfetti, G. A., and Wightman, J. P. (1976). *J. Colloid Interface Sci. 55*, 252.

Pierotti, R. A. (1971). "Physical Adsorption: The Interaction of Gases with Solids." Wiley, New York.

Pierotti, R. A., and Thomas, H. E. (1973). *Trans. Faraday Soc. 70*, 1725.

Ponec, V., Knor, Z., and Cerny, S. (1974). "Adsorption on Solids." Butterworths, London.

Prasad, S. D., and Doraiswamy, L. K. (1977). *Phys. Lett. 60A*, 11.

Radke, C. J., and Prausnitz, J. M. (1972). *AIChE J. 18*, 761.

Rakhlevskaya, M. N., Adler, Yu. P., Gryazev, M. N., and Rumyantseva, G. A. (1971). *Zh. Fiz. Khim. SSSR 45*, 2616.

Regnier, J., Rouquerol, J., and Thomy, A. (1975). *J. Chim. Phys. 72*, 327.

Regnier, J., Thomy, A., and Duval, X. (1977). *J. Chim. Phys. 74*, 926.

Roginsky, S. Z. (1948). "Adsorption and Catalysis on Heterogeneous Surfaces." Izd. Akad. Nauk SSSR, Moscow (in Russian).

Roginsky, S. Z., and Todes, O. M. (1945a). *Acta Physicochim. URSS 20,* 307.

Roginsky, S. Z., and Todes, O. M. (1945b). *Acta Physicochim. URSS 20,* 696.

Roginsky, S. Z., and Todes, O. M. (1946). *Acta Physicochim. URSS 21,* 519.

Ross, S. (1971). *In* "Progress in Surface and Membrane Sci," Vol. 4, p. 386.

Ross, S., and Morrison, J. D. (1973). *Surface Sci. 52,* 103.

Ross, S., and Olivier, J. P. (1964). "On Physical Adsorption." Wiley (Interscience), New York.

Rudziński, W., and Jaroniec, M. (1974). *Surface Sci. 42,* 552.

Rudziński, W., and Jaroniec, M. (1975). *Ann. Soc. Chim. Polonorum 49,* 165.

Rudziński, W., and Wojciechowski, B. W. (1977). *Colloid Polymer Sci. 255,* 870.

Rudziński, W., Oscik, J., and Dabrowski, A. (1973). *Chem. Phys. Lett. 20,* 444.

Rudziński, W., Jaroniec, M., Sokołowski, S., and Cerofolini, G. F. (1975). *Czech. J. Phys. 25B,* 891.

Rudziński, W., Łajtar, L., and Patrykiejew, A. (1977). *Surface Sci. 67,* 195.

Rudziński, W., Waksmundzki, A., Leboda, R., Suprynowicz, Z., and Lasoń, M. (1974). *J. Chromatogr. 92,* 25.

Rusanov, A. J. (1971). *In* "Progress in Surface and Membrane Science," Vol. 4, pp. 57-114. Academic Press, New York and London.

Schay, G. (1956). *J. Chim. Phys. 53,* 691.

Schay, G. (1969). *In* "Surface and Colloid Science" (E. Matijevic, ed.), Vol. 2. Wiley (Interscience), New York.

Schay, G. (1976). *Pure Appl. Chem. 48,* 393.

Schay, G., Fejes, P., and Szathmary, J. (1957). *Acta Chim. Hung.* *12*, 299.

Schay, G., and Nagy, L. G. (1961). *J. Chim. Phys.* *58*, 149.

Schay, G., and Nagy, L. G. (1972). *J. Colloid Interface Sci.* *38*, 302.

Schuchovitzky, A. (1938). *Acta Physicochim. URSS* *8*, 531.

Sircar, S., and Myers, A. L. (1970). *J. Phys. Chem.* *74*, 2828.

Sircar, S., and Myers, A. L. (1973). *Chem. Eng. Sci.* *28*, 489.

Sips, R. (1948). *J. Chem. Phys.* *16*, 490.

Sips, R. (1950). *J. Chem. Phys.* *18*, 1024.

Šiškova, M., and Erdos, E. (1960). *Coll. Czech. Chem. Commun.* *25*, 1729, 2599.

Sloan, E. D., and Mullins, J. C. (1975). *Ind. Eng. Chem. Fundamentals* *14*, 347.

Smutek, M. (1976). *Czech. J. Phys.* *B26*, 699.

Snagovsky, S. (1975). *Kinetika Kataliz* *17*, 1435.

Sokołowski, S., Jaroniec, M., and Cerofolini, G. F. (1975). *Surface Sci.* *47*, 429.

Stebbins, J. P., and Halsey, G. D. (1964). *J. Phys. Chem.* *68*, 3863.

Steele, W. A. (1961). *Adv. Chem. Ser.* *33*, 269.

Steele, W. A. (1963). *J. Phys. Chem.* *67*, 2016.

Steele, W. A. (1973a). *Surface Sci.* *36*, 317.

Steele, W. A. (1973b). *Surface Sci.* *39*, 149.

Steele, W. A. (1974). "The Interaction of Gases with Solid Surfaces." Pergamon Press, Oxford.

Steele, W. A. (1976). *CRC Crit. Rev. Solid State Sci.* *6*, 223.

Steele, W. A., and Ross, M. (1961). *J. Chem. Phys.* *35*, 850.

Szabo, Z. G., and Kall, D. (Editors), (1976). "Contact Catalysis," Vol. 1. Akadémiai Kiadó, Budapest.

Thomy, A., and Duval, X. (1969). *J. Chim. Phys.* *66*, 1966.

Thomy, A., and Duval, X. (1970a). *J. Chim. Phys.* *67*, 286.

Thomy, A., and Duval, X. (1970b). *J. Chim. Phys.* *67*, 1101.

Tiemkin, M. J. (1975). *Kinetika Kataliz* *17*, 1461.

Tompkins, F. C. (1950). *Trans. Faraday Soc. 46,* 569.

Tompkins, F. C., and Young, D. M. (1951). *Trans. Faraday Soc. 47,* 88.

Toth, J. (1970). *Acta Chim. Hung. 63,* 67 and 179.

Toth, J. (1974). *J. Colloid Interface Sci. 46,* 38.

Toth, J., Rudziński, W., Waksmundzki, A., Jaroniec, M., and Sokołowski, S. (1974). *Acta Chim. Hung. 82,* 11.

Tsien, F., and Halsey, G. D. (1967). *J. Phys. Chem. 71,* 4012.

Van Dongen, R. H. (1973). *Surface Sci. 39,* 341.

Van Dongen, R. H., and Broekhoff, J. C. P. (1969). *Surface Sci. 18,* 462.

Van Ness, H. C. (1969). *Ind. Eng. Chem. Fundamentals 8,* 464.

Volmer, M. (1925). *Z. Physik. Chem. 115,* 253.

Waksmundzki, A., Jaroniec, M., and Suprynowicz, Z. (1975). *J. Chromatogr. 110,* 381.

Waksmundzki, A., Sokołowski, S., Jaroniec, M., and Rayss, J. (1975). *Vuoto (Italy) 8,* 113.

Young, D. M., and Crowell, A. D. (1962). "Physical Adsorption of Gases," pp. 247-276. Butterworths, London.

STRUCTURE OF THE BOUNDARY LAYERS OF LIQUIDS AND ITS INFLUENCE ON THE MASS TRANSFER IN FINE PORES

B. V. Derjaguin
N. V. Churaev

Institute of Physical Chemistry
Academy of Sciences of the USSR
Moscow, USSR

PROGRESS IN SURFACE AND MEMBRANE
SCIENCE, VOL. 14

ISBN 0-12-571814-4

I. THEORETICAL CONSIDERATION

The structure of boundary layers is different from that of the bulk liquid. Changes in the specific enthalpy, viscosity, and the solute concentration in the boundary layers give rise to such phenomena as thermoosmosis, capillary osmosis, and filtration anomalies, even in wide pores. In thin liquid interlayers, the boundary layers with a modified structure overlap one another. Therefore, in fine porous bodies the phenomena of mass transfer are substantially complicated, which will be considered below.

The fact that the structure of liquid layers close to the solid substrate does change is in itself beyond debate. Computer solutions, obtained even for simple liquids on the basis of the model of rigid spheres with a spherically symmetrical potential of intermolecular forces, are indicative of the existence of structurally modified boundary layers. The latter are usually characterized by density oscillations that are damped with distance from the solid surface (1-5). In this case, variations in density extend to distances of several molecular diameters.

Unfortunately, sufficiently exact calculations are so far impossible for polar liquids. The theory of polar and especially associated liquids (such as water) is insufficiently developed. The molecular dynamical calculations of Stillinger and Rahman (6,7) carried out for bulk water are indicative only of the existence of a random network of strongly deformed intermolecular hydrogen bonds. The latter were taken into account by introducing (besides the Lennard-Johnson potential) an additional electrical interaction between the 16 pairs of point charges modeling the hydrogen bond. Thus, one succeeds in obtaining good agreement with the known experimental data for water.

It is evident that the formation of H-bonds between molecules of water and the active groups of the solid substrate must distort the network of H-bonds existing in the bulk water. In this case, owing to the directional action of the hydrogen bonds, the structural changes must be accompanied by the appearance of anisotropy and extend at a considerable distance from the substrate.

Calculations (6,7) show, also, that the processes of transfer in water are connected with the collective motion of neighboring molecules rather than with the molecules jumping from one state to another, as in the case of simple liquids. Surface forces can influence this collective motion, and also change flow conditions close to the substrate.

The influence of the surface is expressed still more markedly in the case of anisotropic liquids, i.e., liquid crystals. Calculations carried out on the basis of the macroscopic theory of molecular forces (8) enable one to draw the conclusion that there exists a certain critical thickness ℓ_c of the interlayer of a liquid crystal between solid surfaces: when the thickness of the interlayer ℓ is smaller than ℓ_c, the axes of molecules are oriented parallel to the surface, whereas at $\ell > \ell_c$ they are perpendicular to it. These boundary structures are attributed to the short range forces of interaction of the molecules of a liquid crystal with the substrate surface (as in the case of the H-bond). At a considerable thickness of the interlayer $\ell > 5.10^{-4}$ cm ($\ell >> \ell_c$), the theory predicts the existence of transient boundary layers having thickness δ; within the limits of transient layer the orientation of molecules changes from the surface orientation that is preset by short-range forces to the bulk one. It will be important to note that the values of ℓ_c and δ, which are characteristic of the theoretical extension of the boundary layers of a liquid crystal, are on the order of 10^{-5} to 10^{-4} cm. Experimental examinations of liquid crystals give thicknesses of the boundary layers on the order of 10^{-4} to 10^{-3} cm (9-11). The thickness of the boundary layers somewhat decreases under the effect of the viscous flow (11).

Thus, we have two limiting cases, namely, an isotropic simple liquid and an anisotropic liquid crystal that very easily changes its structure under the effect of weak fields. For these limiting cases, the thickness of the boundary layers changes from ~10 Å to 10^4-10^5 Å. It can be supposed that for the intermediate case of polar and associated liquids possible values of the thickness of the boundary layers are found within that range. As will be shown here, a reasonable assessment are values of the order of scores, and in a number of cases, possibly many hundred Å.

The absence of a strict theory for boundary layers of polar liquids and the large variety of experimental results obtained for them, result in there being no generally accepted point of view, not only relating to the thickness, but also to the significance of boundary layers in the physical chemistry of surface phenomena (12-21).

In our opinion, the scatter of experimental data is primarily attributable to the extreme sensitivity of boundary structure to the lyophilic state of the solid substrate. This has recently been brilliantly demonstrated by Kitchener and Pashley (22). Moreover, one does not often take into account that different methods have different sensitivity and, hence, are suitable to reveal only some part of the boundary layer.

Now, let us consider the surface forces that are responsible for the formation of the boundary liquid layers having a modified structure. These are van der Waals' forces, electrostatic forces, and hydrogen bonding of the molecules of the liquid with the atoms of the solid substrate.

The influence of dispersion forces on the structure of simple liquids and solutions was examined by using statistical mechanical methods by Kuni *et al.* (23). The interactions of molecules with one another and with solid substrates were described by means of effective pair potentials. This allowed them to obtain the variations in the density of liquid in thin interlayers of different forms and dimensions, as well as the distribution of the molecules of the components of solutions. The density of the boundary layer, irrespective of whether it is increased or decreased, depends on the values of the effective potentials. Thus, the possibility of the existence of boundary layers of different structures was shown. Nabutovsky *et al.*, have solved the same problem on the basis of the Lifshitz macroscopic theory (24). They obtained qualitatively similar solutions for the stress tensor and the density distribution near flat, spherical, and cylindrical interfaces. It was shown that the effective pair potentials that are used by Rusanov *et al.* (23), depend on the shape of the interface and cannot be considered as "constants." The macroscopical approach (24) includes, aside from pair potentials, the spectral characteristics of the interacting bodies.

Ninham *et al.* (25,26) improved to a considerable extent the procedure of such calculations by combining the equations of the theory of the liquid state with the macroscopic theory of molecular forces, taking into consideration the force field of the substrate. The calculations, which are still somewhat qualitative, show the following: wetting liquids form boundary layers of an increased density, while nonwetting liquids form boundary layers of a decreased density when compared to that of the bulk liquid phase. Although this approach does not take hydrogen bonding into consideration, it was shown that the forces arising owing to structural effects exceed the molecular attraction of the surfaces when the interlayer thickness decreases to about 10-20 molecular radii. It is evident that the thickness of a lone boundary layer must exceed these values.

The electrical field of a solid surface also makes a certain contribution to the formation of a modified structure for the boundary layers. Thus, taking into consideration the interaction of polar molecules with the monolayer coverage of dipoles that are oriented normally to the surface leads to the appearance of additional attractive forces that become noticeable at a distance between the surfaces smaller than 20-30 Å (27). Distler *et al.* (28,29) have demonstrated, with the use of the electron microscope, the influence of charged areas of the surface on the structure of thin solid films formed on it. Rusanov and Pshenitsin (30) observed by means of ellipsometry the formation of the modified boundary layers of hexane and octane under the effect of the electrical field of the cleavage surface of an ionic LiF crystal. No boundary layers have been formed on the same uncharged surfaces. This is in agreement with the well-known influence of a surface charge on the wetting phenomena (31,32).

In water and aqueous solutions, the influence of the electrical field on the structure of the boundary layers is not the main effect. This is attributed to the fact that the properties of bulk water change but slightly even in relatively strong external electrical fields. According to Sanfeld (33), polarization effects may be neglected when considering the forces of the electrostatic interaction of charged surfaces through thin interlayers of water and aqueous electrolyte solutions. It is also known that the viscous-electric constant of bulk water is relatively small (about 10^{-17} m^2/v^2).

A considerably greater contribution to the structural changes of the boundary layers should be made by the effects produced by deformation of the hydrogen bonds in water under the influence of the substrate, that is, by some sort of epitaxial effect. In this case, the peculiarities of the boundary layer structure must, to a high degree, depend on the nature and the number of the surface atoms or groups that are able to form H-bonds with the water molecules.

The determining role of the surface H-bonds in the formation of the structure of the adsorption layers of water on a quartz surface was examined in detail by Kiselev *et al.* (34), as well as by Zettlemoyer *et al.* (35,36).

Theoretical investigations of the structure of the boundary layers of water, however, are so far only planned; therefore, let us pass over to the discussion of experimental data.

II. STRUCTURE AND PROPERTIES OF THE BOUNDARY LAYERS OF LIQUIDS

In analyzing the experimental data available, we shall take into account both the sensitivity of the method applied and the state of the solid substrate.

The spectral methods (e.g., nuclear magnetic resonance, infrared spectroscopy, x-ray structure analysis), though they provide direct information on a change in structure, do not exhibit a high sensitivity (as compared with other methods that will be discussed later); they reveal only the highly modified part of the boundary layer. Therefore, these methods are usually applied to the examination of the adsorption monolayers undergoing direct contact with the surface.

In view of this, the NMR method is suitable for detecting changes occurring only in those molecular layers of water that are closest to the hydrophilic surface (19,37-41). It was also detected that water molecules in the first two layers are dissociated to a markedly greater extent than those of the bulk water (15,42). This is of considerable importance in the examination of electrokinetic phenomena in fine-pored bodies, and the resultant surface conductivity.

Boundary layers of a greater thickness (up to 50-100 Å) are detected by the NMR method only in cases where the differences in the structures between the bulk phase and the boundary layers are large; namely, in the investigation of the nonfreezing interlayers of water between the surfaces of an ice phase and its solid substrate (43,44). Here, the boundary phase remains in a liquid state, whereas the bulk phase is crystalline. For these systems, a substantial influence of the surface hydrophility on the thickness of the nonfreezing interlayers is detected, and is attributed to the influence of the number of hydrogen bonds formed per unit area of the surface.

IR spectroscopy methods allow one to obtain the shift of the adsorption band, which results from the disturbing action of intermolecular hydrogen bonding on the OH stretching. However, no appreciable variations are detected in the IR spectra for water interlayers 1-2 μm thick between different materials, including hydrophilic materials (45).

In this connection an attempt was made to investigate still thinner interlayers of water (about 0.7-0.1 μm thick) between two polished quartz plates. Quartz of special purity (containing less than 5.10^{-5}% impurities, and less than 0.12% bulk hydroxyls) and a two-beam IR spectrometer were used (46). Even for the thinnest layer of water, the value of the detected band shift (toward the long-wave region) is equal to 20-25 cm^{-1}.

In order to eliminate the influence of the bulk water which constitutes the main portion of water in the interlayers, further measurements were carried out with thin (30-50 Å) polymolecular adsorption layers of water on a flat quartz surfaces. Since the modified structure of the boundary layers is affected by the influence of the solid substrate, it is possible to suppose that the structure of polymolecular water films differs slightly from that of the boundary layers that pass over into the bulk water phase.

For such spectral measurements we need to have a water layer on the order of 10^{-5} cm thick. For this purpose, the IR radiation was made to pass through 40 flat parallel polymolecular water films on 20 thin quartz plates arranged in an evacuated glass chamber (47). Different relative vapor pressures, p/p_s, were maintained in the chamber by the aid of saturated solutions of different salts. The beam was switched in for 2 sec for each wavelength in order to preclude any heating of the films. Between the measurements there were interruptions of about half an hour. The total thickness of adsorption layers on 20 plates was equal to ~0.072 μm (at $p/p_s = 0.975$) and 0.145 μm (at $p/p_s \simeq 1$). An identical second chamber with identical quartz plates was placed in the other beam, but at $p/p_s = 0$.

As compared with bulk water, a shift of the absorption band maximum by about 100 cm^{-1} into the long-wave region was detected.

This result may qualitatively be interpreted as the strengthening of intermolecular H-bonds in thin water films. A quantitative description of structural changes according to these data is insufficiently reliable and requires using certain water structure models.

The observed band shift was not attributable to the formation of a surface gel layer. A prepared layer of silicic acid gel of the same total thickness influenced the absorption spectra in a different manner (48). Now, after hydrothermal treatment of the quartz plates for 1 hr (under a vapor pressure of about 2 atm and a temperature of 120°C), a gel film about 20 Å thick formed at their surface. This gel layer could be removed by a hot chrome mixture, so that the initial state of the quartz surface could be restored. The electrokinetic measurements carried out in thin capillaries of the same quartz have shown that a surface gel layer forms even at room temperature, but only after several weeks of contact with water (49).

Ellipsometric methods were also applied to the investigation of polymolecular adsorption films of water at the quartz surface as a model of the boundary layer. In view of Pashley's publication (50) on the influence of the surface roughness on the ellipsometric measurements, we have compared the results of measurements of thickness by the ellipsometric method with those obtained by the IR absorption. In the latter case, the thickness of a single film is determined by division of the total thickness of the layer by the doubled number of quartz plates. The bulk water absorption coefficient was used to calculate the total layer thickness. This could not have introduced a major error for the absorption spectra differ only slightly from one another.

The differences in thickness obtained by these two methods did not exceed 5 Å, and the isotherms $h(p/p_s)$ present qualitatively the same form (47). Thus, in the case of water films on a polished quartz surface (having the height of roughness of ~100 to 120 Å), the ellipsometric method gives results that are in good agreement with IR absorption methods (these do not depend on the presence of such roughnesses).

Among the more interesting results obtained by the ellipsometric method, we feel the most interesting is the relationship established between the film thickness and the hydrophility of the quartz surface. Thus, the film thickness (at equal values of $p/p_s = 1$) $h_0 = 53$ Å corresponds to the contact angle (measured by the sessile drop method) $\theta = 9°$; the thickness $h_0 = 43$ Å corresponds to $\theta = 20°$, and $h_0 = 40$ Å to $\theta = 30°$. After treating the quartz surface in a glow discharge, that enhanced its hydrophility, the thickness of films (at $p/p_s = 1$) increased up to 70-80 Å (12,51). On the hydrophobized quartz plates the film thickness proved to be commensurable with the ellipsometric measurement error that amounted to about 3 Å.

As was shown by Pashley and Kitchener (22), a further increase in the hydrophility of the quartz surface owing to its very thorough cleaning leads to an increase in the thickness of water films up to 200-300 Å at $p/p_s = 0.975$-0.98. Methylation of the quartz surface or its pollution by hydrophobic impurities drastically decreased the thickness of a water film on the quartz surface. These investigations resulted in a marked increase in the assessment of the range of the structurization effect of surface forces as compared with the first elipsometric measurement data (12).

It will have to be noted here that the formation of thick polymolecular water films with a modified structure (which was revealed by the formation of a finite contact angle with droplets) was for the first time observed by Bangham (52).

Applying ellipsometric methods, Adamson (53) obtained the thickness of polymolecular adsorption layers of different liquids on different solid substrates and also established their relationship with the contact angles.

Calorimetric measurements of the specific heat capacity of polymolecular layers of water on the quartz surface also reveal their structural changes at thickness smaller than 80-100 Å and their dependence on the surface hydrophility (54).

Thus, different methods corroborate the relation between the number of hydroxyl groups per unit surface of quartz (determining the degree of its hydrophility) and the thickness of the modified layer of water.

If the structural changes of the boundary layers of water are actually determined by the rebuilding of the network of H-bonds, then the temperature must exert a substantial influence on their thickness. As is known, the intensification of the thermal motion causes a destruction of intermolecular H-bonds, which makes the behavior of water approach that of simple liquids. Therefore, it is to be expected that an increase in temperature should decrease the thickness of the boundary layers of water.

Figure 1 represents the results of ellipsometric measurements of the thickness of polymolecular films of water on flat quartz

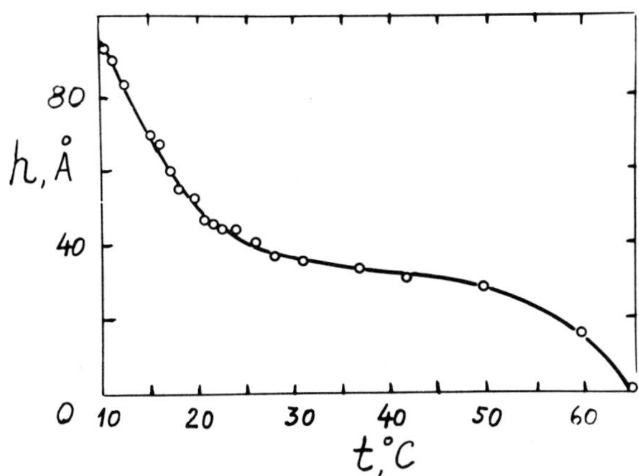

Fig. 1. Dependence of the thickness of polymolecular adsorption films of water (at $p/p_s \simeq 1$) at the quartz surface on temperature.

surface at different temperatures (55,56). The thicknesses of films as measured under the same relative vapor pressure $p/p_s \simeq 1$ decrease as the temperature increases. Thus, it is the weakening and destruction of the network of H-bonds that causes a decrease in the film thickness, though the surface hydrophility remains invariable.

Until now we have primarily discussed the results that were obtained for polymolecular adsorption layers of water on hydrophilic surfaces. Let us now investigate in what relationship those data are with the results of the investigation of the boundary layers found between a solid substrate and a bulk liquid.

Different methods were applied to the investigation of the boundary layers, these methods can be subdivided into equilibrium and nonequilibrium ones. The first ones comprise, for example, of dilatometric methods of measuring the density of liquids and their thermal expansion in fine pores, as well as the methods of investigation of the optical anisotropy of the boundary layers. Equilibrium methods also include the measurement of interaction forces resulting from the overlapping of the boundary layers of particles or flat surfaces, that is, the measurement of the structural component of the disjoining pressure of thin interlayers of liquids (14,21).

Nonequilibrium methods include viscosity investigations of thin liquid films or interlayers from the measurements of the flow rates or molecular diffusion.

Let us first consider the results obtained with the use of equilibrium methods. Precision measurements of the density of nonpolar liquids in contact with graphitized carbon black powders ($S \simeq 10$ m^2/g) allowed calculation of the autoadsorption value $\Gamma > 0$ per unit surface (57,58). It has been shown that as temperature decreases and approaches the melting point, the values of Γ increase in proportion to the number of carbon atoms in the molecule. The formation of dense boundary layers (about 10-20 Å thick) is caused by the orientation of molecules parallel to the surface. The oriented layers gradually pass over into the bulk structure. Thus, even in the case of nonpolar liquids there is observed a departure from the often-used concepts of short-range action on the solid substrate. Similar effects of the formation of polymolecular boundary layers were then detected for nonpolar liquids at the surface of $NiCl_2$ crystals (59).

In the case of water, the analogous measurements are complicated by a possibility of the penetration of water molecules into the solid phase. Thus, for instance, measurements of the density of water in the clays and in boundary layers near silicate surfaces produce conflicting results. In fine clay pores the density of water proves to be slightly decreased (60-62), yet similar measurements for water in contact with the surfaces of glass, quartz, and polymeric fibers, give increased values of density as compared with bulk water (63,64). The degree of the structural modification, as in the aforesaid cases, proves to be stronger as the temperature decreases.

The difference in the boundary layers structure close to hydrophilic and hydrophobic surfaces remains as a topic for further discussion. In the opinion of Tait and Franks (65), close to a hydrophilic surface the structure of the boundary layer approaches that of ice-1 (which must correspond to a decreased density); whereas close to the hydrophobic surface it approaches the structure of water in gas-hydrates (which corresponds to an increased density). Gurikov (66) considers that under the influence of the electric field of a charged surface the nearest layers of water acquire a denser packing. It is replaced by a "disordered" structure passing over into the bulk liquid phase as the distance from the surface increases. The theory of Gurikov, which has been developed on the basis of Samoilov's two-structural model of water, provides a qualitative explanation of the three-zone model of the boundary layer. The latter was suggested earlier (67) for explaining the anomalies of the thermomechanical effect (the appearance of a difference in temperatures owing to the flow of the boundary layers under the effect of an applied pressure drop and then developed in the work of Drost-Hansen (68).

Figure 2 represents the results of measuring the thermal expansion of water in a highly dispersed silica powder (69,70). Here the ordinate axis plots the relative changes in the volume of water in pores, while the abscissa axis plots the temperature. The value ΔV is equal to $V-V_0$, where V_0 is the value of V at $t = 0^\circ$C. The known thermal expansion for bulk water is shown by the dashed line. It has been plotted in such a way that the $\Delta V(t)$ curves coincided in the range $t > 10^\circ$C, when the values of the water density in fine pores and in the bulk approach each other because of the destruction of the network of H-bonds.

For reducing the size of pores between silica particles, the powder was compacted inside a vessel made of the alloy "Invar" (the coefficient of thermal expansion is smaller than 10^{-6}) under the pressure of 1000 atm. The pores thus formed (having the mean radius of about 50 Å) are filled with water after evacuation of the compacted powder up to 10^{-4} Torr. Water was previously degassed by multiple cycles of freezing and melting with simultaneous evacuation.

Figure 2 shows that the thermal expansion of water in fine pores occurs monotonically (at $t > 0^\circ$C), in distinction from bulk water. The density of water in fine pores is increased: curve 1 passes below curve 2 for bulk water. Quantitative evaluations, however, are complicated due to possible differences in the density of water in pores and in bulk at $t = +10^\circ$C. Because of this, this temperature may be considered only approximately as the point of convergence of plots 1 and 2.

Changes in the properties of water in fine pores were not attributable to the dissolution of silica. This follows from the fact that water extruded from the powder retained its bulk properties.

Similar changes in the thermal expansion of water (H_2O and D_2O) were detected by the same method in thin pores (~300 Å)

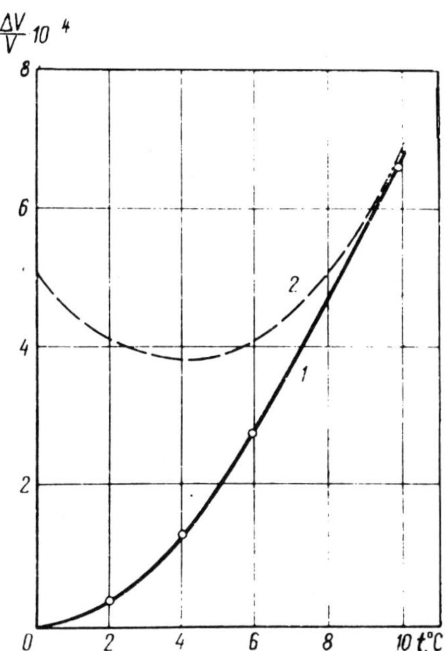

Fig. 2. Thermal expansion of water in the silica pores (r ≃ 50 Å) (curve 1) and bulk water (curve 2).

of the compacted TiO_2 powder (71). As is known, the solubility of TiO_2 in water is much lower than that of SiO_2, and surface gel layer formation is excluded.

All this proves that the structure of the boundary layers is very diverse, and depends strongly on the nature and the state of the solid substrate. Even on such different surfaces as quartz and teflon, nonfreezing interlayers are detected, though their thickness noticeably differs (43,44). It is probable that there may also be such cases when no structural changes of the boundary layers of water take place. This possibly occurs in the case of AgI crystals (72,73).

In a number of systems, the boundary layers exhibit an apparently anisotropic character. As is known, such well-ordered structures are inherent to the boundary layers of liquid crystals (9,10). These are also observed for a few polar liquids. In the case of water, this was experimentally demonstrated by measuring the birefringence of thin aqueous interlayers between platelets of montmorillonite (74).

Permittivity is one of the important characteristics of the structure of polar liquids. Examination of the permittivity of thin interlayers of water between hydrophilic surfaces of clay particles actually detected its marked decrease as compared with the bulk value (75,76).

In this connection, it would be interesting to note the research done by Snitkowskij (77,78) on the electrical properties of thin interlayers of organic liquids, in particular fatty acids. The application of 60 Hz alternating field using gold electrodes revealed that the behavior of the interlayers was similar to that of seignettoelectrics possessing domain structure. If the thickness of the interlayers is smaller than several microns, conductivity is revealed. This disappears when a certain critical value of thickness is exceeded. These phenomena are worth further investigation.

New information on the structure and extension of the boundary layers results from the observation of the stability of thin liquid interlayers and films on different substrates, as well as from the experimentally obtained dependences of their thickness h on the value of disjoining pressure Π. As is well known (79), the value of Π is equal to the difference between the pressure P acting on the interlayer surface, and the pressure P_0 in the bulk phase of liquid, from which the interlayer has been formed. The disjoining pressure determines the value of the force of repulsion

($\Pi > 0$) or attraction ($\Pi > 0$) of the surfaces that are separated by a thin liquid interlayer.

The total value of the disjoining pressure is a sum of several independent (to a first approximation) components such as; Π_m, caused by the dispersion forces; Π_e, resulting from the overlapping of diffuse ionic layers of charged surfaces; and the structural component Π_s. It is just this last component that will be discussed here, since its sign and value are determined by the phenomena accompanying the overlapping of the modified boundary layers.

Such overlapping (taking place when the interlayer is thinning out) is accompanied by the destruction (or, on the contrary, reinforcement) of the peripheral parts of the boundary layers. This gives rise to a change in the Gibbs free energy of the system, $F_s(h)$, which results in the appearance of disjoining pressure (14,80)

$$\Pi_s = -(\partial F_s/\partial h)_T$$

The method of finding the isotherm of the structural component of the disjoining pressure $\Pi_s(h)$ consists of the following: the values of $\Pi_m(h)$ and $\Pi_e(h)$ that are calculated theoretically are to be subtracted from the experimentally found values of the total disjoining pressure $\Pi(h)$. For determining the $\Pi_m + \Pi_e$ values, the known DLVO theory is used (81,82).

In order to illustrate this method, let us consider, for example, the isotherm $\Pi(h)$ for water films on a quartz surface (14). This isotherm exhibits an s-shaped form (curve I, Fig. 3a) corresponding to the case of incomplete wetting (the contact angle $\theta = 10 - 20^\circ$). Within the region of thick metastable β-films ($h > 500$ Å), the experimental curve $\Pi(h)$ is in good agreement with the theoretical one for $\Pi_e(h)$ (Fig. 3b). For thick water films, this result is also well corroborated by other measurements (22, 83). In this case, the contribution of Π_m component is small as compared with that of Π_e.

Within the region of thin α-films corresponding to the state of the full thermodynamic stability of the system, neither Π_e component nor Π_m component, nor their sum is able to explain the form of the experimental isotherm $\Pi(b)$ (curve I). The theoretical de-

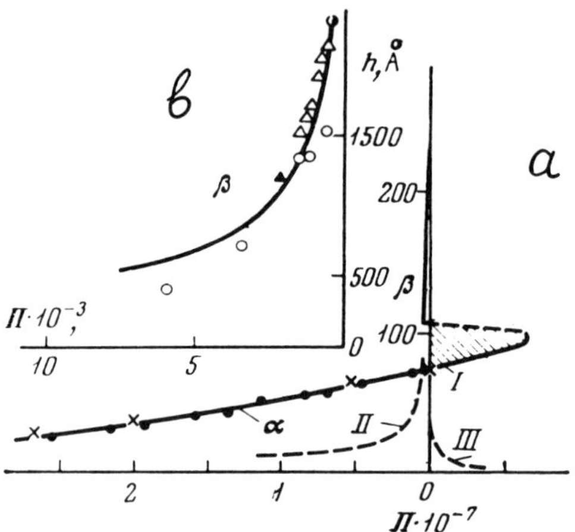

Fig. 3. Isotherm of the disjoining pressure Π(h) of water films at the quartz surface (a), and β-part of this isotherm (b). (Π, dyne/cm^2; h, Å).

pendences of $\Pi_e(h)$ (curve II) and $\Pi_m(h)$ (curve III) are shown in Fig. 3a.

Thus, the lower α-branch of the isotherm in Fig. 3a closely corresponds to that of the $\Pi_s(h)$ isotherm alone, reflecting the dependence of the structural repulsion ($\Pi_s > 0$) or attraction ($\Pi_s < 0$) forces on the water film thickness. As shown in Fig. 3a, these forces prevail when the film thickness is smaller than ~100 Å. The values of Π_s (in $\Pi_s > 0$ region) increase exponentially as the film thickness decreases, which is in good agreement with the theory of Marčelja and Radic (84). The exponential form of the $\Pi_s(h)$ isotherms was also obtained by Adamson (53) for α-films of other liquids on different solid substrates.

Pashley and Kitchener (22) have carried out similar calculations of the $\Pi_s(h)$ isotherm. They showed that the thickness of α-films increases as the procedure employed for cleaning the quartz surface improves and as its hydrophility increases. This means that the structural repulsion forces exert their influence at still greater distances, up to 200-300 A. The whole isotherm $\Pi(h)$ is shifted toward the region of positive values of Π and does not intersect (in the case of the most perfect hydrophility) the axis of thicknesses. As is known, this condition corresponds to the complete wetting of the surface by the liquid (31,85).

Many surface pollutants can cause marked changes in the radius of action of structural forces and their dependence on distance. As was shown by Schulze (86), the value and sign of Π_s of the wetting films of water are strongly influenced by the adsorption of even very small amounts of surfactants, changing the structure of the boundary layers.

The effect of the forces of structural repulsion is also revealed by direct measurements of the interaction of the surfaces of mica and quartz across a thin liquid interlayer. Thus, Israelachvili and Adams (87,88) state that the forces of repulsion of crossed cylinders (coated with mica platelets) in dilute electrolyte solutions surpass the values of $\Pi_e + \Pi_m$ forces at a distance smaller than 40-75 Å. The magnitude of such excess forces was somewhat different for different samples of mica, and decreased with distance according to a law that is close to the exponential one (as in the case of α-films!). The authors (87,88) put forward a supposition that these forces are attributable to the structural changes in the boundary layers of water occurring close to the mica surface.

The repulsion forces that were connected with the effect of the structural component on the disjoining pressure were also detected by Peschel *et al.* (89-93) for water and electrolyte solutions as well as for cyclohexane, benzene, and monohalogenobenzene interlayers between the surfaces of the quartz lens and the quartz plate. For aqueous systems, the dependence of Π_s (with h = const) on temperature exhibited several maxima. This cannot yet be explained, and is in contradiction with other known results. For nonaqueous systems, the structural disjoining pressure was found to exist close to the melting point.

In the first series of work of Peschel *et al.*, the radius of action of the structural forces was evaluated as several hundred angstroms. Quite recently (94), the values of $\Pi_s(h)$ have been determined by subtracting the theoretically calculated values of Π_m and Π_e from the experimentally obtained values of $\Pi(h)$. In this case, the Π_s forces radius has been reduced to 60-80 Å, which is close to other known estimations.

Approximately the same estimations of the distance at which the contribution of the Π_s component dominates follow from analysis of the stability of many colloid systems (95/99). Thus, for example, without the structural component of the disjoining pressure it would be difficult to explain such facts as colloid stability at an isoelectric point and a decrease in the stability as temperature increases. Both effects may be explained by the destruction of boundary water layers.

The boundary layers of nitrobenzene forming close to the surface of glass or fused quartz, which is purified (or/and activated) by a glow discharge or hydrogen flame, are of a special interest (100,101).

First of all, Popovskij (100) showed that the specific heat capacity of nitrobenzene in layers less than 0.1 μm thick on the activated surface of a glass powder is 8% lower than its bulk value, and does not depend on thickness. Now, the specific heat capacity of similar nitrobenzene layers on the nonactivated surface of the glass powder does not differ from the bulk value, yet the heat capacity of adsorption layers (having the thickness of several molecules) exceeds the bulk value.

The specific heat capacity of nitrobenzene layers that are found beyond a certain distance from the substrate is equal to the bulk value. Thus, one may speak about modified boundary phases of nitrobenzene, which, in distinction from usual phases, have a thickness h_c that is a definite function of temperature. It is the measurement of the heat consumed in heating the nitrobenzene layers of different thicknesses in the work of Popovskij and Derjaguin (100) that allowed, not only the dependence of h_c on temperature, but also the specific heat of transition from the boundary phase into the bulk phase to be determined. The latter is an increasing function of temperature. Figure 4 shows a plot of the boundary phase thickness (curve 2) and the specific transition heat (curve 1) versus temperature. The points that are differently indicated in curve 2 correspond to the experiments carried out with nitrobenzene films of different initial thicknesses on the glass powder.

A further research of Silenko (101) concerning polarized light passing through a nitrobenzene interlayer between two pieces of

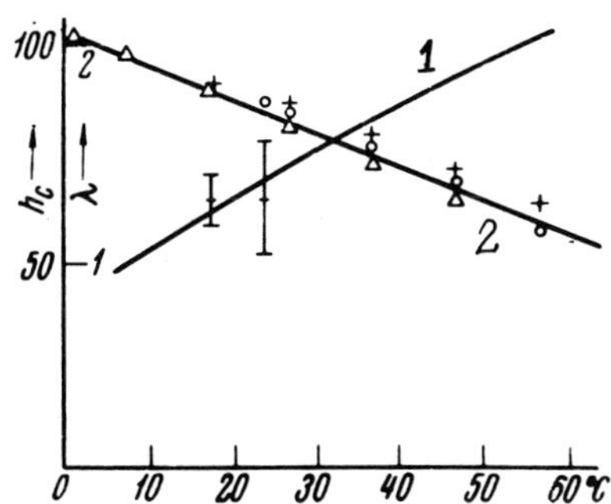

Fig. 4. The heat of phase transition, λ(cal/g), the boundary phase, the bulk nitrobenzene (1) and the thickness, h_c(nm) of the nitrobenzene boundary layer (2) versus temperature. Curve 1 is calculated from the dependences obtained earlier (101) for the difference in specific heats of the bulk and boundary nitrobenzene phases.

black glass (in particular with the use of the integral, waveguide optics calculation methods) revealed the following: The boundary layers of nitrobenzene possess birefringence with the difference in refraction indices Δn for light rays having a polarization plane that is parallel and normal to the layer plane of the order of $\Delta n \simeq 0.0004$. Simultaneously, the evaluation of the thickness of the boundary layers of nitrobenzene based on calorimetric measurements was corroborated.

Examination of the wetting films of nitrobenzene provide still more corroboration of the influence of the treatment and cleaning of glass on the thickness of the boundary layers. On the nonactivated surface, the thickness of the wetting film is much smaller than 10^{-5} cm, and the contact angle of the bulk phase is equal to $\sim 20°$. On the activated surface of glass, the thickness of the wetting films within the limits of measurement accuracy coincides with that of the boundary layer of nitrobenzene $\sim 10^{-5}$ cm, which has been determined from the calorimetric measurements, and the contact angle of a nitrobenzene droplet reduces to a value of 1-2°. Hence, on the basis of contact angle theory (85), it follows that, in this case, nitrobenzene layers thinner than 10^{-5} cm reveal a positive value of the structural component of disjoining pressure. In accordance with Derjaguin (85), for great thicknesses either the disjoining pressure is negative, or it changes in a jumpwise fashion because of the jumpwise character of the transition boundary-bulk phase.

Popovskij *et al.* substantiated the existence of a positive disjoining pressure for the boundary layers of nitrobenzene by examining the stabilizing effect of glass particles suspended in nitrobenzene (100). Thus, the following may be considered to be established: Under the influence of the surface the appearance of a special structure with a definite orientation of nitrobenzene molecules may be observed in layers a few hundred molecules thick.

Specific relay (or quasiepitaxial) mechanism of the boundary layer formation is also demonstrated by experiments (100) in which polymolecular boundary layers of benzene form on the surface of the activated glass under the influence of an adsorped nitrobenzene monolayer. These adsorption layers form when ~0.2-0.3% by volume of nitrobenzene is dissolved in the benzene.

Zorin investigated the adsorption of benzene vapor on mercury (102) and revealed the ability of nonpolar benzene to form structurally modified polymolecular layers. It was found that adsorption layers of two different thicknesses can coexist in a durable manner on a mercury surface under a certain pressure of benzene vapor. Thus, two isotherms of the adsorption of benzene on the same surface exist. This phenomenon can be explained by assuming that both modifications of polymolecular adsorption layers differ from each other by different orientations of benzene rings, parallel or perpendicular to the surface of mercury. Let us note that the thickness of adsorption layers increases jumpwise as the benzene vapor approaches the saturation point. Sheludko (103) examined the corresponding equilibrium-thicker β-films by using the known gas bubble pressing-on method. The disjoining pressure isotherms thus obtained tally well with Zorin's data on α-films. The measurements were carried out on identical samples of mercury and benzene.

Researches into boundary layers of nitrobenzene are of special interest in that these layers occupy an intermediate place between the boundary layers of water and the layers of liquid-crystalline substances.

Nonequilibrium methods were mainly applied to obtain information on the viscosity of the boundary layers of liquids. A great number of data were obtained by the blow-off method (104-106). The basis of the method consists in blowing a wetting liquid film off a well-polished solid surface by a constant stream of air. The film profile $h(y)$ obtained after blowing off, reflects the viscosity distribution $\eta(x)$, where y is the distance along the surface, and x is the distance from the substrate. The local viscosity values are proportional to the derivative dh/dy, assuming that the film flow is one-dimensional and laminar.

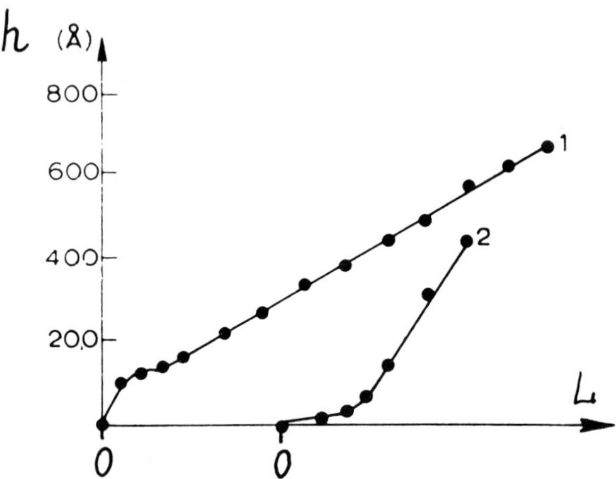

Fig. 5. Thickness of the liquid films after blowing off as a function of the distance L from the wetting line. (1) amylsebacate; (2) dibutylphthalate (a); (3) vaseline oil (b).

Figure 5 represents profiles $h(y)$ that are obtained for several nonvolatile liquids, such as amylsebacate (curve 1), dibutylphthalate (curve 2), and especially purified vaseline oil (curve 3) on a steel surface. For polar esters, the viscosity can be shown to be changed in the boundary layer: It is increased at $h \leq 100$ Å (curve 1) or decreased at $h \leq 20$-30 Å (curve 2). This is connected with the different orientation of polar molecules (normal or parallel) relative to the surface. At a certain distance from the substrate, the viscosity takes its bulk values. For the nonpolar vaseline oil (curve 3), the obtained profile $h(y)$ corresponds to the exactly constant viscosity and the absence of any boundary layer with a modified structure.

The boundary layers of a considerably greater thickness (up to several microns) exhibiting an increased viscosity are detected by the blow-off method in the case of the polymers of vinyl esters and their solutions in oils (107). This is connected with the greater length of the chains of polymeric molecules that are oriented perpendicular to the surface of the metal substrate. Very interesting results were obtained in the examination of the boundary viscosity of some polymer series (108) by applying the blow-off method.

Now the lower homologues of this series, for example, tetraethoxyhexane, reveal a lower viscosity in layers up to 0.5 μm thick, yet higher homologues (beginning with tributoxy-butane) exhibit the viscosity that has been increased markedly in layers up to 1 μm thick. The profile of these films retains the shape that indicates a change in local viscosity as a function of the distance from the substrate during the entire blow-off period, even when the slope of the film surface decreases to a value of the order of 10^{-4}-10^{-5}. Under these conditions, the influence of the gradients of disjoining pressure and other surface effects is so small that it may be neglected.

In the work of Bascom and Singleterry (109), an attempt was made to cast doubts on the data obtained by the blow-off method: this was done by carrying out experiments, in which such conditions were chosen, under which the influence of either spreading beyond the boundary of the perimeter of the film or of the poor wetting was enhanced drastically. In the paper of Derjaguin *et al.* (110), the conclusions of the work by Bascom and Singleterry (109) were refuted.

Recently, the application of the blow-off method was also extended to volatile liquids, requiring development of a corresponding theory (111). Application of this method to hexadecane has demonstrated that at thicknesses smaller than 200 Å, the viscosity is lower than the bulk value. This may be explained by the horizontal orientation of molecules in polymolecular layers.

Another method of examination of the viscosity of thin liquid interlayers is based on measuring the flow rates through ultrathin quartz capillaries ~10-0.03 μm in radius (112,113). One end of a capillary having length of ~5-7 cm is immersed into a vessel containing the liquid to be investigated, while the second end communicates with a high-pressure chamber. The mean viscosity of liquid in the capillary is calculated from the Poiseulle equation:

$$\eta = r^2 \cdot P/8\, \ell v,$$

where v is the meniscus shift velocity in the capillary; $P = P_0 - P_c$ the difference in the gas pressures applied to the ends of the liquid column having length ℓ; P_0 the excess pressure of gas (nitrogen) in a chamber; P_c the meniscus capillary pressure. The latter is determined according to the intersection point between plot $v(P)$ and the pressure axis. At $P_0 > P_c$, the meniscus retreats in the capillary ($v < 0$); whereas at $P_0 < P_c$, the meniscus advances ($v > 0$).

The radius of the capillary r was determined from the meniscus maximum pressure (112). For water, benzene, and CCl_4 it practically coincided with the capillary pressure $P_c = 2\sigma/r$, which corresponds to complete wetting. In the case of water, this required the measurement to be carried out on the retreating meniscus.

The measurements of the meniscus shift velocity were carried out on a very small part of the capillary as compared with the liquid column length, which assumed ℓ = const. Under this condition, the viscosity values are determined by the slope of the plots $v(P)$, which were strictly linear.

Figure 6 shows the obtained dependence of η/η_0 (η_0 is the viscosity of the bulk liquid) on the radius of capillaries at 20°C. The mean viscosity of water increases as the radius of the capillary decreases. In contrast to this, nonpolar benzene and CCL_4 in the same capillaries do not show any change in viscosity.

Assuming different models of the boundary layer (with the stepwise or exponential distribution of viscosity), it will be possible to determine its effective thickness h_c and the viscosity values (114). Using the data presented in Fig. 6, the values of h_c were found to be equal to 150-200 Å, while the viscosity close to the capillary walls was increased about 1.5-1.8 times, depending on the model adopted. This result does not contradict the afore-

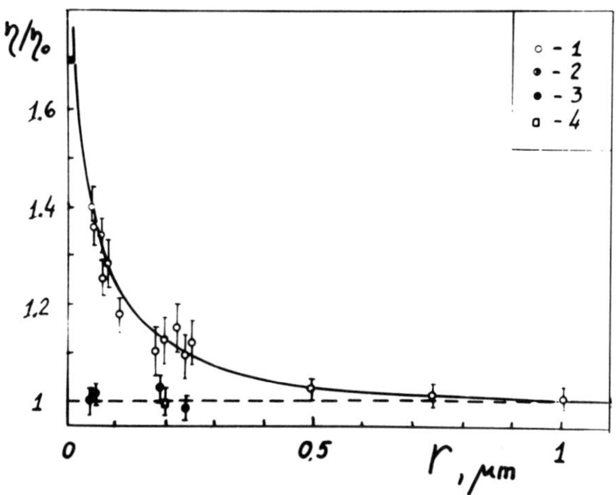

Fig. 6. Dependence of the relative viscosity of liquids η/η_0 on the capillary radius. (1) water; (2) water, the data of Tovbina (124); (3) CCl_4; (4) benzene.

mentioned evaluations of the thickness of the boundary layers of water, since the viscosity measurements involve also the less modified outer part of the boundary layers.

In the case of dilute electrolyte solutions ($C \leq 10^{-3}$ *N* KCl) exactly the same dependences of $\eta/\eta_0(r)$ were obtained as those for pure water. This makes it impossible to explain the obtained results by the electroviscous effect. In the latter case, just at $C < 10^{-3}$ *N* the electroviscous counterflow should depend very strongly on the electrolyte concentration (115).

In quartz capillaries the surface gel layers form after a few weeks of contact with water (49). When surface gel layers are present, considerably higher values of the viscosity may naturally be obtained, and even the yield stress can be detected (49). This,

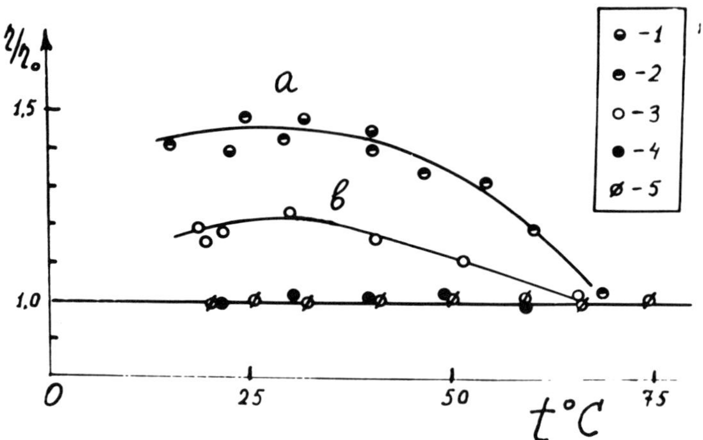

Fig. 7. Influence of temperature on the values of relative viscosity of water (1-3), benzene (4) and CCl_4 (5) in thin capillaries. (a) r = 0.05 μm; (b) r = 0.17 μm.

however, will not be connected with the boundary layers of water proper.

As shown in Fig. 7, an increase in temperature will decrease the difference between the mean viscosity of water in thin capillaries and in the bulk. As in the other cases, this is connected with the thermal destruction of the boundary layers. In our opinion, the results shown in Fig. 7 provide the most convincing proof of that effect. As is known, the electrical potential of the quartz surface depends only slightly on temperature. Therefore, an explanation of the decrease in viscosity values owing to electroviscosity is here completely excluded. The possible formation of the gel layers by heating should increase rather than decrease the measured values of viscosity.

The data shown in Fig. 7 are in good agreement with the measurement of the thickness of polymolecular films (Fig. 1) as a function of the thermal expansion, with the stability of some colloids and, as will be shown here, with the thermal dependence of the filtration and thermoosmosis rates in fine-pored bodies.

It would be more difficult to interpret the results of the measurement of viscosity in fine-pored bodies owing to the complex geometry of the pore space. In a number of cases, however, it is possible to obtain sufficiently reliable, at least qualitatively, results.

Figure 8 shows the results of examining water filtration rates v through a porous glass membrane having a mean radius of pores of about 10 Å at different temperatures (116). The experimental data are independent of whether the pressure drop P increases or decreases. The filtration dependences $v(P)$ pass through the origin. This means that water, even in very fine pores, behaves like a Newtonian liquid. Within the limits of experimental error (~0.1 dyne/cm^2) no yield stress is detected.

The slope of $v(P)$ dependences is proportional to the viscosity of the liquid in pores. Though its absolute values cannot be determined from these experiments, it is possible to examine a variation in the value of ratio η/η_{20} (where the η_{20} value relates to 20°C) by comparing the slopes at different temperatures. It has been shown that the values of η/η_{20} for water in pores decrease faster as temperature increases, than the corresponding values of the ratio η/η_{20} of bulk water. Hence, destruction of the modified structure of water occurs even in fine pores as the temperature increases.

Viscosity measurements were carried out also for nonpolar CCl_4 in a porous glass having wider pores (r = 30-40 Å) (117). The viscosity of nonpolar liquids in such pores remains practically equal to its bulk value. Hence, these measurements may be used as a reference point for the viscosity measurement of water in the same glass. Comparison of the plots 1 and 2 in Fig. 9 shows that the viscous resistance of water in fine pores is approximately 1.5 times greater than that for CCl_4, though the bulk viscosity of water is only about 3% higher.

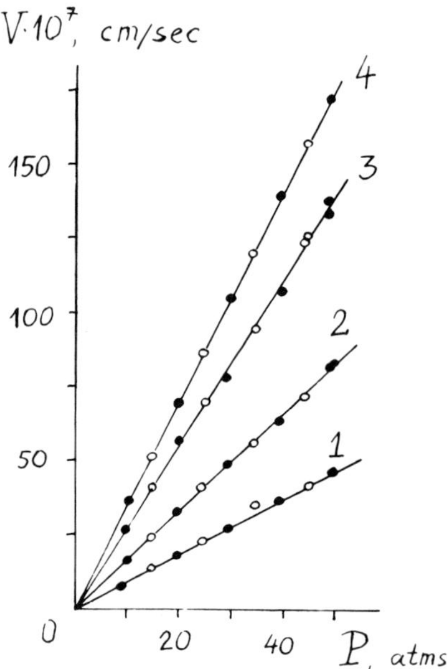

Fig. 8. Influence of temperature on the filtration dependences v(P) in the fine-porous glass. $t = 20°C$ *(1);* $40°C$ *(2);* $60°C$ *(3);* $70°C$ *(4).*

An increase in the electrolyte concentration makes the plots $c(P)$ for KCl solutions approach the plot for CCl_4. This means that the boundary layers are destroyed under the effect of a high electrolyte concentration. At $C \geq 10^{-2}$ N, a noticeable destruction is observed, which is in good agreement with other known data (118, 119). At $C = 2$ N, the differences between the plots 1 and 5 are already comparable with the differences in bulk viscosity between CCl_4 (0.97 cP) and KCl solution (1.02 cP).

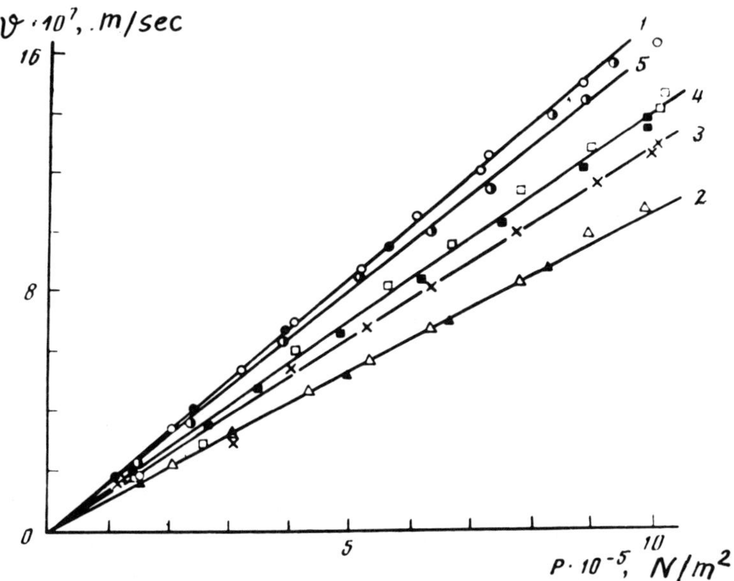

Fig. 9. Influence of the concentration C of KCl electrolyte on the water filtration rate in the fine-porous glass. CCl_4 (1); C = 0% (2); C = 0.1% (3); C = 1% (4); C = 20% (5).

Viscosity measurements that are based on the examination of the thinning out of the interlayers of water and aqueous solutions between two polished silica surfaces appear to be less reliable. The height of the roughness of polished surfaces (at least 100-150 Å) is of the same order of magnitude as the thickness of the zone having a modified viscosity. In this case, viscosity calculations are very complicated. Therefore, it is not surprising that conflicting results were obtained even for similar objects (120,121). On the surface of fused quartz capillaries, the height of the roughness as measured by the use of the electron microscope does not exceed 3-5 Å.

Low (122) obtained increased values of the viscosity of water in Na-montmorillonite using data of viscous flow of water at different temperatures, as well as the data of water self-diffusion (from neutron scattering and the isotope tracer technique). An increase in viscosity is detected at an interlayer thickness smaller than 100 Å.

An attempt to explain these results only by the effect on the polarization of water in a field of surface charges (123) can hardly be considered as worthwhile. In order to obtain agreement with the data of Low, a viscoelectric constant for water increased by two orders of magnitude had to be used.

Similar results were obtained earlier by Tovbina (124) using NMR techniques and by measuring the diffusion rate of different molecules through the porous silica with different mean radii (from 100 to 10 Å). Already in pores of about 70-80 Å radius, the values of viscosity have been increased approximately two times. An increase in temperature or in the concentration of electrolyte makes the viscosity values decrease. At 70-80°C, as in the experiments carried out with quartz capillaries, the differences between the pore and bulk viscosity would disappear.

For a number of systems no appreciable changes in viscosity are detected even in thin layers. Thus, in fine cylindrical pores from 30 to 400 Å in radius, which have been etched out in a thin mica plate, small changes in viscosity of 0.1 *N* KCl solution are observed only in pores $r < 50$ Å (125). It is probable that the etched pore walls do not possess sufficient hydrophility in distinction from the cleavage (basal) mica surface.

Recently, we have succeeded in measuring the viscosity of thin (about 50-100 Å thick) nonfreezing interlayers of water between the molecular smooth surface of quartz capillaries and the ice formed in them (126). For this purpose, the shift velocities of an ice column in strictly cylindrical capillaries under the effect of the gas pressure drops were measured. This allows the ratio of the interlayer thickness h to its viscosity η to be obtained. Then, using the thicknesses of the nonfreezing water interlayers obtained by the NMR method (43,44), values of η are determined as a function of the temperature below zero. The viscosity of the nonfreezing interlayers many times exceeds that of the supercooled bulk water at the same temperature. The viscosity values obtained are, however, much lower than the corresponding first data of Jellinek (127).

III. MASS TRANSFER IN FINE PORES

The modified structure of the boundary layers of liquids influences their flow rate in fine pores. As has been shown above, the effective thickness of the boundary layers is sensitive to the state of the solid surface, as well as the composition and concentration of the solution, and the temperature. Therefore, it is impossible to indicate any definite critical value of the radius of pores, in which structuring effects are acting. Thus, in the case of simple liquids, the appreciable anomalies of flow will probably be detected only in pores having a radius of the order of scores of molecular diameters. The thickness of peculiar boundary water layers near to the quartz surface is within the range of 10^{-6}-10^{-5} cm.

Let us first of all consider the influence of changes in the rheological properties of the boundary layers of liquids on filtration. Furthermore, we shall also examine the influence of the electric field, the concentration and temperature drops on the mass transfer in fine pores.

Figures 8 and 9 show that only an increase in the Newtonian viscosity of water is detected in the fine pores of glass membranes, without violating the Darcy law:

$$v = K_f P \ , \tag{1}$$

where v is the filtration rate, P the pressure gradient, and K_f the filtration coefficient.

Now, in a number of cases, deviations from that law were observed; the deviations being accompanied by the nonlinear run of the dependence v versus P and sometimes by the appearance of the threshold pressure gradient P_0 (128-132). In view of this, another form of the empirical filtration equation was suggested (131):

$$v = K_f P - K_f P_0 [1 - \exp(-P/P_0)] \ . \tag{2}$$

At $P >> P_0$, it transforms into an equation that is frequently used in practice:

$$v = K_f (P-P_0) \ . \tag{3}$$

Here it is assumed that at $P \le P_0$, no flow of liquid takes place.

The physical nature of the threshold pressure gradient P_0 remains still to be discussed. Generally, it is connected with a small, but finite value of the yield shear stress ($\tau_0 \simeq 10^{-2} - 10^{-3}$ dyne/cm^2) of the bulk water, or with increased value of τ_0 in the boundary layers that are considered as a plastic fluid (128). It is not excluded, however, that these effects could have been caused by the influence of the capillary osmosis counterflow or by the presence of colloidal particles in the pore moisture, as well as by the influence of swelling. Now, in the nonswelling porous bodies the filtration obeys, as a rule, the Darcy law (132-134), which also follows from our experiments (Figs. 8 and 9).

Jackson (135) showed that the dissolution of sodium and potassium silicates during the water filtration across porous ceramics plays an important part. In a number of cases, the influence of the incomplete wetting of the flow-rate measuring capillary is exerted. This effect was examined by Novak (136).

The nonlinearity of dependences $v(P)$ was most frequently caused by a reversible (and sometimes even an irreversible) change in the mutual position and orientation of the particles of a porous body under the effect of pressure drop P. The nonlinear dependences

$v(P)$ are usually observed in swelling systems (130,132,137,138), where such changes in the porous structure are possible. Here, the dependences $v(P)$ are qualitatively the same as in the case of the non-Newtonian flow through a nondeformable porous body.

If pores exhibit the wide and the narrow parts, the nonlinearity of the dependences $v(P)$ may be attributed to a change in the velocities field when P and, accordingly, the flow rate vary. In this case, a circulation flow of different intensities may arise in the wide parts of pores. This is both experimentally (139,140) and theoretically (141,142) well corroborated.

The filtration measurements are often complicated by the time instability of the flow. The instability may be caused by the bacterial activity (143,144), by the evolution of dissolved air (128), by the slow alteration of the porous structure under the influence of changes in moisture content, composition, and the concentration of dissolved substances (138,145). The influence of dust particles in wide pores and colloidal particles in fine ones is also possible (146,147).

When only the Newtonian viscosity of the liquid changes, calculation of the flow rates must take into account the dependence of viscosity η on distance x from the pore wall. Such solutions are derived on the basis of Navier-Stokes equations for different types of function $\eta(x)$ and for different shapes of cross sections (114,148). The variable viscosity $\eta(x)$ of liquid within the region of electrical double layer (EDL) was taken into account in the new theory of the electrokinetic phenomena in capillaries (149).

The deviations from the Darcy law may result from the gradients of concentration (induced capillary osmosis) and electrical potential (electroviscous effect), which arises when solutions flow through fine pores. In the case of wide pores, both these effects have an insignificant effect on the liquid flow rate. Their influence becomes significant only in the case where the diffuse layers of ions or molecules in sufficiently thin pores overlap one another.

The theory of the electroviscosity in fine pores where the EDLs overlap each other, is developed to a sufficient extent (150-152). It has been shown that the maximum relative decrease in the filtration rate occurs at $\kappa h \sim 1$, where κ is the inverse Debye radius, and h the width or the radius of a pore. As the radius of pores decreases further ($\kappa h < 1$), the convective flow of ions decreases faster than the electrical conductivity of the solution in a pore. This causes a decrease in the streaming potential values and a reduction in the effect of the electroviscosity.

Under steady-state flow conditions, the influence of electroviscosity cannot cause any deviations from the Darcy law; for the relative decrease in the flow rate does not depend on the pressure gradient, and is determined only by the values of κ. The deviations may be observed only in the case where the streaming potential has not yet been established. In this connection, it will have to be noted that the streaming potential relaxation times may be considerable (153).

The situation is quite different when a solution concentration gradient arises in filtration. In this case, the capillary-osmosis counterflow arises. This effect is connected with the formation of a concentration drop, when the solution flows through fine pores.

If the pores are filled with a pure liquid, then, as has already been shown above, changes in the structure of the boundary layers may occur under the influence of surface forces. This leads to local changes in the liquid density ρ and viscosity η, which become functions of the distance from the pore surface $\rho(x)$, $\eta(x)$. In fine pores, the boundary layers overlap one another. This causes additional structural changes within the overlapping zone, which so far cannot be predicted theoretically for water and other associated liquids.

If the pores are filled by the solution, then, besides the aforementioned changes in the solvent structure, a nonuniform distribution of the solute molecules occurs throughout the pore cross section. The concentration of the solute molecules also becomes a function of the distance from the pore walls, $C = C(x)$. The function $C(x)$ is determined by the value and sign of the forces acting between the solute molecules and the pore surfaces across the solution interlayers.

When the repulsion forces are predominant, the solution concentration close to the surface is decreased, corresponding to negative adsorption. Such a situation occurs for many aqueous solutions: This was interpreted earlier as the existence of "nonsolving volume" or as the content of "bound" water in disperse systems (154). The effect of decreasing the solute concentration is also applied to the separation of solutions by using the reverse osmosis method (155-157).

When the attraction forces are predominant, positive adsorption takes place. Unlike the concepts adopted earlier, positive adsorption (like negative adsorption) is not limited to the formation of the adsorption monolayer. The adsorption layers of molecules (like the EDLs) have a diffuse structure, which has been shown for the first time by one of the authors in the experimental investigation of the capillary osmosis (158). As is known, capillary osmosis is possible only when the mobile part of an adsorption layer is present. The adsorption value determines the intensity, while the adsorption sign (negative or positive) determines the direction of the capillary osmosis flow.

Distribution of the solute concentration across the pore space may be found from Boltzmann's equation

$$C(x,z) = C_0 \exp[-\ \phi(x,z)/\bar{\bar{K}}T] \ . \qquad (4)$$

Here $\phi(x,z)$ is the energy of interaction of the solute molecules with the pore surfaces; x and z designate the coordinates in the plane of the pore cross section; $\bar{\bar{K}}$ the Boltzmann constant; T temperature $^\circ\underline{\underline{K}}$; C_0 the concentration of the bulk solution outside the pore or in that point of the pore (if the pores are wide), where $\phi = 0$.

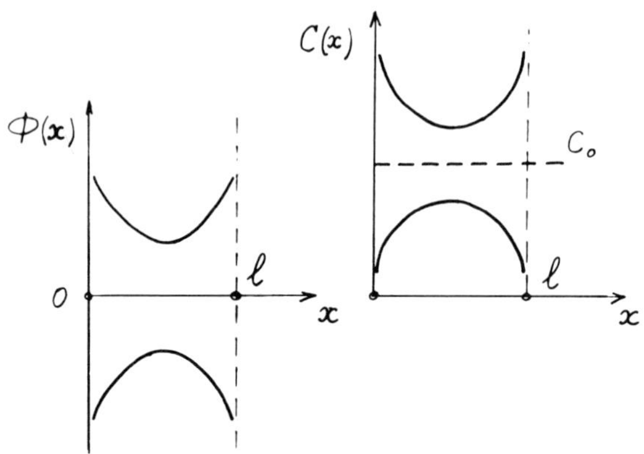

Fig. 10. Potential energy $\phi(x)$ (curves 1 and 1') of solute molecules in a thin slit pore and the distribution of concentration $C(x)$ corresponding to the former (curves 2 and 2'). 1 and 2, positive adsorption; 1' and 2', negative adsorption.

Figure 10 represents the schematic form of the function $\phi(x)$ and the corresponding distribution of solute concentration $C(x)$ across a slitlike pore when the values of ϕ and C depend only on one coordinate x. Here and subsequently, ℓ is the width of the slit.

The values of $\phi(x)$ are determined by the contributions of different components of the surface forces, and depend on the composition of the solution and on the properties of the substrate. In the case of nonaqueous systems, the main part is played by dispersion forces. In this case, the function $\phi(x)$ may be calculated on the basis of the known equations of the macroscopic theory of molecular forces (159-161). For low solute concentrations the calculation is essentially simplified. In this case, changes in the solvent structure may be neglected, and it may be considered that the value of $\phi(x)$ is an additive function of the energy interaction of the solute molecules with both surfaces of the slit (162):

$$(x) = \frac{A}{KT}\left[\frac{1}{x^3} + \frac{1}{(\ell - x)^3}\right] . \qquad (5)$$

Here A is the constant of molecular forces, which is equal to (160)

$$A = \frac{\hbar}{16\pi^2}\int_0^\infty \frac{(\varepsilon_2 - \varepsilon_1)}{(\varepsilon_2 + \varepsilon_1)} \frac{(\partial\varepsilon_m/\partial C)}{\varepsilon_1}\, d\xi \qquad (6)$$

where $\hbar$ is the Planck constant; $\varepsilon(i\xi)$ are the dielectric permeabilities of the materials along the imaginary frequency axis $i\xi$. Here, the indices 1 and 2 relate to the solvent and the solute substances, respectively. In this case, the dielectric permeability of solution is assumed to be equal to

$$\varepsilon_m = \varepsilon_1 + (\partial\varepsilon_m/\partial C)C .$$

This expression is valid at a low solution concentration: the value of the derivative is taken at $C \to 0$.

In the case of aqueous solutions, the modified structure of the boundary layers and, hence, the changed thermodynamical potential value produce the main effect on the solute concentration in thin hydrophilic pores.

The strict calculation of $\phi(x)$ is complicated and frequently impossible. Thus, in solving the problems of the solution flow through fine-pored bodies, one may limit oneself to the introduction of the effective potential

$$\phi_\alpha = \left(\frac{1}{\ell}\right)\int_0^\ell \phi(x)dx ,$$

which has been averaged throughout the pore cross section. Here index α relates to a definite component of the solution. The solution flow problem may be reduced to one-dimensional by using for an isotropic fine-pored body the averaged values of concentration C_α and diffusion coefficient D_α.

The steady-state flux of component α (in the absence of an electrical field) may be found from the equation of the convective diffusion of the solute molecules in the field of surface forces:

$$F_\alpha = C_\alpha \cdot v - D_\alpha\left(\frac{\partial C_\alpha}{\partial y} + C_\alpha \frac{\partial \phi_\alpha}{\partial y}\right) = \text{const} \tag{7}$$

where v is the filtration rate, and y the coordinate in the direction of the solution flow.

This equation was applied to the problem of membrane separation, when an intense agitation of the solution was effected at the membrane inlet. In this case, a laminary sublayer is formed close to the membrane surface whose thickness H is a decreasing function of the agitation intensity. The ratio of the concentration of the supplied solution C_0 to the concentration of the outflowing solution C_∞, is obtained to be equal to

$$\frac{C_0}{C_\infty} = 1 + [\exp(\phi_\alpha) - 1][1 - \exp(-vh/D_\alpha)]\exp(-vH/D_0) \ , \tag{8}$$

where h is the membrane thickness, and D_0 is the bulk diffusion coefficient.

As appears from this equation, at $\phi_\alpha > 0$ the concentration of the outflowing solution C_∞ is decreased, which corresponds to the negative adsorption ($A < 0$); whereas at $\phi_\alpha < 0$, C_∞ is increased, which corresponds to the positive adsorption ($A > 0$). In the equilibrium state ($v = 0$), the concentrations are leveled out as a result of diffusion through the membrane, and $C_\infty = C_0$. The same situation also occurs at $\phi_\alpha = 0$, when there is no adsorption ($A = 0$), and when the concentration of the solution in the pores does not differ from that of the bulk solution.

In the case of cellulose acetate and the thin porous glass membranes, the effective potential values are approximately equal to $\phi_\alpha = 0.5 - 1\ \bar{\bar{K}}T$ for nonaqueous solutions and to $\phi_\alpha = 4 - 5\ \bar{\bar{K}}T$ for aqueous ones. The high values of ϕ_α for aqueous solutions are determined by the modified structure of water in thin pores. This leads to the high selectivity of such membranes: $\phi = 1 - (C_\infty/C_0) = 0.96 - 0.98$.

The filtration of the solution through fine pores (at $\phi_\alpha \neq 0$) leads to the formation of a concentration drop and, therefore, to the development of the capillary osmosis counterflow. This retards the filtration, which results in deviations from the Darcy law. The solution flow rate in fine porous bodies is equal to (163)

$$v = K_f \cdot P - K_f \cdot RT\Delta C[1 - \exp(-\phi_\alpha)] \ , \tag{9}$$

where P is the pressure gradient causing the solution flow, R the gas constant, and ΔC the concentration difference at the ends of membrane pores. This concentration difference, however, is not equal to $C_0 - C_\infty$ owing to the phenomenon of concentration polarization, which causes an increase in the concentration at the membrane inlet as compared with C_0. Calculation of the value of ΔC, which is a function of the flow rate v and the solution agitation intensity, was carried out by solving Eq. (7) and by determining the distribution of concentration $C(y)$ across the membrane and beyond it, i.e., in bulk solution zones.

This calculation gives the following dependence $v(P)$, which has been plotted in Fig. 11 for the particular case of a fine-pored body having the mean radius of pores $r = 20$ Å and $K_f = r^2/8\eta h = 2.5.10^{-9}$ cm^3/dyne·sec. In this calculation we have adopted $h = 0.02$ cm, $\phi_\alpha = 3.3\ kT$, $D_\alpha = 5.10^{-6}$ cm^2/sec, $D_0 = 10^{-5}$ cm^2/sec, and solution viscosity $\eta = 0.01\ P$.

Figure 11 shows that the influence of the capillary-osmotic flow leads to appreciable deviations from the Darcy law. Thus, the nonlinearity of filtration dependence $v(P)$ may not be at-

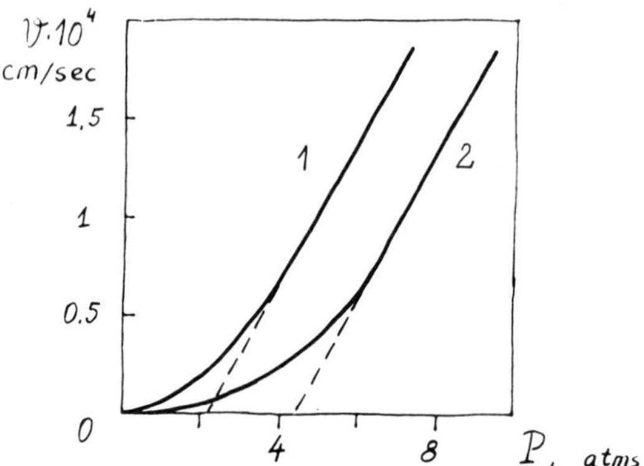

Fig. 11. Influence of the capillary osmosis on the solution flow rate (with concentration C_0) through a fine-porous body under the effect of different pressure drops P. ($\eta = 10^{-2}$ poise, $T = 300$ °K, $H = 10^{-3}$ cm); (1) $C_0 = 10^{-2}$ mole/ℓ; (2) $C_0 = 2.10^{-2}$ mole/ℓ.

tributed only to the deformation of a porous body and to changing porous structure.

Now, at high filtration rates the dependence $v(P)$ becomes linear, yet it does not pass through the origin of coordinates, intersecting the pressure axis at $P = P_0$. The value of P_0 is equal to the osmotic pressure of an ideal semipermeable membrane (for which $\phi_\alpha = \infty$ and $C_\infty = 0$). This pressure $P_0 = RTC_0$ may erroneously be interpreted as some yield shear stress, if the capillary osmosis phenomenon is not taken into account.

It will be interesting to note that Eq. (9), taking into account the influence of the capillary osmosis on the filtration of solutions, coincides in its form with empirical Eq. (2), which was earlier suggested by Swartzendruber (131) for describing the experiments carried out with fine-pored bodies.

An important piece of information on the structural changes in liquids in fine pores may be obtained by examining the thermoosmotic flow. One of the authors (164) was the first to have detected the phenomenon of thermoosmosis and explained it by the difference in the enthalpy per unit volume of the boundary liquid from the bulk values ΔH. On the basis of thermodynamics of irreversible processes, the following expression for the thermoosmosis coefficient was derived (164):

$$\chi = -\int_0^{\infty} \frac{\Delta H(x)\,x\,dx}{\eta(x)}\ \text{cm}^2/\text{sec}\ , \tag{10}$$

which determines the thermoosmotic flow rate:

$$j_{\mathrm{T}} = \chi\,\frac{\text{grad } T}{T}\ \text{cm/sec}\ . \tag{11}$$

Equation (10) is written for wide pores, when the thickness of the boundary layers is much smaller than the pore width. The coordinate x is read off the normal from the solid wall; grad T is the temperature gradient.

In fine pores where the boundary layers overlap one another, the following simplifying assumption may be used for integrating Eq. (10) [inasmuch as the exact form of functions $\Delta H(x)$ and $\eta(x)$ is unknown]. Introducing the values of excess enthalpy ΔH and viscosity η, which are averaged throughout the pore cross section, the following expression was derived for a slitlike pore (165):

$$\chi = -\frac{2\ell^2 \Delta H}{3\eta}\ , \tag{12}$$

where ℓ is the slit width.

This equation enables one to obtain the averaged ΔH values, knowing the rate of thermoosmotic flow j_T, the sizes of pores ℓ, and the mean viscosity of liquid η.

It has been of interest to examine how the values of ΔH vary at different degrees of overlapping of the boundary layers. For this purpose, experiments were carried out with the porous bodies of the same composition (eg., porous glass), but having different mean radii r of pores (165,80). Samples of porous glass had the form of disks 2.5 cm in diameter and 0.1 cm thick. The technique of the measurements was described earlier (165).

Figure 12 shows the results of measuring the thermoosmosis rates j_T, as a function of the temperature gradient for five samples of porous glass (r = 45 Å - 1.5 μm). The experimental dependences j_T(grad T) are in good agreement with theory [Eq. (11)]. It follows from the obtained data that the value and the sign of the thermoosmosis coefficient χ prove to be dependent on the pore size. In the case of wide pores, where the boundary layers do not overlap (linear plot 5), the thermoosmosis flow is directed toward the hot side ($\chi > 0$). This corresponds to a decreased enthalpy of single boundary layers of water ($\Delta H < 0$), i.e., an increased energy of intermolecular hydrogen bonds.

In finer pores, the thermoosmotic flow is directed toward the cold side ($\chi < 0$), which corresponds to an increased mean enthalpy of water in fine pores. The negative values of χ and, accordingly, the positive values of ΔH increase as the pore radius further decreases. This result may be explained by the progressing destruction of the intermolecular bonds in water within the overlapping zone. The weakening of the intermolecular bonds of water molecules in thin pores of cellulose acetate membranes and of montmorillonite is confirmed by IR spectroscopy (166) and NMR (167) studies.

The water layers nearest to the surface probably possess negative ΔH values, independent of the radius of pores. Here, the intermolecular bonds are stronger, as in the case of polymolecules adsorption layers of water on quartz surfaces (47). The absolute values of ΔH decrease as the distance from the surface increases. Then they change the sign and become positive in the overlapping zone.

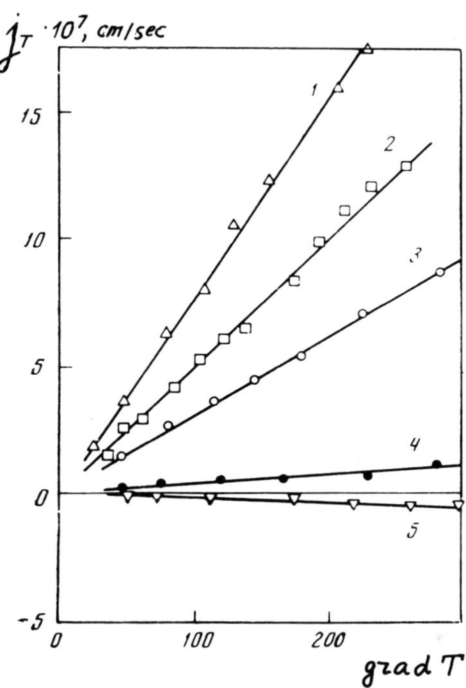

Fig. 12. Dependence of the thermoosmotic flow rate on the temperature gradient in the porous glass having the mean radius of pores: r = 45 Å (1); 83 Å (2); 100 Å (3); 550 Å (4); 1.5 μm (5); t = 20°C (grad T, degree/cm).

The ΔH values that are established in the experiments made with porous glass amount to $\sim 10^{-3}$-0.1 cal/mole, depending on the pore radius. They are very small as compared, for example, with the specific melting heat of ice I, equal to 1440 cal/mole. This means that thermoosmotic measurements open up the possibility of detecting even very small changes in the structure of water. Therefore, the thermoosmosis method may be considered as one of the most sensitive methods of registering structural changes.

As shown by the data thus presented, this method enables one to detect the overlapping of the boundary layers already in pores about 550 Å in radius. Thus, the most sensitive method gives also the largest thickness of the boundary layer of water.

As in the case of water viscosity in fine capillaries (Fig. 7), the values of j_T decrease as the mean temperature increases (Fig. 13); this is indicative of the thermal destruction of the modified structure of water in fine pores. At temperatures above 65-70°C, the thermoosmotic flux is not registered ($j_T = 0$). At this temperature, the specific enthalpy of water in fine pores does not differ from the specific enthalpy of bulk water. The same result ($j_T = 0$) is obtained in the hydrophobized porous glass.

When the pores are filled with electrolyte solutions, the thermoosmosis may be accompanied by electroosmosis arising under the effect of the thermodiffusion potential (168,169). In the aforementioned experiments made with the porous glass and pure water, this effect was absent. The liquid flow rate was not changed by the short-circuiting of electrodes.

In the general case, the influence of the surface charge is able to make an additional contribution to the effect of the thermoosmosis proper. Polarization of the water molecules in the electrical field of surfaces would also cause a change in the specific enthalpy of liquid in the region of EDL. However, this contribution to ΔH, which was calculated earlier (170), is considerably smaller than the ΔH values experimentally obtained (164, 165).

In experiments carried out with fine-pored glass ($r < 550$ Å), the negative values of χ for water were on the order of 10^{-6}-10^{-7} cm^2/sec. Similar values of $\chi = -10^{-7}$ cm^2/sec were obtained for pure water in cellulose acetate membranes (171), when the thermodiffusion potential exerted no influence either. For electrolyte

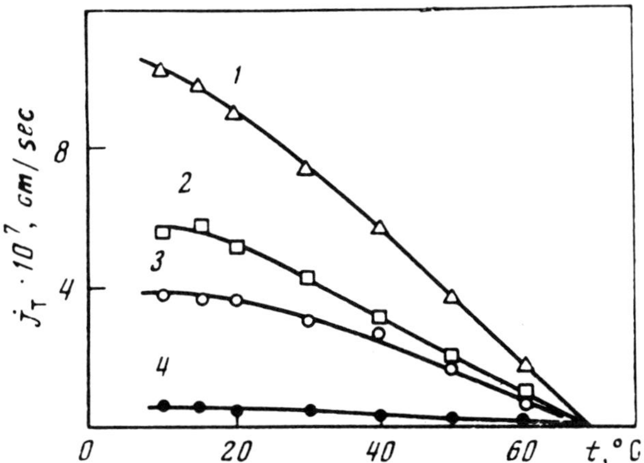

Fig. 13. Influence of mean temperature t on the thermoosmosis flow rate j_T (at grad T = 120°/cm) in the porous glass having the mean radius of pores: r = 45 Å (1); 84 Å (2); 100 Å (3); 550 Å (4).

solutions in thin pores, as well as for ion-exchange membranes, the values of χ vary within a very broad range: Their sign and value depend on the composition and concentration of the solution, as well as on the temperature gradient (171,172).

In the case of clays, the interpretation of thermoosmotic measurements is complicated not only by the presence of electrolytes, but also by the nonlinearity of filtration dependence $v(P)$ (173), which is probably attributable to the porous structure deformation effect. In many systems, even in the fine-pored ones, no thermoosmosis is detected, if the surface of the system is insufficiently hydrophilic to provide boundary layers of an appreciable size and with a sufficient degree of structural change.

Another method of examination of thermoosmosis consists in noting that the changes in the profile of a thin liquid layer are observed under the effect of the temperature gradient directed along the external normal to the three-phase contact line disposed in the substrate plane (174). The surface tension gradient having been caused by the gradient of temperature, its influence was similar to that of the air stream in the blow-off method: it turned the film to a sloping wedge.

As has been shown, in blowing off a silicone polymer film, the equidistant interference bands are revealed, which are indicative of the viscosity being constant at all the points of the layer. Now the shape of the film forming under the influence of the temperature gradient is different: The equidistant bands are observed only where thickness exceeds a micron. At smaller thicknesses, the interference bands draw closer together, which is attributable to the existence of another flow mechanism, namely, thermoosmotic flow. That the entire layer slips as a whole under the influence of thermoosmotic effect toward the hot side and that a resistance is offered to the motion of the wetting perimeter to the outside explain an increase in the profile steepness close to that perimeter. Simultaneously, this also shows that the thermoosmotic effect is more sensitive to the structure of the boundary layers than viscosity.

The application of this method in examining the incomplete hydrated benzontrone presents a point of still greater interest. In this case, the zone of more narrow interference bands is separated from that of equidistant bands by the zone, in which the distances between the bands are increased, exceeding their distance in the latter zone (of maximum thickness). This implies the inversion of the direction of thermoosmotic slip with distance from the substrate, thus confirming the earlier assumption of the inversion of the sign of the excess of ΔH (the specific enthalpy) at a certain distance from the substrate.

IV. CONCLUSION

At present, in view of the absence of liquid state theory, no strictly quantitative prediction of the structure of the boundary layers is possible. From the experimental facts, however, it is clear that the strongest structural changes must occur close to the most hydrophilic surfaces free from hydrophobic impurities.

The state of the boundary layers of water close to the hydrophobic surfaces requires some special consideration. It may be supposed that here the structural state of boundary water is different, whether from the bulk state or from the state close to hydrophilic surfaces. This modified structure is responsible for the so-called hydrophobic interaction that has been studied to a sufficient extent for molecules (175), and is probably revealed also in the interaction of macroscopic bodies (176,177). Close to a hydrophobic surface, the slip of water may occur (178), which is similar to that of mercury in glass capillaries (179). This is a proof of the modified state of water layers adjacent to the hydrophobic surface.

Increasing electrolyte concentration in the boundary layers of water and raising temperature destroy the modified structure of the boundary layers of water close to the hydrophilic substrates. The overlapping of the boundary layers in fine pores or in thin interlayers between the approaching particles causes additional structural changes in the overlapping zone. This effect necessitates further careful examination.

Water as well as some other polar liquids form boundary layers (essentially polymolecular). Without this assumption, it would be impossible to explain the aforementioned facts. However, a peculiar situation has arisen. The first proofs of the thesis thus formulated were obtained about 40 years ago, parallel with the experimental research lying at the basis of the development of the theory of stability of lyophobic colloids (180,81).

These researches have since been further systematically developed and extended. However, they were almost invariably either ignored or sceptically considered, or attempts were made to refute some of the proofs. The basis of this attitude is the presupposed scepticism with regard to all that is new and not found in the main current of the development of the science of surface phenomena and stimulated in the first place by the successes in investigation and statement of the important role of monolayers. These successes are associated with the names of Langmuir, Harkins, Adams, and others: They have given rise (essentially without sufficient substantiation) to the concept that the radius of surface effects is limited to a few monolayers.

The first breach was made by researches into the electrostatic and molecular interactions between macroobjects. However, a further perception of a broader concept of the surface long-range effect of a different mechanism was delayed for a long time. It would be unfair not to note here the views of an eminent British physicochemist, W. Hardy, whose views had been well in advance of his time. He had formulated a concept on diachysis, some sort of a relay of surface effects over great distances from the substrate into the liquid (181). Unfortunately, an attempt at substantiating this hypothesis by direct experiments proved to be erroneous, so that a greater success has been achieved by Soviet scientists.

Measurement of the viscosity of liquid interlayers by Bastow and Bowden (182) may be cited as an opposite example of scepticism corroborated even by the misleading attempts. The authors of that work did not detect any appreciable deviations of viscosity from the bulk value. But they had applied the interlayer thickness measurement method, which gave, however, an error of ~0.2 μm, the error being much greater than the thickness at which a variation in the viscosity of interlayers of the liquids investigated might be noticed. These authors had come to a conclusion that there could be no such variation in viscosity at all, which conclusion had also been joined by many followers. This reminds us of the case where one person had to face five witnesses of a crime alledgedly committed by him. The accused replied that he could have named scores of persons that did not witness the crime ascribed to him. What has been said does not mean that certain categories of work pretending to prove the special properties of the boundary layers are sufficiently convincing, or that they cannot be criticized.

At present, there are noticeably some changes in the attitude to structural effects and forces. This is proved by the experimental results as given in the present paper. Probably, the clearest example of the manifestation of structural effects is the membrane separation of aqueous solutions by reverse osmosis.

A further examination of structural effects by the aid of model systems, under controlled conditions, will enable one to obtain new information that will be of importance for the development of physical chemistry of surface phenomena and its numerous practical applications.

References

1. C. A. Croxton, "Liquid State Physics." Cambridge Univ. Press, Cambridge, Massachusetts, 1974.
2. B. Borstnik and A. Azmar. *Mol. Phys. 30,* 1565, 1975.
3. D. Henderson, F. F. Abraham, and J. A. Barker. *Mol. Phys. 31,* 1291, 1976; E. Waisman, D. Henderson, and J. L. Lebowitz. *Mol. Phys. 32,* 1374, 1976.
4. S. Toxvaerd and E. Praestgaard. *J. Chem. Phys. 67,* 5291, 1977; *Ann. Isr. Phys. Soc. 2,* 495, 1978.
5. D. E. Sullivan and G. Stell. *J. Chem. Phys. 69,* 5450, 1978.
6. A. Rahman and F. H. Stillinger. *J. Chem. Phys. 55,* 336, 1971.
7. F. H. Stillinger and A. Rahman. *J. Chem. Phys. 57,* 1281, 1972; *60,* 1545, 1974.

8. E. I. Katz. *Zhurn. Eksper. Teoretich. Fisiki 70*, 1394, 1976.
9. J. E. Proust, L. Ter-Minassian-Saraga, and E. Guyon. *Solid State Comm. 11*, 1227, 1972; J. E. Proust and L. Ter-Minassian-Saraga. *Colloid Polymer Sci. 254*, 492, 1976.
10. J. E. Perez, J. E. Proust, and L. Ter-Minassian-Saraga. *Colloid Polymer Sci. 256*, 784, 1978.
11. C. K. Yuu and A. G. Fredrickson. *Phys. Fluids 16*, 1, 1973.
12. B. V. Derjaguin and Z. M. Zorin. *Zhurn. Fisich. Khimii 29*, 1755, 1955; *Proc. II Int. Congr. Surf. Activ., V. 2*, p. 145. Butterworths, London, 1957.
13. B. A. Pethica. *Spec. Disc. Farad. Soc.*, No. 1, 7, 1970.
14. B. V. Derjaguin and N. V. Churaev. *J. Colloid Interface Sci. 49*, 249, 1974.
15. R. Parsons. *Croat. Chem. Acta 48*, 597, 1976.
16. B. V. Derjaguin. *Chem. Ser. 9*, 97, 1976; *Pure Appl. Chem. 48*, 387, 1976.
17. J. Lyklema. *J. Colloid Interface Sci. 58*, 242, 1977; *Croat. Chem. Acta 50*, 77, 1977.
18. J. A. Kitchener. *Chem. Britain 13*, 105, 1977.
19. J. Clifford. *In* "Water in Disperse Systems,: p. 75. Plenum Press, New York and London, 1975.
20. J. N. Israelachvili and B. V. Ninham. *J. Colloid Interface Sci. 58*, 14, 1977.
21. B. V. Derjaguin. *Disc. Faraday Soc.*, No. 65, 306, 1978.
22. R. M. Pashley and J. A. Kitchener. *J. Colloid Interface Sci. 71*, 491, 1979.
23. A. I. Rusanov, F. M. Kuni and E. N. Brodskaya. *Zhurn. Fisich. Khimii 44*, 553, 1231, 1970; *Kolloidn. Zh. 32*, 836, 1970; *Res. Surf. Forces*, ed. B. V. Derjaguin, V. 4, p. 240. Cons. Bureau, New York, 1975.
24. V. M. Nabutovskij, V. R. Belosludov, and A. M. Korotkih. *Kolloidn. Zh. 41*, 722, 876, 1979.
25. D. J. Mitchell, B. W. Ninham, and B. A. Pailthorpe. *J. Colloid Interface Sci. 64*, 194, 1978; *J. Chem. Soc. Faraday Trans. II 74*, 1116, 1978.

26. D. Y. C. Chan, D. J. Mitchell, B. W. Ninham, and B. A. Pailthorpe. *Mol. Phys. 35,* 1669, 1978.
27. G. A. Martynov and V. P. Smilga. *Kolloidn. Zh. 27,* 250, 1965.
28. G. I. Distler, V. M. Gerasimov, and V. G. Obronov. *Thin Solid Films 10,* 195, 1972.
29. G. I. Distler *et al.* "Decoration of the Surface of Solids." Published in Russian by "Nauka," Moscow, 1976.
30. V. I. Pshenitsyn and A. I. Rusanov. Collection "Surface Forces in Thin Films." Published in Russian by "Nauka," Moscow, p. 61, 1979.
31. A. N. Frumkin. *Zhurn. Fisich. Khimii 12,* 337, 511, 1938.
32. A. Kitahara, M. Fujiwara, T. Ogawa, and T. Ishibashi. *J. Colloid Interface Sci. 51,* 540, 1975.
33. A. Sanfeld, C. Devillez, and M. Lubelski. *Electrochim. Acta 13,* 1937, 1968.
34. A. V. Kiselev and V. I. Lygin. *Usp. Khimii 31,* 351, 1962; A. V. Kiselev. *Disc. Faraday Soc. 52,* 14, 1971.
35. A. C. Zettlemoyer, F. J. Micale, and K. Klier. *In* "Water in Disperse Systems,: p. 249. Plenum Press, New York and London, 1975.
36. A. C. Zettlemoyer and H. H. Hsing. *J. Colloid Interface Sci. 55,* 637, 1976; *58*, 263, 1977.
37. J. Hongardy, J. M. Serratoza, W. Stone, and H. van Olphen. *Spec. Disc. Faraday Soc.* No. 1, 187, 1970.
38. J. Clifford, J. Oakes, and G. J. T. Tiddy. *In* "Thin Liquid Films and Boundary Layers," p. 175. Academic Press, New York and London, 1971.
39. V. I. Kvlividze and A. V. Krasnushkin. *Doklady AN SSSR 222,* 388, 1975.
40. Yu. I. Tarasevich and F. D. Ovcharenko. "Adsorption on Clay Minerals." Published in Russian by "Naukova Dumka," Kiev, 1975.
41. E. Almagor and G. Belfort. *J. Colloid Interface Sci. 66,* 146, 1978.

42. J. J. Fripiat. *Bull. Cl. Sci. Acad. R. Belg. 56,* 1188, 1970.
43. A. B. Kurzaev, V. I. Kvlividze and V. F. Kiselev. *Biophysica 20,* 544, 1975; Collection "Bound Water in Disperse Systems," V. 4, p. 156. Published in Russian by the Moscow State University, 1977.
44. V. I. Kvlividze and A. B. Kurzaev, Collection "Surface Forces in Thin Films," ed. B. V. Derjaguin, p. 211. Published in Russian by "Nauka," Moscow, 1979.
45. G. V. Yukhnevich. "Infrared Spectroscopy of Water." Published in Russian by "Nauka," Moscow, 1973.
46. G. F. Ershova and N. V. Churaev. *Kolloidn. Zh. 39,* 1151, 1977.
47. G. F. Ershova, Z. M. Zorin, and N. V. Churaev. *Kolloidn. Zh. 41,* 19, 1979.
48. G. F. Ershova and N. V. Churaev. *Kolloidn. Zh. 41,* 1176, 1979.
49. N. V. Churaev, I. P. Sergeeva, V. D. Sobolev, and B. V. Derjaguin. *J. Colloid Interface Sci.* (to be published).
50. R. M. Pashley. *Surf. Sci. 71,* 139, 1978.
51. A. C. Hall. *J. Phys. Chem. 74,* 2742, 1971.
52. D. H. Bangham. *J. Chem. Phys. 14,* 352, 1946.
53. A. W. Adamson. *J. Colloid Interface Sci. 27,* 180, 1968; M. E. Tadros, P. Hu, and A. W. Adamson. *J. Colloid Interface Sci. 49,* 184, 1974; P. Hu and A. W. Adamson. *J. Colloid Interface Sci. 59,* 605, 1977.
54. M. N. Plooster and S. N. Gitlin. *J. Phys. Chem. 75,* 3322, 1971.
55. G. F. Ershova, Z. M. Zorin, and N. V. Churaev. *Kolloidn. Zh. 37,* 208, 1975.
56. B. V. Derjaguin, Z. M. Zorin, N. V. Churaev, and V. A. Shishin. *In* "Wetting, Spreading, and Adhesion," p. 201. Academic Press, London, 1977.
57. D. H. Everett. *Isr. J. Chem. 14,* 267, 1975.
58. H. Kern, W. V. Rybinski, and G. M. Findenegg. *J. Colloid Interface Sci. 59,* 301, 1977.

59. M. Grosse-Rhode and G. H. Findenegg. *J. Colloid Interface Sci. 64,* 374, 1978.

60. D. M. Anderson and P. F. Low. *Soil Sci. Soc. Amer. Proc. 22,* 99, 1958.

61. W. F. Bradley. *Nature 183,* 1614, 1959.

62. C. T. Deeds and H. van Olphen. *Adv. Chem. Ser.,* No. 33, 332, 1961.

63. G. E. van Gils. *J. Colloid Interface Sci. 30,* 272, 1969.

64. B. V. Zheleznij. *Izv. AN SSSR, Ser. Khimich.,* No. 6, 1276, 1972.

65. M. J. Tait and F. Franks. *Nature 230,* 91, 1971.

66. Yu. V. Gurikov. *Collection* "Surface Forces in Thin Films," ed. B. V. Derjaguin, p. 76. Published in Russian by "Nauka," Moscow, 1979.

67. B. V. Derjaguin, Yu. Šutor, S. V. Nerpin, and M. A. Arutyunyan. *Doklady AN SSSR 161,* 147, 1965.

68. W. Drost-Hansen. *Chem. Phys. Letters 2,* 647, 1968; *Ind. Eng. Chem. 61,* 10, 1969.

69. V. V. Karasev, B. V. Derjaguin, and E. N. Efremova. *Kolloidn. Zh. 24,* 471, 1962.

70. V. V. Karasev, B. V. Derjaguin, and E. N. Chromova. *Res. Surf. Forces,* V. 3, p. 25. Conc. Bureau, New York, London, 1971.

71. B. V. Derjaguin, V. V. Karasev, and E. N. Chromova. *Kolloidn. Zh.* (In press).

72. B. H. Bijsterbosch and J. Lyklema. *Adv. Colloid Interface Sci. 9,* 147, 1978.

73. D. D. Eley, M. J. Hey, and C. K. Rix. *J. Colloid Interface Sci. 65,* 592, 1978.

74. D. Green-Kelly and B. V. Derjaguin. *Doklady AN SSSR 153,* 638, 1964; *Trans. Faraday Soc. 60,* 449, 1964.

75. B. V. Derjaguin, N. A. Krylov, and V. F. Novik. *Doklady AN SSSR 193,* 126, 1970; Collection "Surface Forces in Thin Films and the Stability of Colloids," p. 164. Published in Russian by "Nauka," Moscow, 1974.

76. S. S. Dukhin and V. N. Shilov. "Dielectrical Phenomena and Double Layer in Disperse Systems and Polyelectrolytes." Published in Russian by "Naukova Dumka," Kiev, 1972.
77. M. M. Snitkovskij. Collection "Surface Forces in Thin Films and the Stability of Colloids," p. 38. Published in Russian by "Nauka," Moscow, 1974.
78. B. V. Derjaguin and M. M. Snitkovskij. *Kolloidn. Zh. 37*, 254, 1975.
79. B. V. Derjaguin and N. V. Churaev. *J. Colloid Interface Sci. 66*, 389, 1978.
80. B. V. Derjaguin and N. V. Churaev. *Croat. Chem. Acta 50*, 187, 1977.
81. B. V. Derjaguin and L. D. Landau. *Acta Physicochim. URSS 14*, 633, 1941.
82. E. J. W. Verwey and J. Th. G. Overbeek. "Theory of the Stability of Lyophobic Colloids." Elsevier, New York and Amsterdam, 1948.
83. T. D. Blake and J. A. Kitchener. *J. Chem. Soc. Faraday Trans.*, p. I, *68*, 1435, 1972.
84. S. Marcelja and N. Radic. *Chem. Phys. Lett. 42*, 129, 1976.
85. B. V. Derjaguin. *Zhurn. Fisich. Khimii 14*, 157, 1940; B. V. Derjaguin and L. M. Shcherbakov. *Kolloidn. Zh. 23*, 40, 1961.
86. H. J. Schulze. *Colloid Polymer Sci. 254*, 438, 1976; *256*, 1037, 1978.
87. J. N. Israelachvili, G. E. Adams, *J. Chem. Soc. Faraday Trans.*, p. I, *74*, 975, 1978.
88. J. N. Israelashvili. *Faraday Disc. Chem. Soc.*, No. 65, 20, 1978.
89. G. Peschel and K. H. Adlfinger. *Naturwissenschaften 54*, 614, 1967.
90. K. H. Adlfinger, R. Schnorrer, and G. Peschel. *Z. angew. Phys. 29*, 136, 1970.
91. G. Peschel and K. H. Adlfinger. *Z. Phys. Chem. (BRD) 63*,

150, 1969; *Z. Naturforsch 26a*, 707, 1971; *Z. angew. Phys. 31*, 137, 1971.

92. G. Peschel and R. Schnorrer. *Z. Phys. Chem. (BRD) 75*, 97, 1971.
93. G. Peschel, K. H. Adlfinger, and G. Schwarz. *Naturwissenschaften 61*, 215, 1974.
94. G. Peschel and P. Belouschek. *Z. Phys. Chem. (BRD) 108*, 145, 1977.
95. Zh. G. Blashctuk and Yu. M. Glazman. *Res. Surf. Forces*, V. 4, p. 72. Cons. Bureau, New York, London, 1975.
96. A. A. Baran, I. M. Solomentzova, V. V. Mank, and O. D. Kurilenko. *Doklady AN SSSR 207*, 363, 1972; *Kolloidn. Zh. 37*, 219, 1975.
97. R. D. Harding. *J. Colloid Interface Sci. 35*, 172, 1971.
98. L. M. Barclay and R. H. Ottewill. *In* "Thin Liquid Films and Boundary Layers," p. 138. Academic Press, New York, London, 1971.
99. Yu. M. Chernobereshskij, T. F. Girfanova, L. M. Labunetz, and E. B. Golikova. Collection "Surface Forces in Thin Films," p. 67. Published in Russian by "Nauka," Moskva, 1979.
100. Yu. M. Popovskij and B. V. Derjaguin. *Doklady AN SSSR 159*, 897, 1964; *175*, 385, 1967; Yu. M. Popovskij. Collection "Surface Forces in Thin Films and the Stability of Colloids." Published in Russian by "Nauka," Moscow, 1974, p. 242; 1979, p. 81.
101. B. V. Derjaguin, Yu. M. Popovskij, and G. P. Silenko. *Doklady AN SSSR 207*, 1153, 1972; *239*, 828, 1978.
102. Z. M. Zorin. *Res. Surf. Forces* V. 2, p. 134. Cons. Bureau, New York, 1966.
103. A. Sheludko and D. Platikanov. *Doklady AN SSSR 138*, 2, 1961.
104. B. V. Derjaguin and V. V. Karasev. *Doklady AN SSSR 62*, 761, 1948; *101*, 289, 193.
105. B. V. Derjaguin, V. V. Karasev, N. N. Zakhavaeva, and V. P. Lazarev. *Zhurn. Tekhnich. Fisiki 27*, 1976, 1957; Wear *1*, 277, 1958.

106. B. V. Derjaguin, V. V. Karasev, I. A. Lavygin, I. I. Skorokhodov, and E. N. Chromova. *Res. Surf. Forces* V. 4, p. 227. Cons. Bureau, New York, London, 1975.

107. B. V. Derjaguin and N. N. Zakhavaeva. Collection "Research in Polymer Compounds," p. 223. Published in Russian by AN SSSR, Moscow-Leningrad, 1949; *Res. Surf. Forces V. 1,* ed. B. V. Derjaguin, p. 110. Cons. Bureau, New York, 1963.

108. N. N. Zakhavaeva, B. V. Derjaguin, A. M. Chomutova, and S. V. Andrejev. *Res. Surf. Forces* V. 2, p. 156. Cons. Bureau, New York, 1966.

109. W. D. Bascom and C. R. Singleterry. *J. Colloid Interface Sci. 66,* 559, 1978.

110. B. V. Derjaguin, V. V. Karasev, and E. N. Chromova. *J. Colloid Interface Sci. 66,* 573, 1978.

111. B. V. Derjaguin, V. V. Karasev, V. M. Starov, and E. N. Chromova. *J. Colloid Interface Sci. 67,* 465, 1978.

112. N. V. Churaev, V. D. Sobolev and Z. M. Zorin. *In* "Thin Liquid Films and Boundary Layers," p. 213. Academic Press, New York, London, 1971.

113. B. V. Derjaguin, B. V. Zheleznyi, Z. M. Zorin, V. D. Sobolev, and N. V. Churaev. Collection "Surface Forces in Thin Films and the Stability of Colloids," ed. B. V. Derjaguin, p. 90. Published in Russian by "Nauka," Moscow, 1974.

114. O. A. Kiseleva, V. D. Sobolev, V. M. Starov, and N. V. Churaev. *Kolloidn. Zh. 41,* 245, 1979.

115. J. T. Davies and E. K. Rideal. "Interfacial Phenomena," p. 126. Academic Press, New York, 1961.

116. N. V. Churaev. Proc. Symp. "Water in Heavy Soils," V. 1, p. 8. Bratislava, 1976.

117. V. I. Lashnev, V. D. Sobolev, and N. V. Churaev. *Teor. Osnovy Khimich. Technol. 10,* 926, 1976.

118. B. V. Derjaguin, N. N. Zachavaeva, and A. M. Lopatina. *Res. Surf. Forces,* V. 1, p. 141. Cons. Bureau, New York, 1963.

119. F. D. Ovcharenko, M. I. Medvedeva, A. G. Brekhunets, and V. V. Mank. *Kolloidn. Zh. 38,* 286, 1976.

120. G. Peschel and K. H. Adlfinger. *Naturwissenschaften 56,* 558, 1969; *J. Colloid Interface Sci. 34,* 505, 1970.

121. A. D. Roberts. *J. Appl. Phys. D4,* 433, 1971.

122. P. F. Low. *Soil Sci. Soc. Amer. Proc. 40,* 500, 1976; *43,* 651, 1979.

123. R. J. Hunter and J. V. Leyendekkers. *JCS Faraday Trans.* 1, *74,* 450, 1978.

124. Z. M. Tovbina. *Res. Surf. Forces, V. 3,* ed. B. V. Derjaguin, p. 20. Cons. Bureau, New York, London, 1971.

125. J. L. Anderson and J. A. Quinn. *JCS Faraday Trans.* 1, *68,* 744, 1972.

126. S. S. Barer, N. V. Churaev, B. V. Derjaguin, O. A. Kiseleva, and V. D. Sobolev. *J. Colloid Interface Sci. 74,* 173, 1980.

127. H. H. G. Jellinek. *J. Colloid Interface Sci. 25,* 192, 1967.

128. N. F. Bondarenko. "Physics of the Motion of Underground Water." Published in Russian by "Hydrometeoizdat," Leningrad, 1978.

129. S. P. Li. *Soil Sci. 95,* 410, 1963.

130. R. I. Miller and P. F. Low. *Soil Sci. Soc. Amer. Proc. 27,* 605, 1963.

131. D. Swartzendruber. *Soil Sci. 93,* 22, 1962.

132. D. A. Russell and D. Swartzendruber. *Soil Sci. Soc. Amer. Proc. 35,* 21, 1971.

133. E. C. Childs and E. Tzimas. *J. Soil Sci. 22,* 319, 1971.

134. H. W. Olsen. *Soil Sci. Soc. Amer. Proc. 29,* 135, 1965; *Water Resources Res. 2,* 287, 1966.

135. R. D. Jackson. *Soil Sci. Soc. Amer. Proc. 31,* 713, 1967.

136. V. Novak. *Vodohop. Casopis 20,* 213, 1972.

137. H. W. Olsen. *Clays Clay Minerals 11,* 131, 1962.

138. V. Novak. *J. Soil Sci. 23,* 248, 1972.

139. V. A. Nelidov. *Inzh.-Fisich. Zhurnal 21,* 1017, 1971.

140. R. Skawinski and A. Lasowska. *Bull. Acad. Polonaise Sci., Ser. Sci. Techniques 22,* 235, 307, 1974.

141. L. M. Simuni. *Pzicladnaja Math. Teor. Fisika,* No. 6, 106, 1965.

142. V. S. Golubev. *Doklady AN SSSR 238,* 1318, 1978.
143. R. P. Gupta and D. Swartzendruber. *Soil Sci. Soc. Amer. Proc. 26,* 6, 1962.
144. A. Paulovassilis. *Soil Sci. 113,* 81, 1972.
145. N. V. Churaev. *Kolloidn. Zh. 25,* 718, 1963.
146. G. A. Nikitin. *Izv. Vuzov, Aviats. Tekhnika* No. 4, 38, 1965.
147. A. T. J. Hayward and J. D. Isdale. *Brit. J. Appl. Phys. D2,* 251, 1969.
148. K. J. Mysels and S. P. Frankel. *J. Colloid Interface Sci. 66,* 166, 1978.
149. V. M. Starov and N. V. Churaev. *Kolloidn. Zh. 41,* 297, 1979.
150. N. V. Churaev and B. V. Derjaguin. *Doklady AN SSSR 169,* 396, 1966.
151. C. L. Rice and R. Whitehead. *J. Phys. Chem. 69,* 4017, 1965.
152. S. S. Dukhin. "Electrical Conductivity and Electrokinetic Properties of Disperse Systems. "Published in Russian by "Naukova dumka." Kiev, 1975.
153. P. P. Zolotarev and N. V. Churaev. *Kolloidn. Zhurn. 32,* 56, 1970.
154. A. V. Dumanskij "Lyophility of Disperse Systems." Published in Russian by AN USSR, Kiev, 1960.
155. N. Lakshminarayanaian. "Transport Phenomena in Membranes." Academic Press, New York, 1969.
156. S. Sourirajan. *Pure and Appl. Chem. 50,* 593, 1978.
157. Yu. I. Dytnerskij. "Reverse Osmosis and Ultrafiltration." Published in Russian by "Khimija," Moscow, 1978.
158. B. V. Derjaguin, G. P. Sidorenkov, E. A. Zubaschenko, and E. V. Kiseleva. *Kolloidn. Zhurn. 9,* 335, 1947.
159. I. E. Dzyaloshinskij, E. M. Lifshitz, and L. P. Pitajevskij. *Zh. Eksp. Teor. Fiz. 37,* 229, 1959; *Advan. Phys. 10,* 165, 1961.

160. B. V. Derjaguin, I. E. Dzyaloshinskij, M. M. Koptelova, and L. P. Pitajevskij. *Disc. Faraday Soc. 40,* 246, 1965.
161. B. V. Derjaguin and M. M. Koptelova. *Res. Surf. Forces,* V. 4, p. 182. Cons. Bureau, New York, London, 1975.
162. B. V. Derjaguin and N. V. Churaev. *Dokl. Akad. Nauk SSSR 222,* 554, 1975; *J. Colloid Interface Sci. 62,* 369, 1977.
163. G. A. Martynov, V. M. Starov, and N. V. Churaev. *Kolloidn. Zhurn. 42,* No. 3, 489; No. 4, 607, 1980; B. V. Derjaguin, N. V. Chuzaev, G. A. Maztynov, *J. Colloid Interface Sci.* (in press).
164. B. V. Derjaguin and G. P. Sidorenkov. *Doklady AN SSSR 32,* 622, 1941.
165. P. A. Voznyj and N. V. Churaev. *Kolloidn. Zhurn. 39,* 264, 438, 1977.
166. C. Toprak, J. N. Agar, and M. Falk. *JCS Faraday Trans. I 75,* 803, 1979.
167. V. V. Mank, Z. E. Sujunova, Yu. I. Tarasevich, and F. D. Ovcharenko. *Doklady AN SSSR 202,* 117, 1972.
168. C. W. Carr and K. Sollner. *J. Electr. Soc. 109,* 616, 1962.
169. Y. Kobatake and H. Fujuta. *J. Chem. Phys. 41,* 2963, 1964.
170. N. V. Churaev, B. V. Derjaguin, and P. P. Zolotarev. *Doklady AN SSSR 183,* 1139, 1968.
171. M. P. Dariel and O. Kedem. *Isr. Atom. Energy Commis.* N° 1262, 105, 1971; *J. Phys. Chem. 79,* 336, 1975.
172. J. W. Lorimer and S. H. Chan. *Prep. Int. Symp. Macromol.,* Helsinki, V. 3, Sec. 2, p. 219, 1972.
173. R. C. Srivastava and A. K. Jain. *Indian J. Chem. 12,* 1276, 1974.
174. V. V. Karasev, B. V. Derjaguin, and E. N. Chromova. *Res. Surf. Forces* V. 2, p. 153. Cons. Bureau, New York, 1966.
175. C. Tanford. "The Hydrophobic Effect." Wiley-Interscience, New York, 1973.
176. V. A. Pchelin. *Doklady AN SSSR 194,* 621, 1970.
177. V. V. Yaminskij and V. A. Pchelin. *Doklady AN SSSR 210,* 157, 1973; *Kolloidn. Zhurn. 37,* 412, 1975.

178. E. Schnell. *J. Appl. Phys. 27,* 1149, 1956.

180. B. V. Derjaguin and M. M. Kussakov. *Acta Phys.-Chim. URSS 10,* 25, 1939; B. V. Derjaguin. *Trans. Faraday Soc. 36,* 203, 1940; *Acta Phys.-Chim. URSS 12,* 314, 1940.

181. W. B. Hardy. "Collected Scientific Papers," p. 827. Cambridge Univ. Press, 1936.

182. S. Bastow and F. P. Bowden. *Proc. R. Soc. (London) A151,* 220-233, 1935.

COUPLING OF IONIC AND NONELECTROLYTE FLUXES IN ION SELECTIVE MEMBRANES

O. K. Stefanova
Department of Physical Chemistry
Leningrad State University
Leningrad, USSR

M. M. Shultz
Institute of Silicate Chemistry
Academy of Sciences of the USSR
Leningrad, USSR

I. INTRODUCTION

This chapter is concerned with the problems of transport phenomena in ion-conducting media. The analysis is carried out by the method of thermodynamics of irreversible processes permitting a general description of the phenomena of mass and energy transfer under the influence of various forces. Among these most complex phenomena, of particular interest is the interaction of some mate-

ISBN 0-12-571814-4

rial fluxes and, primarily, the coupling of fluxes of charged and neutral particles. Such a dynamic interaction is found to occur in all electrolyte-participating transport phenomena, irrespective of the nature of forces inducing fluxes of separate components. The easiest way to gain information about such an interaction is to investigate the processes in which it is expressed in a pure form. Thus one may create in a system a motive force that would directly affect only some of the components. Any transfer of other components of the system under the influence of this force would indicate a dynamic interaction of appropriate components. The transfer of neutral components in the electric field as well as the displacement of electrical charges due to nonequilibrium distribution of nonelectrolyte components in the system provide a vivid example of such processes.

This work is devoted to the analysis of such kind of phenomena: the coupling of fluxes of neutral and charged particles in the membrane potential. In what follows, we shall consider some general principles of thermodynamics, which will be then specified and illustrated by experimental materials using as an example comparatively simple systems, primarily membranes with fixed charges, provided they show high selective permeability for definite species of ions. In conclusion we shall turn to processes in membranes with neutral carriers regarding induced ion transport in these systems as a limiting case of coupling of ionic and nonelectrolyte fluxes.

II. GENERAL CONSIDERATION OF THE COUPLING EFFECT OF IONIC AND NONELECTROLYTE FLUXES IN A MEMBRANE POTENTIAL FROM THE VIEWPOINT OF THE THERMODYNAMICS OF IRREVERSIBLE PROCESSES

The principal benefit in applying the thermodynamics of irreversible processes to the study of transport phenomena is that it permits to take into account, in the most general terms, the interaction and interdependence of some material and energy fluxes.

This is achieved by introducing an equation for definite species of fluxes with some members that reflect the influence, not only of the motive force proper, but also that of all other efficient forces in the system (Haase, 1963; Denbigh, 1951; de Groot, 1963; Prigogine, 1961). Thus in the linear approximation the flux of particles i in the direction x under isobar - isothermal conditions is expressed as

$$I_i = \sum_j L_{ij}X_j = \sum_j L_{ij} \frac{d\bar{\mu}_j}{dx} ; \qquad j = 1,\ldots,i,\ldots,n \tag{1}$$

where X_j is the force representing a gradient of electrochemical (chemical in the case of neutral species) potential of j particles. L_{ij} is the phenomenological coefficient reflecting (at $i \neq j$) a dynamic interaction of i and j particles and depending only on local properties of the system (i.e. not on their gradients). Equation (1) is true for any medium up to and including a membrane separating two solutions. We shall restrict ourselves to consideration of such membrane systems. Therefore, without touching on the difficulties that selection of a characteristic frame of reference used in studies of transfer in free solutions, we shall read flows in the membrane relative to its matrix. The choice of this reference system is rational for membranes with high hydrodynamic resistance. In these membranes, the flows show low sensitivity to a possible change of hydrostatic pressure within the range of the usual experimental conditions.

Equation (1) forms the basis for a quantitative description of transfer phenomena where the electrical field and the chemical potential gradients of components serve as motive forces.

The interesting effect of coupled fluxes of neutral and charged particles in a membrane potential arises when the membrane contains nonequilibrium neutral components. Flows of relevant particles induced by gradients of their chemical potentials, interact with mobile charges in the membrane. As a result, chemical potential gradients of these neutral components also become a motive force toward charged particles (acting along with the "own" forces, namely, the external electrical field, gradients of electrochemical potentials of charged particles, etc.). Consequently, they must affect ion fluxes or displacement of charges about each other if the electric circuit is disconnected. This influence is dependent both on the degree of nonequilibrium distribution of neutral components (gradients of their chemical potentials) and on the values of corresponding cross phenomenological coefficients in Eq. (1).

Since in the general case the displacement of charges (and as a result, the appearance of an electrical potential gradient in the membrane) occurs not so much because of or due to their being involved in fluxes of nonelectrolytes as due to the nonequilibrium distribution of electrolyte components in the system, for a quantitative consideration of the proper effect of coupled fluxes of neutral and charged particles in the membrane, one must isolate relevant addends in the expression for membrane potential.

If diffuse regions are absent in solutions separated by the membrane, these members arise only at the expense of the membrane diffusion potential. The latter is defined according to the conventional procedure: Eq. (1) is put down for all the species of charged particles, and the resulting system is solved in relation to the electrical potential gradient on condition that there is no current in the system. As a consequence, we obtain:

$$\frac{d\phi}{dx} = - \sum_i \sum_j z_i L_{ij} \frac{d\mu_j}{dx} / F \sum_i \sum_j z_i z_j L_{ij} \tag{2}$$

It would be rational to transform Eq. (2) by introducing reduced transfer numbers determined by the relation

$$\tau_i = \frac{I_i}{\sum_i z_i I_i} \quad \text{if} \quad \frac{d\mu_j}{dx} = 0 \quad i = 1, \ldots, n \tag{3}$$

Here n is the number of particle species that are mobile in the membrane. As seen from the definition, τ_i is the flux of i particles (in moles or gram-ions) induced by the only motive force, the electrical potential gradient and related to the common flow of charged particles in the system. The reduced transfer number τ_i is connected with the common transfer number t_i, meaningful only with respect to charged particles and representing a portion of current carried by ions of the species via the relation

$$t_i = Z_i \tau_i \tag{4}$$

Unlike the ionic transfer numbers, the reduced transfer numbers may be of different signs. Further, we shall consider the transfer number , as positive in the case where the direction of motion of relevant particles (of species) coincides with that of positive charges. If in (3) fluxes are expressed according to Eq. (1), the reduced transfer number may be presented in the following way:

$$t_i = \left(\sum_j Z_j L_{ij} \right) \Big/ \left(\sum_i \sum_j Z_i Z_j L_{ij} \right) \tag{5}$$

Hence it follows that, upon fulfillment of the reciprocity relation

$$L_{ij} = L_{ji} \tag{6}$$

the diffusion potential gradient is defined by values of the reduced transfer numbers:

$$\frac{d\phi}{dx} = -\frac{1}{F} \sum_i \tau_i \frac{d\mu}{dx} \tag{7}$$

Here, the summation includes all species of particles that are mobile in the diffuse layer. With due regard to our stated objectives, it seems only reasonable to divide addends relating to neutral and charged components.

Let us introduce the index K for neutral molecules and the index ℓ for ions. Then the expression for diffusion potential takes the form

$$\phi_{\text{diff}} = -\frac{1}{F}\sum_K \int_{'}^{''} \tau_K d\mu_K - \frac{1}{F}\sum_\ell \int_{'}^{''} \frac{t_\ell}{z_\ell} d\mu_\ell$$

$$= -\frac{RT}{F}\sum_K \int_{'}^{''} t_K d\ln a_K - \frac{RT}{F}\sum_\ell \int_{'}^{''} \frac{t_\ell}{z_\ell} d\ln a_\ell \qquad (8)$$

where τ_K denotes the reduced transfer number of neutral molecules, t_ℓ the ionic transfer numbers, μ_K, μ_ℓ, a_K, and a_ℓ the chemical potentials and activities of relevant components.

Summation is over all species of neutral (K) and charged (ℓ) particles that are mobile in the diffuse layer. The integration limits correlate with the boundaries of this layer.

The first addend in the equation for the diffusion potential (8) is the effect of coupled fluxes of ions and neutral components. It is known that the diffusion potential is a thermodynamically ambiguous value. Therefore, the question of a rigorous identification and experimental determination of the effect of coupled fluxes may be solved only by considering the complete equation for the emf of a galvanic cell comprising a membrane. Besides the diffusion potential in the membrane, the emf of such a cell must contain potential jumps at the membrane-solution boundary and potentials of auxiliary electrodes reversible to one of the ions in the solution. Standard transformations commonly made to obtain an equation for the emf of the cell with a membrane are as follows: The transfer number of one of the charged particles distributed in equilibrium between membrane surface layers and contacting solutions is eliminated (in accordance with the condition $\sum_\ell t_\ell = 1$), and the potential jump at the membrane-solution boundary is obtained from the terms of equality for the electrochemical potentials of this particle in media under investigation. As a result, it is found that the integrals comprising ion transfer numbers may be combined so that the integrand expressions would contain only activities of the electrolytes as a whole. In this instance, the activities of electrolytes relating to the membrane may be eliminated from the final expressions as soon as allowances are made for their equilibrium distribution between the membrane surface layers and relevant solutions.

In a particular case, when coions do not penetrate the membrane, the reduction of emf to electrolyte activities make allowances for equilibrium conditions of exchange of two (or a few) counterion species between the solution and the membrane.

The above scheme may be used for obtaining an equation for the emf of a galvanic cell with a membrane containing any number of electrolytes of any valence type. The number and stoichiometry of electrolytes determine only the number of addends in the expression for emf and the values of constant coefficients that reflect the electrolyte valence type. Therefore, for our further description, we shall restrict ourselves to conclusive results using a comparatively simple system as an example. Let us assume that the system contains two species of cations (M^{z_M} and L^{z_L}), two species of anions (X^{z_X} and Y^{z_Y})[1] and an arbitrary number of nonelectrolytes. An assumption is made that all components of the solution can penetrate the membrane and are sufficiently mobile in it. If the galvanic cell is patterned as

		′	″	
X	solution (1) M^{z_M}, L^{z_L} X^{z_X}, Y^{z_Y} nonelectrolytes (K)	membrane	solution (2) M^{z_M}, L^{z_L} X^{z_X}, Y^{z_Y} nonelectrolytes (K)	X

(9)

[1]*We do not stipulate here for a correlation between cationic and anionic species in solution since in this case only conditions of electrical neutralites are to be met.*

after the above described transformations, the emf may be presented by the relation

$$E = \frac{RT}{\nu_{MX}F} \ln \frac{a_{MX}^{(1)}}{a_{MX}^{(2)}} - \frac{RT}{\nu_{LX}F} \int_{'}^{''} t_L \, d \ln a_{LX}$$

$$+ \frac{RT}{\nu_{MX}F} \int_{'}^{''} t_L \, d \ln a_{MX} + \frac{RT}{\nu_{MX}F} \int_{'}^{''} t_X \, d \ln a_{MX}$$

$$+ \frac{RT}{\nu_{MY}F} \int_{'}^{''} t_Y \, d \ln a_{MY} - \frac{RT}{F} \sum_K \int_{'}^{''} \tau_K \, d \ln a_K \qquad (10)$$

where the bottom indices L, X, and Y symbolize relevant ions and the indices MX, LX, and MY the electrolytes, without reflecting their stoichiometry. ν_{MX}, ν_{LX}, and ν_{MY} indicate the number of equivalents in the molecule of a given electrolyte. The upper indices (1) and (2) relate to solutions. The integration limits ' and " correspond to the membrane surfaces.

Since, by the above condition, all components of the system are distributed in equilibrium between surface layers of the membrane and the neighboring solutions, the integration limits are to be related to these solutions as well. The integration variable should be interpreted as the activity of a corresponding electrolyte in the solution, which is in equilibrium with some intermediate membrane layer. In this case, the expression for emf of the cell would be as follows:

$$E = \frac{RT}{\nu_{MX}F} \ln \frac{a_{MX}^{(1)}}{a_{MX}^{(2)}} - \frac{RT}{\nu_{LX}F} \int_{(1)}^{(2)} t_L \, d \ln a_{LX}$$

$$+ \frac{RT}{\nu_{MX}F} \int_{(1)}^{(2)} t_L \, d \ln a_{MX} + \frac{RT}{\nu_{MX}F} \int_{(1)}^{(2)} t_X \, d \ln a_{MX}$$

$$+ \frac{RT}{\nu_{MY}F} \int_{(1)}^{(2)} t_Y \, d \ln a_{MY} - \frac{RT}{F} \sum_K \int_{(1)}^{(2)} \tau_K \, d \ln a_K \tag{11}$$

Thus, in harmony with the above, the final expression for the emf includes only activities of molecules of nonelectrolytes and neutral formations containing ions of different signs, i.e., thermodynamically definite and experimentally detectable values.[1]

Therefore, the effect of coupled fluxes of ions and neutral components may be found from an equation of type (II) on condition that membranes reflecting the ionic transfer in the membrane are independently determined.

However, it is less labourous to define this value in conditions when all electrolyte components are distributed in equilibrium along the system. In this instance, regardless of the degree of membrane permeability to ions, the effect of coupled fluxes is equal to the emf of a cell with a membrane:

$$E_{coupl.} = - \frac{RT}{F} \sum_K \int_{'}^{''} \tau_K \, d \ln a_K = - \frac{RT}{F} \sum_K \int_{(1)}^{(2)} \tau_K \, d \ln a_K \tag{12}$$

[1] *It is noteworthy that the expression for the emf of a cell (9) containing four ion species incorporates activities of three electrolytes composed of these ions. The results are in agreement with the fact that the activities of an electrolyte, which is a fourth possible combination of ions of different signs, may be expressed through values of the other three.*

An analysis of the properties of this value first poses the question of whether it is an unambiguous function of the boundary conditions (integration limits) or whether it depends on the concentration profiles of the membrane. For a solution of the problem, one must know the number of independent variables that determine the state of each membrane layer (assumed to be locally in equilibrium). This number depends on the number of components in the system. Difficulties arise, however, due to the presence of a matrix of the membrane. No matter how complex the composition and the structure of the latter may be, it is nonetheless an entity which interacts with the mobile components of the membrane, and, in the general case, should be regarded as a component of the system. In connection with the nature of mobile components and the range of their concentrations, the energy state of the matrix, and consequently, its chemical potential, can alter to a different extent along the membrane. If the chemical potential of the matrix is practically constant, this matrix may be ignored when the number of degrees of freedom is accounted for. Where electrolyte components are also distributed in equilibrium along the system (see above), in isobar - isothermal conditions, the number of degrees of freedom is one less than the number of mobile neutral components of the membrane. This means that in the case when there are two such components and the energy state of the matrix in the membrane undergoes no change, the system is monovariant, and the coupling effect in the membrane potential is unambiguously determined by boundary conditions. Although such a case is unlikely, it appears that it may apply to systems with nonelectrolyte components of kindred nature (differing, for instance, only in their isotope composition). If the matrix behaves as a "variable" component, the coupling effect should be dependent on the concentration profile. Therefore, under given limiting conditions, a definite value for the potential is established only when stationary concentration profiles are attained for all nonequilibrium components along the membrane.

Since these speculations concern systems with strictly equilibrium distribution of electrolytes, the establishment of membrane matrix properties based on dynamic potential data under conditions of nonelectrolyte activity changes seems problematic.

To prevent the disturbance of the equilibrium distribution of the electrolyte components through nonelectrolyte fluxes (different for different profiles and changing until a steady state is attained), the electrolytes should be highly mobile so as to be redistributed almost immediately (as compared to nonelectrolytes) along the membrane. In the opposite case, equilibrium for electrolytes (if it is provided by the compositions of separated solutions) is attained simultaneously with the steady state for neutral components. Under these conditions for systems containing two nonelectrolytes, the dependence of coupling effect on the concentration profile (evidenced, among other things, by slow attainment of the potential with changing boundary conditions for nonelectrolyte) is not evidence that the membrane matrix behaves as a component with a chemical potential gradient.

If there are more than two nonelectrolyte components in the system, the effect of coupled fluxes, regardless of properties of the membrane matrix should be susceptible to the form of the concentration profile.

It appears that theoretical calculations of integrals in Eq. (12) for the given form of the concentration profile will require knowledge of the laws of dynamic interaction of all mobile membrane ions and neutral components that are mobile in the membrane. Recent studies, based on concepts of friction forces between fluxes of separate components, still yield insufficient data for calculating the effects of coupled fluxes. Systems with specifically interacting ions and neutral components seem to be more promising in this respect.

III. EFFECT OF IONIC AND SOLVENT FLUXES COUPLING IN THE POTENTIAL OF A MEMBRANE WITH FIXED CHARGES

We shall not further concern ourselves with classical capillary systems that are considered primarily when electrokinetic phenomena are discussed, but shall instead confine ourselves to quasihomogeneous membranes with charges fixed on a matrix. The fact that these membranes may be used as electrodialytic diaphragms and are connected with the advent of selective electrodes for ionometry accounts for the interest they arouse. Most of the extensive studies were made in the past two decades when a great deal of work devoted to the investigation of their electrochemical properties (including electrode properties inclusive) appeared (see, i.e., monographs and reviews: Helfferich, 1962; Schlögl, 1964; Lakshminarayanaiah, 1969; Bergsma and Kruissink, 1961).

However, as the investigators did not find the high selective permeability necessary to define species of ions for such objectives (with the exception of glass electrodes, with their matchless properties of pH-metric pick-up), interest in these membranes as objects of potentiometric studies has notably decreased in recent years. This is largely due to the detection of other ion-exchange systems that are more promising for ionometric research, primarily of liquid membranes with extracting agents, solid-crystal membranes (Moody and Thomas, 1971; Koryta, 1975; Lakshminarayanaiah, 1976; Cammann, 1977).

Nevertheless, the main defect of the majority of ion-exchange membranes with fixed charges (a weak discrimination of different species of counterions) does not adversely affect the "attractiveness" of these membranes as objects for investigating coupled fluxes of definite species of ions and neutral components, since the variation of their permeability to ions with different signs (counterions and coions) may be great.

For such kinds of membranes, conditions may be created when conductance would be performed by ions of one species and, it would be possible to eliminate the difference character of the value of the coupling effect, which in the general case is due to the presence of several species of mobile ions. The results of such model investigations may be used for an evaluation of these phenomena in membranes of a similar but more selective nature (but much more selective), as well as for more complicated conditions when several ionic species are mobile. Most of the known ion-exchangers with fixed charges more or less swell in solutions. Therefore, the problem of coupled fluxes of ions and nonelectrolytes is concerned largely with the interaction of charged particles with solvent.

In membrane systems for which membrane potentials are commonly measured, solutions are not identical. Therefore, in the overwhelming majority of cases, there is no complete equilibrium with a solvent, and a term that reflects its transfer must be presented in the equation for membrane potential. Thus for the simplest galvanic cell patterned as

$$\mathrm{X} \left| \begin{array}{l} \text{solution (1)} \\ \mathrm{MX}, \\ \text{solvent} \end{array} \right|' \quad \text{membrane} \quad {}''\left| \begin{array}{l} \text{solution (2)} \\ \mathrm{MX}, \\ \text{solvent} \end{array} \right| \mathrm{X} \qquad (13)$$

the emf is expressed by the relation:

$$E = \frac{RT}{F} \ln \frac{a_{\mathrm{MX}}^{(1)}}{a_{\mathrm{MX}}^{(2)}} + \frac{RT}{F} \int_{'}^{''} t_{\mathrm{X}} \, d \ln a_{\mathrm{MX}} - \frac{RT}{F} \sum_{K} \int_{'}^{''} \tau_K \, d \ln a_K \qquad (14)$$

In this case summation is made using all components of the solvent. For a pure solvent, the picture becomes simpler not only because the effect of coupling is expressed by one addend, but also because for such systems the only cause of

τ_K inconstancy is the difference between the electrolyte concentrations in solutions. Hence the possible degree of inconstancy of this value is less than that for more complex multisolvent systems. For these most simple systems, the method of direct determination of the coupling effect on the strength of emf cannot be used, since its value differs from zero only in the case of a nonequilibrium distribution of electrolytes. Therefore, to determine it from emf data, one must undertake an independent investigation of ion transfer in the system and calculate the first integral addent in Eq. (14). We are acquainted with such works concerned only with water systems (Graydon and Steward, 1955; Lorimer *et al.*, 1956; Steward and Graydon, 1957; Hills *et al.*, 1961; Lakshminarayanaiah and Subrahmanyan, 1964; Lakshminarayanaiah, 1966; Tombalakina and Graydon, 1966). They were primarily made to mostly elucidate the cause of the discrepancy between the EMF of a cell with an ion-exchange membrane and the Nernst dependence from the activities of a corresponding electrolyte in solution. It was found that the addends in emf determined by water transfer in some instances reach several millivolts and, consequently, may be the cause of notable deviation from the counterion electrode function (Graydon and Steward, 1955; Steward and Graydon, 1957; Lakshminarayanaiah and Subrahmanyan, 1964; Tombalakian and Graydon, 1966). This conclusion is in agreement with the results of other authors (Kamo *et al.*, 1971; Kamo and Kobatake, 1971).

For more complex systems containing several solvents (or any other nonelectrolyte in the general case), we find essential deviations from the systems discussed above. First, greater differences occur in the states of some solvent components at different sides of the membrane. Second, the effect of coupling (even if electricity in the membrane is carried by particles of one species) is an algebraic sum of several members that may mainly eliminate each other. That is why in this instance one may expect considerable effects only for systems where there is a notable selectivity of transfer of definite components. In this connection, we may speak about potentiometric investigations of the systems under study as one of the possible methods for detecting such selective interactions. There are works that provide an example of detection of rather fine effects of such kind (Greyson, 1967a,b). These publications contain data on the effects of solvent for cation-exchanger and anion-exchanger membranes separating alkali metals and tetramethylammonium halides solutions in common and heavy water. For a cation-exchanger membrane in NaCl and KCl solutions, the effects produce 1.2 and 1.5 mV, respectively. Surprisingly, for anion-exchanger membranes they are greater (about 3 - 6 mV) growing upon transition from chlorides to bromides and iodides. The sign of the emf obtained indicates that the energy effect of the transfer of common water is predominant. By making an assumption about the constancy of activity coefficients and mobilities of common and heavy water with the isotope composition, the authors relate the observed effects to differences in the τ_{H_2O} and τ_{D_2O} values in the membrane surface layers in contact with H_2O and D_2O, respectively. This conclusion is in agreement with the literature evidence that H_2O participates preferentially (as compared to D_2O) in the formation of hydrate sheaths (Ionov and Bystrov, 1970; Samoylov and Yastremsky, 1971) and also with data on the preferential absorption of H_2O (Mjaghkoy *et al.*, 1966). It must be pointed out that the increase of the effect of coupling upon transition from Cl^- to I^- observed by Greyson is rather unexpected.

Greater effects of coupling in the membrane potential have been revealed in systems that contain solvents differing significantly in their properties (Shultz *et al.*, 1972; Illjashik *et al.*, 1972a, b).

A study was made of systems comprising membranes with a variety of fixed ions (SO_3^-, COO^-, $-\overset{|}{\underset{|}{N}}^{+}-$) and solutions of electrolytes (HCl, NaCl) in water and 96% water ethanol.

The emf was defined for elements designed according to the scheme

$$\mathrm{Ag}\;\left|\begin{array}{c}\text{solution (1)}\\ \mathrm{AgCl,\ MCl}\\ \mathrm{H_2O,\ C_2H_5OH}\end{array}\right|' \begin{array}{c}\text{membrane}\\ x\rightarrow\\ \mathrm{M{=}H,Na}\end{array} ''\left|\begin{array}{c}\text{solution (2)}\\ \mathrm{MCl,\ AgCl}\\ \mathrm{H_2O}\end{array}\right|\;\mathrm{Ag} \qquad (15)$$

In the majority of the experiments, the HCl and NaCl concentrations in solutions separated by the membrane were chosen so that they were isoactive with respect to the electrolyte.[1] In accordance with the above, the emf of the cell in this case is a direct effect of solvent. The sum of two addends as determined by the transfer of water and alcohol across the membrane:

$$E_{\mathrm{coupl}} = -\frac{1}{F}\int_{'}^{''}\tau_{\mathrm{H_2O}}\,d\mu_{\mathrm{H_2O}} - \frac{1}{F}\int_{'}^{''}\tau_{\mathrm{C_2H_5OH}}\,d\mu_{\mathrm{C_2H_5OH}}$$

$$= -\frac{RT}{F}\int_{'}^{''}\tau_{\mathrm{H_2O}}\,d\ln a_{\mathrm{H_2O}} - \frac{RT}{F}\int_{'}^{''}\tau_{\mathrm{C_2H_5OH}}\,d\ln a_{\mathrm{C_2H_5OH}} \qquad (16)$$

Notable effects of solvent were detected in all the systems studied. The maximum values obtained in a series of experiments for each of the systems are summarized in Table 1. They relate to electrolyte concentrations in solutions for which the highest permeability of the membrane toward the counterion was obtained under our experimental conditions. The transference numbers of counterions in the membrane that is in equilibrium with the relevant water and alcohol solutions (which characterize these values) are also given in the table.

[1] *The equality of the emf of elements composed of electrodes that are reversible to M^+ and Cl^- ions and contain as solvents water and 96% water ethanol, respectively, was used as an index of isoactivity.*

TABLE 1. Effects of Coupled Transfer of Solvent in the Membrane Potential for Membranes with Fixed Ions of Various Nature

Electrolyte	Fixed ion	H_2O	C_2H_5OH	E_{coupl}[a]
HCl	$-SO_3^-$	1.00	1.00	-34
	$-COO^-$	0.9	0.7	-10
	$-\overset{\vert}{\underset{\vert}{N}}{}^+-$	1.00	1.00	43
NaCl	$-SO_3^-$	1.00	0.90	-45
	$-COO^-$	0.99	0.81	-47
	$-\overset{\vert}{\underset{\vert}{N}}{}^+-$	0.98	1.00	32

[a]*The sign of emf corresponds to that of silver chloride electrode immersed in solution (2). See scheme (15).*

Let us first consider the signs arising from the obtained effects of coupled transfer of solvent and conclusions that may be drawn.

As indicated above, the effect of solvent in the system under analysis is given by the differential value determined by Eq. (16). In the region of complete counterion function of the membrane, the signs of transference numbers of neutral ions are defined as positive for cation-exchangers and negative for anion-exchangers. Moreover, signs denoting changes in chemical potentials of the solvent components are also determined, in our case toward the axis x $d\mu_{H2O} > 0$ and $d\mu_{C2H5OH} < 0$ (see scheme (15)). Therefore, based on the sign of the effect of solvent, we may draw conclusions about the decisive role of one of the integral addends of Eq. (16), conditioned by the transfer of some components of the solvent.

In this case, the effect of solvent is negative for cation-exchangers and positive for anion-exchangers. The combination of all the listed conditions results in the inequality

$$\left| \int_{'}^{''} \tau_{H_2O} \, d\mu_{H_2O} \right| > \left| \int_{'}^{''} \tau_{C_2H_5OH} \, d\mu_{C_2H_5OH} \right| \qquad (17)$$

Thus it may be said with certainty that in the region of high selectivity of all membranes to counterions, the predominant contribution to the solvent effect, is that due to the coupled water transfer.

If the participation of a coion in the transfer of electricity is not to be neglected, one cannot predict the direction of motion of nonelectrolytes along the membrane under the influence of the electric field. Consequently, the sign of the transference number of nonelectrolyte grows indefinite, being dependent on the extent that the molecules of the electrolyte are involved by oppositely charged ions. Under such conditions, one cannot draw the conclusion about relations of some addends in Eq. (16).[1] These data, however, enable us to gain some information about the signs of transference numbers of neutral components, i.e., to reveal the directions of coupled transfer of nonelectrolytes that cannot be realized in the system under study. Let us illustrate this by an example.

Table 1 presents data on carboxyl membranes in the hydrogen form where coions transfer a notable portion of electricity. For systems under study, the auxiliary experiments on the determination of the effect of solvent were performed in much more concentrated solutions, in which membranes with intensively dissociating groups lose selectivity to counterions. In this instance, the absolute values of the effects are found to decrease, but their signs do not change. This means that, under these conditions the effects of solvent are negative for cation-exchanger membranes and are positive for anion-exchanger ones. The direction of changes of the chemical potentials of water and alcohol in the membrane remains as $d\mu_{H_2O} > 0$, $d\mu_{C_2H_5OH} < 0$.

In accordance with Eq. (16), we must acknowledge as impracticable values for the transference numbers of nonelectrolytes, which

[1] *Adhering to the position of pure thermodynamics, we ignored here the well-known structural ideas of a varying tendency for solvation of ions of diverse chemical nature developed in modern investigations of electrolyte solutions.*

would have given positive (for cation-exchangers) and negative (for anion-exchangers) values of both integral addends simultaneously. Consequently, an unambiguous conclusion should be made for cation-exchanger membranes that the conditions $\tau_{C_2H_5OH} > 0$ and $\tau_{H_2O} < 0$ cannot be performed simultaneously, i.e., no assumption can be made that water is involved mainly by anions and alcohol by cations.

As regards the absolute values of the effects of solvent given in Table 1, the difference between them for strongly and weakly dissociating forms of ion-exchanger is striking. The principal cause for such difference is that these membranes are to a various extent permeable to cations.

In other cases the effects are similar. This seems rather unexpected as there is a special motion mechanism for hydrogen ions in solutions, and the degrees of hydration of chloride and sodium ions are different.

There is evidence, however, that in water - alcohol solutions the participation of hydrogen ions in conductance, based on a relay-race prototriphic transfer mechanism, may be less than in pure water solutions (Erdey-Grúz, 1974a; Erdey-Grúz and Majthenyi, 1958). The introduction of alcohol into the system may affect, not only the ability of hydrogen but also the ability of other ions to interact with water molecules. Thus the negative solvation (hydration) that is mainly characteristic of water solutions in the presence of organic substances, commonly decreases (Erdey-Grúz, 1974b; Engel and Hertz, 1968). This implies that after the introduction of alcohol, the strength of the bonds between Cl^- ions and water increase. The result of all these phenomena may be the levelling off of the transfer of the solvent components by ions of various nature.

However, no matter how strongly the above phenomena are expressed in the system under analysis, comparatively small differences in the effects of solvents may be accounted for by the differential nature of this value since, upon transition from one system to another, changes in each addend in Eq. (16) are able to considerably compensate for each other.

All these general regularities in the expression of the effect of coupled transfer of ions and nonelectrolytes in a fixed charge membrane potential may be applied to glass electrodes, however in this instance, the phenomenon is more complicated due to the presence of a coarse glass bulk containing counterions of definite species which are introduced during the synthetic process.

Due to this thick layer, a true stationary state is difficult to establish along the membrane. Each glass surface interacts with its "own" solution, and relevant surface layers differing, among other things, in the composition of a solvent that penetrates them, are also spatially separated. The general thermodynamic relations of the membrane potential are still efficient: integrals in Eqs. (8), (10)-(12), and (14) may be put down within the bounds corresponding to the glass membrane surfaces or compositions of appropriate solutions.[1] But it must be remembered that this occurs in the case of separate interactions of ions with solvents contained in solution on both sides of the membrane. Thus, the physical picture differs from that for a membrane treated through with a solvent. For solvent effects on the membrane potential, this difference is larger; the greater the dynamic interaction of ions with a given nonelectrolyte component. This, in turn, depends on the presence of other nonelectrolyte components. The presence of a thick coarse glass layer is also essential since the glass membrane commonly contains definite species of counterions (introduced during synthesis), regardless of what solutions it makes contact with.

In this context, we must discriminate between two cases. If the electrode works in the region of its "proper" ionic function (i.e., potential-defining are counterions contained in the original glass), the mechanism of solvent effect on the electrode potential should be basically the same as that for ion-exchanger membranes in contact with a solution of one of the electrolytes at sufficiently

[1]*In this case inner coarse layers will not contribute to the integral value.*

low concentrations such that the coion sorption from solutions is negligibly small.

Small effects of solvent (10 - 20 mV) were obtained for some (comparatively "loose") glasses in the region of the electrode function proper (Ivanovskaya *et al.*, 1970; Eisenman, 1965). For a great number of systems, however, no effect was observed (Eisenman, 1965; Ivanovskaya *et al.*, 1970; Ivanovskaya and Shultz, 1968; Shultz and Parfenov, 1958; Lowe and Smith, 1974). This may imply that solvent does not penetrate glass due to the structural density and stability of the latter. Another possible explanation is that solvent, for some reason or other, is distributed in the glass surface layer without a chemical potential gradient. This occurs if penetration of solvent into glass is restricted at the first stage of its interaction with the crude inner layer, and the resultant layer is easy to diffuse through. Solvent spreads along the surface, penetrating in depth of the glass along its frontage, and, in the treated layer, it is found to be distributed practically in equilibrium.

In cases where there is an effect of solvent in the region of the glass electrode function proper, the solvent must be distributed in the glass surface layer with a chemical potential gradient. The facts that the chemical potentials of glass components and also the mobility of ions (electricity carriers) will not by themselves induce significant auxiliary effects in the membrane potential, unlike the effects those already discussed. The reason for this is that glass contains the only species of mobile ions (naturally, the transference number is equal to unity). In this connection, the only addend in the diffusion potential corresponding to the ion transfer in the membrane (see Eq. (8)) turns into a logarithmic member containing a potential-defining ion activity at the boundary of the diffusion region. After allowances are made for an interphase potential jump, the activities are eliminated in the glass surface layer in equilibrium with the solution. The ion activity in the depth of the glass is also eliminated upon transition to the membrane potential because of potential jumps in the glass surface layers facing another solution. As a consequence, only the terms corresponding to the effect of coupled fluxes of ions and solvent components are retained in the solution.

When the glass electrode exhibit an "exchange" function (in this case potential-defining ions were not originally contained in the glass) the surface layer always comprises a region with ion concentration gradients. If a solvent penetrates the glass, a diffusion potential arising in its surface layers should be determined by the non-equilibrium distribution of ionic and neutral components (of the solvent). Since the quantity and nature of a solvent penetrating the glass in the general case must generally affect both chemical potentials and the mobilities of ions the glass contains, the potential by its nature will double. It is induced not only by a shift of charges owing to their involvement by fluxes of neutral particles but also by the effect of solvent in the "primary" distribution of charges at the action of motive forces "proper." Thus the term "the effect of solvent" has a wider sense than "the effect of coupled fluxes" of ions and solvent.

Considerable solvent effects, in some cases up to 80 - 90 mV, were detected for glass electrodes in the region of the "exchange" (primarily hydrogen) function (Ivanovskaya and Shultz, 1968; Shultz and Ivanovskaya, 1967). Since, in the glass, the coupling effect proper is small for sodium ions, it is hard to suggest that it may be significant for hydrogen ions in the surface ion-exchange layer. This permits to interpret great effects of solvent in the region of hydrogen function is mainly a consequence of the solvent influence on the ionic mobilities and activity coefficients (Ivanovskaya and Shultz, 1968; Shultz and Ivanovskaya, 1967). It appears, however, that the interpretation of a mechanism for the solvent effect on the glass electrode potential is rather complex. For glasses with an ionic concentration gradient in the glass surface layer, the physical pattern of the phenomena is very dependent on the correlation of the rates of the ion-exchange process and on the diffusion of electrolytes. Depending on their rate of penetration into the depth of the glass, regions with gradients of ionic and nonelectrolyte concentrations may superimpose or may be separated spatially. This defines the extent to which the resulting effect of solvent will be determined by coupling of fluxes or by the influence of solvent on the thermodynamic and kinetic parameters of mobile ions.

Irrespective of the physical picture of the phenomena, considerable solvent effects detected by glass electrodes and other ion-exchange membranes would be necessarily displayed in ionometric studies of relevant media. Neglect of them may be the cause of errors when ion activities making up hundreds of percent are defined.

IV. EFFECT OF COUPLING OF FLUXES IN A MEMBRANE POTENTIAL FOR A SYSTEM WITH SPECIFICALLY INTERACTING NEUTRAL AND IONIC COMPONENTS

The question of effects of coupled transfer of neutral and charged particles in the membrane potential for systems in which these components interact specifically arose owing to attention paid in the last few years to membranes based on mobile neutral complexons (Ovchinnikov *et al.*, 1974; Markin and Chizmadzhev, 1974).

High selective permeability to definite species of ions when lipophylic complexons are introduced into membrane (both bilayer and bulk[1]) is accounted for by the fact that these ions are bound selectively by the complexon and that the resulting complexes are mobile.

Thus during functioning of the membrane as a partition permeable to ions, the neutral complex participates in transfer processes, the extent of the participation being dependent on the stoichiometry of corresponding complexes. If, for one reason or another, the complexon is at nonequilibrium along the membrane, this

[1] *In what follows, we shall focus our attention largely on bulk membranes the main practical meaning of which is to function as ion-selective electrodes.*

should be expressed in the membrane potential in the same manner as the nonequilibrium in distribution of any neutral component participating in coupled transport.

Nonequilibrium in the complexon distribution along the membrane may either be produced artificially in the process of membrane formation, or it may arise as a result of contact between membrane surfaces and solutions of various compositions.

All these equations for the thermodynamics of irreversible processes and quasithermostatics, which account for a contribution of coupled fluxes of charged and neutral particles to the membrane potential, must be valid for our case of specifically interacting components, in addition data on the complex composition help us clarify the expression for the transfer number of a neutral component.

It is not unlikely that in this instance one may treat fluxes of complex ions as one of the species of charged particles in the expression for the diffusion potential, i.e., to introduce a specific model of the process at the very first stage of quantitative consideration. Such an approach is often taken when the electrode properties of membranes containing neutral complexons are discussed (Eisenman, 1969; Wuhrmann *et al.*, 1973). Since the complex composition unambiguously determines connections between the transfer numbers of the neutral complexon and a corresponding complex ion, and the conditions of complex formation equilibrium determine the connections between relevant chemical potentials, the final result would be the same, irrespective of at what stage the stoichiometry of complex formation was taken into account.

The complexon concentration in the membrane is commonly small as compared to that of other membrane components (solvent-plasticizer, structure-forming polymer agent). Therefore it may be suggested that, based on the Gibbs-Duhem equation, nonequilibrium in the complexon distribution along the membrane cannot be the cause of significant deviation from the equilibrium distribution of other components. Moreover, if the difference in the states of solvents in solutions separated by the membrane is also small, the effect of coupling in the membrane potential would be due only to the complexon transfer even though there are coupled fluxes of other neutral components.

Let us consider quantitative regularities that must determine the value of coupling effect in systems with specifically interacting components for a very simple case when solutions separated by the membrane contain the only electrolyte MX (Stefanova, 1979).

Let the galvanic cell patterned on the basis of this system assume the form

$$\mathrm{X}\ \left|\ \begin{matrix}\text{solution (1)}\\ \mathrm{MX}\end{matrix}\ \right|\ \begin{matrix}' & & ''\\ & \text{membrane} & \\ c'_{\mathrm{S}} & & c''_{\mathrm{S}}\end{matrix}\ \left|\ \begin{matrix}\text{solution (2)}\\ \mathrm{MX}\end{matrix}\ \right|\ \mathrm{X} \qquad (18)$$

On the strength of the general thermodynamics approach discussed in Section II, one may describe for the emf of this cell an expression that still does not suggest specification of the transfer mechanism of both charged and neutral particles:

$$E = \frac{RT}{F} \ln \frac{a^{(1)}_{\mathrm{MX}}}{a^{(2)}_{\mathrm{MX}}} + \frac{RT}{F} \int_{'}^{''} t_{\mathrm{X}}\, d \ln a_{\mathrm{MX}} - \frac{RT}{F} \int_{'}^{''} \tau_{\mathrm{S}}\, d \ln a_{\mathrm{S}} \qquad (19)^{1}$$

where t_{X} and τ_{S} are the total transfer number of anion and complexon in the membrane; a_{MX} and a_{S} are the activities of respective components.

The last addend in Eq. (19) is a contribution to the membrane potential of the effect of coupled transfer of the neutral complexon *S* and ionic components. It is likely to differ from zero only in one case when the membrane contains the *S* activity gradient. Strictly speaking, the latter must arise at any asymmetry in the system, regardless of whether the compositions of solutions sepa-

[1] *Equation (19) is a particular case of Eqs. (11) and (14).*

rated by the membrane are different or differences in the S concentration in the membrane layers are artificially produced. It appears that in the general case of determining the effect of coupled transfer in accordance with Eq. (19), we should have at our disposal independent evidence of the ion transfer number in the membrane. But for a particular case when the effect of coupling is determined only by nonequilibrium of the complexon distribution and there is a complete equilibrium of the electrolyte MX throughout the system, the emf of the cell (18) is a direct effect of the coupling of ion fluxes and the S component in the membrane, independent of the extent the latter is permeable to ions of one or another species:

$$E_{\mathrm{coupl}} = -\frac{RT}{F}\int_{'}^{''} \tau_{\mathrm{S}}\, d\ln a_{\mathrm{S}} \tag{20}$$

Integration of the last equation is possible provided the mechanism of coupled transport of the complexon S is established. Below we shall confine our analysis of the simplest case practicable for a great number of systems, where the coupling of fluxes of S and of ionic components is expressed only in the motion of complex ions of MS^+ composition.

The maximal effect of coupled fluxes in the diffusion potential for systems with cationic complex formation is likely to be observed in cases when anion fluxes in the membrane can be neglected. In what follows, we shall consider only this approximation, which is consistent with the manifestation of membrane high cationic selectivity. As customary, we shall assume as constant both the activity coefficients of the free complexon and the thermodynamically definite combinations of the activity coefficients of some ions. Then Eq. (20) would take the form

$$E_{\text{coupl}} = -\frac{RT}{F}\int_{'}^{''} \tau_S \, d \ln C_S \tag{21}$$

In this case,

$$\tau_S = t_{MS} = \frac{U_{MS} \cdot C_{MS}}{U_{MS}C_{MS} + U_M C_M} \tag{22}$$

where C_M, C_{MS}, U_M, and U_{MS} are the concentrations and mobilities of relevant particles in the membrane.

If allowances are made for the equilibrium of complex formation in the membrane in the form

$$K_{MS} = \frac{C_M \cdot C_S}{C_{MS}} \tag{23}$$

and the relations of mobilities of all particles participating in the transfer are assumed to be constant, the integration in Eq. (21) is easy to perform. As a result, we obtain

$$E = -\frac{RT}{F} \ln \frac{C_S'' + \frac{U_M}{U_{MS}} \cdot K_{MS}}{C_S' + \frac{U_M}{U_{MS}} \cdot K_{MS}} \tag{24}$$[1]

where $_S'$ and $_S''$ are the concentrations of free complexon in the membrane surface layers.

If one of the concentrations (C_S'') is kept constant, then

[1] *This equation for the effect of coupled transfer of the complexon is more complicated than the logarithmic addend obtained by Ciani et al. (1969), which has the same meaning. This is explained by the fact that, not only complex, but also free cations M^+ are assumed to be transferred across the membrane.*

$$E = E^0 + \frac{RT}{F} \ln\left(C_S + \frac{U_M}{U_{MS}} \cdot K_{MS}\right) \tag{25}$$

Consistent with the last equation, the relation between E and C_S in semilogarithmic coordinates will be described as a curve with two linear plots; the first (in the region $C_S \ll (U_M/U_{MS}) \cdot K_{MS}$) is horizontal and the second (for the concentration range $C_S \gg (U_M/U_{MS}) \cdot K_{MS}$) has the slope coefficient equal RT/F (Fig. 1a).

However since in the experimental conditions one may establish only a total concentration of the complexon for experimental checking of the derived relationships, a transition from the concentration to the total concentration of the complexon is indispensable. The type of connection between these values depends on the laws of membrane absorption of cations and anions from the solution; they may be absorbed in equivalent quantities (pure sorption mechanism) or nonequivalent quantities if the membrane shows some ion-exchange properties.

In the literature there is evidence that on the membranes of the type discussed here, the equivalence in the penetration of ions of different signs is disturbed (Thoma *et al.*, 1977; Yurinskaya *et al.*, 1979a). In the region of mean concentrations of electrolytes (10^{-1} - 10^{-2} m), contributions of sorption and ion-exchange mechanisms are commensurable (Yurinskaya *et al.*, 1979a). In this case, it is particularly difficult to obtain in the general form a connection between the concentration values under discussion. Thus it would be reasonable to first of all consider limiting cases of pure ion-exchange and pure sorption mechanisms of ion penetrating the membrane.

Let us assume that there is a purely ion-exchange mechanism, which means that anion X^- practically does not penetrate the membrane and the general content of cation M^+ is constant and equal to the ion-exchange capacity (C^0). In this case, a relation between the concentration of a free complexon and its total concentration may be established from the relations

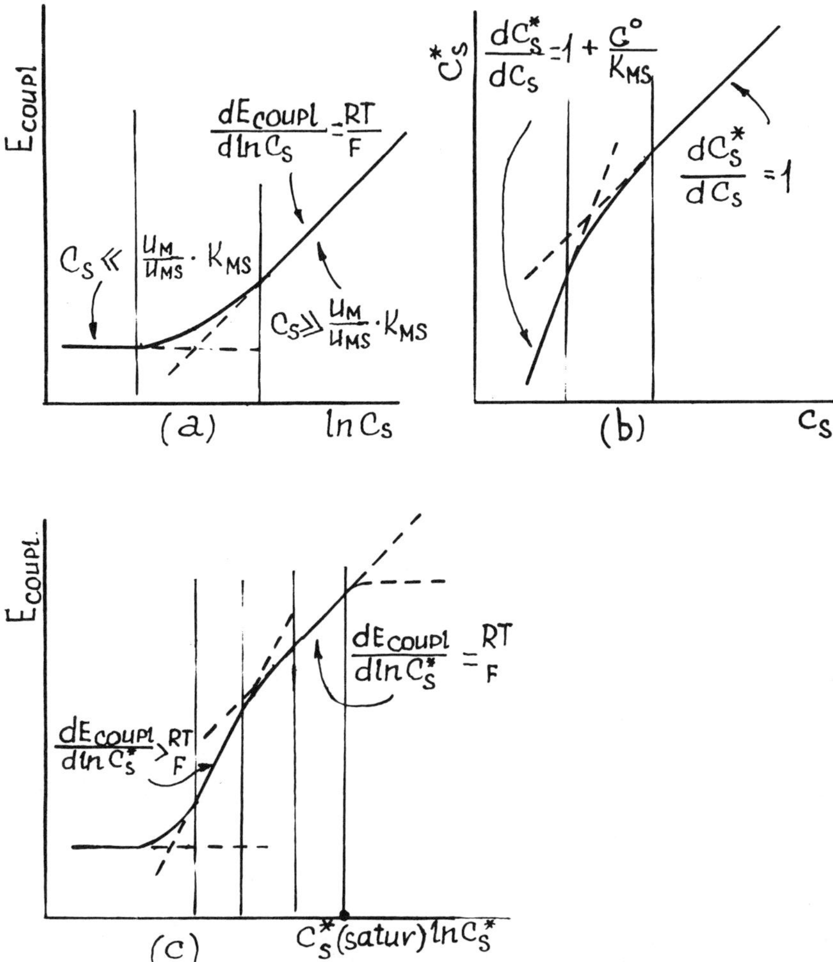

Fig. 1. Effect of coupling of ionic and neutral complexon fluxes in membrane potential (model systems with purely ion-exchange mechanism of cation absorption). (a) Dependence of the effect of coupling (E_{coupl}) on the concentration of a free complexon (C_S); (b) relation between the concentration of a free complexon (C_S) and its total concentration (C_S^*); (c) dependence of the effect of coupling (E_{coupl}) on the total concentration of a complexon (C_S^*); the region $C_S^* > C_{S(satur)}^*$ corresponds to saturation of the membrane with a complexon.

$$C^*_S = C_S + C_{MS}$$

$$C^*_M = C_M + C_{MS} = C^0$$

$$K_{MS} = C_M \cdot C_S / C_{MS} \tag{26}$$

By solving this system of equations we obtain:

$$C^*_S = C_S\left(1 + \frac{C^0}{K_{MS} + C_S}\right) \tag{27}$$

The last relation, upon performance of limiting conditions $C_S << K_{MS}$ or $C^0 << K_{MS} + C_S$, gives a linear dependence of C^*_S from C_S with the slope $1 + (C^0/K_{MS})$ or 1 (Fig. 1b). Correspondingly, the inverse relation $C_S - C^*_S$ in these regions has the slope $K_{MS}/(K_{MS}+C^0)$ and 1, and the second derivative for the intermediate region is more than zero. Consequently, at some mean concentrations of the complexon, C_S grows more rapidly than C^*_S, which should lead to a more drastic dependence of the emf of the cell (18) from C^*_S than from C_S. Thus in the semilogarithmic coordinates $E - \ln C^*_S$ at low S concentrations, one may still expect to find a horizontal linear plot[1] and a growth of emf with the growth of C^*_S (Fig. 1c). If, simultaneously, conditions leading to the linear relation $E - \ln C_S$ with the slope RT/F ($C_S >> (U_M/U_{MS}) \cdot K_{MS}$) and a more rapid growth of C_S as compared to C^*_S ($(d^2C_S/dC^{*2}_S) > 0$) are performed, a region with the slope exceeding the Nernstian coefficient will appear on the dependence curve $E - \ln C^*_S$. With further growth of C^*_S (if the complexon solubility permits it), one may find a linear plot with the Nernstian slope corresponding to the linear connection

[1] *It is easy to prove that its extent is determined by the condition* $C^*_S << \frac{U_M}{U_{MS}} \cdot K_{MS}\left(1 + \frac{C}{K_{MS}}\right)^{-1}$.

$C_S - C_S^*$ which, as soon as it reaches saturation, continues as a horizontal line.

This discussion is concerned with cases when the mechanism of ion penetration from solution is of a purely ion-exchange nature. If, on the contrary, there occurs a sorption of electrolyte as a whole, the following relations for establishing a connection between C_S and C_S^* should be initially chosen:

$$C_M^* = C_X$$

$$C_M^* = C_M + C_{MS}$$

$$C_S^* = C_S + C_{MS} \tag{28}$$

$$K_{MS} = C_M \cdot C_S / C_{MS}$$

$$C_M \cdot C_X = K_{MX} \cdot a_{MX}$$

where K_{MX} is the coefficient of the MX electrolyte distribution between the membrane and solution, and A_{MX} is its activity in the solution.

The solution of this system is as follows:

$$C_S^* = C_S \left[1 + \left(\frac{K_{MX} \quad a_{MX}}{K_{MS}} \right)^{\frac{1}{2}} \frac{1}{(K_{MS} + C_S)^{\frac{1}{2}}} \right] \tag{29}$$

The analysis shows that in this case we also find plots of linear connection between C_S and C_S^* in the region of maximally low and high concentrations, while in the intermediate region there is a nonlinear dependence for which $d^2 C_S / dC_S^{*2} > 0$.

Thus when the mechanism of the electrolyte penetration from solution is purely sorptive, the mode of dependence of the emf of the cell (18) on $\ln C_S^*$ is in qualitative agreement with the case considered earlier of ion-exchange penetration of cations. This allows the suggestion that a similar curve must correspond to the mixed mechanism of membrane sorption of the electrolyte. The extent to which these plots will differ and the correlation between the observed curvature and the Nernstian coefficient value are likely to be the function of parameters of the above equations.

The experimental study of the effect of coupled fluxes in the systems of the type considered here was carried out on membranes of plasticized polyvinyl chloride containing valinomycin as an active agent (Stefanova and Suglobova, 1979a). Nonequilibrial distribution of the complexon along the membrane is accomplished by introducing it in various amounts to the membrane layers during synthesis; KCl was chosen as an electrolyte.

The element was patterned as

$$\text{Ag} \left| \begin{array}{l} \text{AgCl, KCl} \\ \quad\quad (0.01\ \text{m}) \end{array} \right| \begin{array}{l} \text{membrane (1)} \\ C_V \neq 0 \end{array} \vdots \begin{array}{l} \text{membrane (2)} \\ C_V = 0 \end{array} \left| \begin{array}{l} \text{KCl, AgCl} \\ (0.01\ \text{m}) \end{array} \right| \text{Ag} \tag{30}$$

The membrane of this cell consisted of two layers, one of which (membrane (2)) contained no valinomycine, while in the second (membrane (1)) the valinomycine concentration varied over a wide range up to saturation. Both layers were brought into equilibrium with each and the same KCl solution (0.01 m).

To determine the value of a coupling effect corresponding to the given difference between the concentration of valinomycin in membrane layers, the time changes in the emf of the galvanic cell were investigated. Horizontal portions were distinct on the curves plotted with these coordinates. The appropriate emf values were interpreted as effects of coupled fluxes in the membrane potential. It was assumed that regions of constant emf values correspond to a state of the system when the process of "resolution" of contacting surface layers of membranes (1) and (2) is over, but the limiting conditions for valinomycin in the whole combined membrane are unaffected.

The results obtained concerning membranes with different amounts of valinomycin are given in Fig. 2. It is seen that, for the system under study, the mode of dependence of the effect of coupled fluxes on the valinomycin concentration is consistent with the predicted by the above analysis. In those cases where large amounts of the valinomycin were added between layers, the emf values reached hundreds of millivolts. A plot of the greatest curvature has a slope about 75 mV, which notably exceeds the Nernst coefficient.

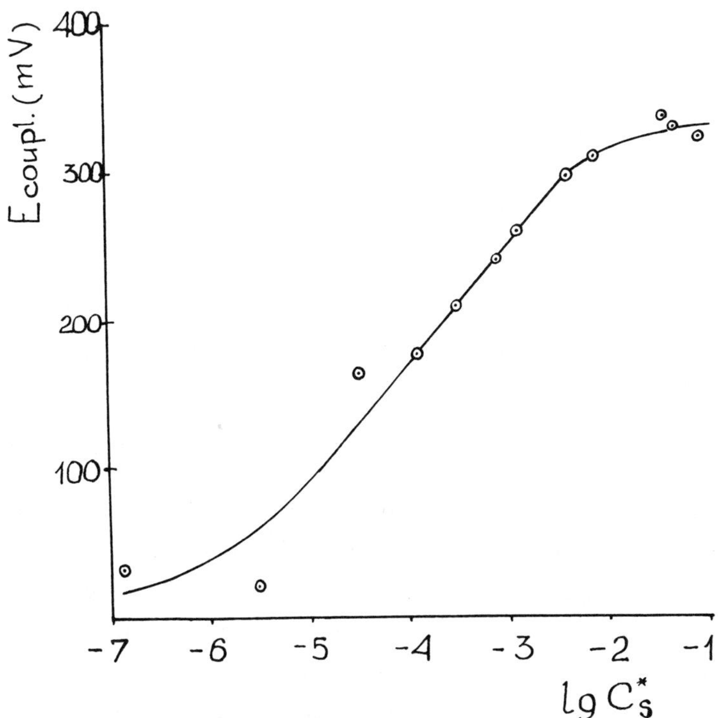

Fig. 2. Effect of coupling of the fluxes in the potential of a membrane containing valinomycin. The membrane is in equilibrium with the KCl solution (0.01 M). The valinomycin concentration (C_S^*) varies at one membrane surface and is equal to zero at the other.

Somewhere between the system and the above models, such a high slope must characterize a region for which the conditions $C_S >> (U_M/U_{MS})K_{MS}$ and $d^2C_S/dC_S^{*2} > 0$ are simultaneously satisfied. It may be demonstrated that in both cases, for ion-exchange and pure electrolyte sorption, the relation $d^2C_S/dC_S^{*2} > 0$ is applicable only if the values of C_S and K_{MS} are commensurable. Consequently, the high slope of the dependence E - ln C_S^* may be indicative of a small value of the relation U_M/U_{MS}. Although this result is hard to interpret in physical terms since a larger ion will be more mobile, it is in agreement with the experimentally supported evidence that, in valinomycin-containing membranes, chloride ions of comparatively small size are practically nonlabile Yurinskaya *et al.*, 1979b).

It appears that the conclusive interpretation of the data will call for a separate scrupulous study of both the complexon-formation processes and the transfer phenomena in the given system. However, irrespective of to what extent the above assumptions may be realized for this system, notable effects of coupled fluxes of potassium and valinomycin and their sensitivity to the complexon concentration exceeding the membrane-potential in some regions, sensitivities to the concentration of charged particles seem noteworthy. The effect under study may be used as a method for analyzing the nature of processes in membranes themselves primarily when the function mechanism of membrane-active agents is to be clarified.

If all these deal with the physical aspect of the phenomena and prospects of its use in physicochemical research, another method of evaluating the result is to consider a "negative" effect of coupling in conditions when it can break comparatively simple regularities in the properties of membrane systems which underly their utilization. This concerns the functioning of membranes with neutral carriers in ionometry as ion-selective electrodes. Although addends corresponding to the effect of coupling are in general terms present, for the potential of a membrane with a neutral carrier (Wuhrmann *et al.*, 1973; Ciani *et al.*, 1969), they are commonly neglected when properties of ion-selective electrodes are discussed. The ground for this taciturn assumption was that the equilibrium of the complexon distribution along the membrane is practically never disturbed. Direct checking of this assumption is difficult and, as far as we know, no research has been yet done along these lines.

Indirect determination of the coupling effect in conditions under test from (19) also presents difficulties. In this case we must have in our disposal independent data to calculate the first integral addend, i.e., carry out an experimental investigation of ion transfer in the membrane depending on the concentration of the solution. We must be certain that the ion-transference numbers are not sensitive to the complexon concentration within the limits which it is assumed to change along the membrane (inducing an effect of coupled fluxes in conditions under discussion). In the opposite case, no independent determination of the first integral addend in Eq. (19) is possible. We are not aware of works aimed at defining the effect of coupled fluxes by this indirect method. At the same time, there is qualitative evidence, although qualitative, permitting to suggest that under functional membrane conditions for the functioning of a membrane with a neutral carrier as an ion-selective electrode, the effect of coupled fluxes of ions and complexon in the membrane potential can be defined (Stefanova and Suglobova, 1979b). This conclusion is drawn from the study of electrode properties of a series of membranes containing different amounts of valinomycin. The membranes are based on polyvinyl chloride plasticized with dibutyl phthalate (DBP). The galvanic cell was patterned as

Ag	AgCl, KCl (satur.)	solution (1) KCl	' membrane "	solution (2) KCl, AgCl 0.01 m	Ag

(31)

With change of the composition of solution (1), the emf of the element alters in accordance with the equation

$$E = E^{0} + \frac{RT}{F} \ln a_{K}^{(1)} + \frac{RT}{F} \int_{'}^{''} t_{\mathrm{Cl}}\, d \ln a_{\mathrm{KCl}}$$

$$- \frac{RT}{F} \int_{'}^{''} \tau_{S}\, d \ln a_{A} \qquad (32)$$

Both integral terms lead to a deviation from the membrane cationic function in the same direction since the activity gradients of KCl and valynomicine (S) in the membrane must have different signs. Difficulties in the precise determination of these terms have already been mentioned. But for a qualitative estimation of the effect of coupled transfer in the membrane potential, it is important to note that the dependence of these addends on the complexon concentration in the membrane may be different.

The first integral term may change (at a given composition of solution (1)) since the amount of membrane-absorbed electrolyte will be dependent on valinomycin concentration. It is hard to predict the extent that the transfer number t_{Cl} will alter in this instance, but there is no reason to believe that changes would be nonmonotonic. Consequently, it may be assumed that at the given composition of the solution the first integral in Eq. (19) will change monotonously, depending on the valinomycin concentration.

In contrast, the second integral term is expected to change nonmonotonously with change of the complexon content. In fact this integral is equal to zero if the complexon is absent from the membrane, or its activity along the membrane is constant. The last condition will be satisfied if the membrane is saturated with respect to this component. In the intermediate range of membrane complexon concentrations, both the first and the second cofactors in the integrand differ from zero, which results in the integral value

$$\int_{'}^{''} \tau_S \, d \ln a_S$$

also differing from zero, and, hence, in a nonmonotonic binding of the coupling effect with the complexon concentration. It is likely that under certain conditions, the effects of coupling would reach such values that this nonmonotonicity would lead to a nonmonotonic deviation from the cationic function depending on the complexon concentration.

Experimental data obtained in the work cited (Stefanova and Suglobova, 1979b) demonstrate such a dependence (see Fig. 3). Breaks in a close-to-Nernst dependence of the emf from the potassium ion activity are observed for both high and low electrolyte concentrations. In both cases they are associated nonmonotonically with the complexon concentration. These results suggest that deviations from the cationic function, in the system under investigation, are determined to a considerable extent by the effect of coupled fluxes of potassium and valinomycin. For more concentrated KCl solutions, this conclusion may be drawn with considerable certainty due to a greater potentials stability. In contrast to this pattern for concentrated solutions, the range of nonmonotonic deviations from the function at larger dilutions is shifted toward low valinomycin concentrations in the membrane. Such shifts seem to be natural, since at lower valinomycin concentrations, notable changes in the complexon activities[1] along the membrane surface layers arise in the region of more diluted KCl solutions. Despite the growth of the degree of S complexity with KCl concentration for such diluted membranes, the extent of their participation in the transfer of free (noncomplex) potassium ions increases and, hence, the transfer num-

[1] *In conformity with the earlier assumptions, the complexon activity may be identified with the concentration of its free molecules.*

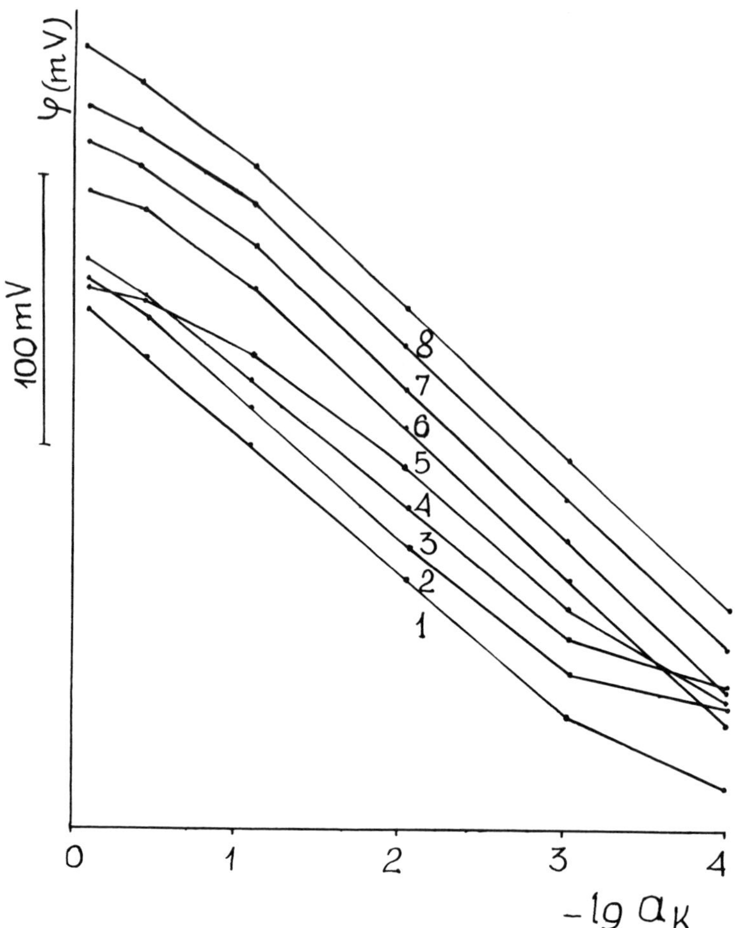

Fig. 3. Potential of the membrane electrodes with the different valinomycin concentration (C_S^*) as a function of the K^+ ion activity in solution. The valinomycin concentrations in the membranes (mol/kg DBP) are as follows: (1) 0; (2) 1.5×10^{-7}; (3) 3.00×10^{-6}; (4) 3.00×10^{-5}; (5) 3.00×10^{-4}; (6) 1.50×10^{-3}; (7) 7.50×10^{-3}; (8) 7.50×10^{-2}.

ber is diminished. As a result, for such membranes the effect of coupling may be insignificant in the range of high KCl concentrations. This is likely to be our case.

It is clear that, in regions of both the high and low concentrations, deviations from the potassium functions may be due to other different reasons (anion transfer, influence of admixture ions). It is impossible to single out factors issuing only from the emf data. In this context, one may speak about a great role of the coupling effect only with respect to membranes where distinct nonmonotonic deviations are found to occur in the valinomycin content.

So far we have restricted ourselves to the analysis of the properties of bulk membranes. For such membranes as shown above, the artificially produced gradient of complexon concentration did permit detection of the essential coupling effects.

The picture was different for thin bilayer membranes (Eisenman *et al.*, 1973; Szabo *et al.*, 1970). No notable deviation in the membrane potential was observed when the authors made an attempt to attain a nonequilibrium distribution of the complexon (macrotetralide) along the membrane by introducing it in various amounts to solutions separated by the membrane. This result is accounted for by the fact that the limiting stage of complexon transfer is located outside the membrane (in a diffuse film of the solution or at the phase boundary), while membrane inner layers are so permeable to the complexon that its concentration in the membrane is independent of the depth of penetration under the experimental conditions. This is why there are no conditions for the effect of coupling of ionic and complexon fluxes in the membrane potential.

V. CONCLUSIONS

It must be pointed out that in this chapter attention is focused on the thermodynamic aspect of the phenomena. To turn from the phenomenological description to consideration of the quantitative regularities of the phenomena at the molecular level, detailed information on ionic - molecular interactions of dynamic nature is needed in such complex systems as polymer membranes. Studies in this direction are based on concepts of friction forces between fluxes of some components of the system (Spiegler, 1958; Meares *et al.*, 1972). Difficulties that arise in a quantitative consideration of the phenomena are mainly due to the necessity to establish a concentration dependence of these values. However, even though the equilibrium aspect of ion solvation in media of differing nature is being experimentally studied and may be interpreted theoretically, our knowledge of the mechanism of transport processes is meager, particularly in relation to interactions leading to coupled transfer of ions and molecules of the solvent. This is especially true for systems with encounter fluxes of nonelectrolytes. However, irrespective of the level of the theory developed, the accumulation of experimental data treated from the viewpoint of thermodynamics of irreversible processes seems to be a necessary stage in understanding of the nature and mechanism of the coupled transport of ions and nonelectrolytes in membranes. Data on coupling effects in the membrane potential not only permit qualitative conclusions about interactions of the appropriate components, but also lead to new approaches of the quantitative evaluation of the properties of complex membrane systems and transport phenomena within.

Thus the presence of a coupling effect offers a considerable scope for direct determination of the nonelectrolyte activities by the emf method. In practice, the problem is with systems with strongly pronounced complex formation. For these systems, a situation may be real when the transference number of nonelectrolytes is constant. If a complex of the definite composition is the only electricity carrier across the membrane, Eq. (12) is easy to integrate, and the effect of coupling appears to be clearly related with the complexon activity.

On the other hand, the differential form of Eq. (12) may be used to obtain evidence of membrane transference numbers of neutral components from experimental data of the emf dependence upon the activities of these components.

Finally, the potential coupling method seems to be efficient and economical at the first stages of the investigation of the complex formation between ions and neutral ligands in insufficiently studied systems, when the complex charge sign or a set of selectivities characterizing the interaction between the ligand and the different cations or anions can be detected.

This method may also be useful in elucidating the mechanism of action of the membrane modifiers when there is some question whether a given agent is a motive carrier or it forms canals of selective conductivity.

REFERENCES

Bergsma, F., and Kruissink, C. A. (1961). *Adv. Polymer Sci. 2*, 307.

Cammann, K. (1977). "Das Arbeiten mit ionenselektiven Elektroden." Springer, Berlin.

Ciani, S., Eisenman, G., and Szabo, G. (1969). *J. Membrane Biol. 1*, 1.

Denbigh, K. G. (1951). "The Thermodynamics of the Steady State." Methuen, London.

Eisenman, G. (1965). *Adv. Anal. Chem. Instrum. 4*, 213.

Eisenman, G. (1969). In "Ion-Selective Electrodes" (R. A. Durst, ed.), p. 1. NBS Spec. Publ. 314, Washington.

Eisenman, G., Szabo, G., Ciani, S., and Krasne, S. (1973). In "Progress in Surface and Membrane Science" (J. F. Danielli, M. D. Rosenberg, and D. A. Cadenhead, eds.), Vol. 6, 139. Academic Press, New York.

Engel, G., and Hertz, H. G. (1968). *Ber. Bunsenges. Phys. Chem. 72*, 808.

Erdey-Grúz, T. (1974a). "Transport Phenomena in Aqueous Solutions." §4.2.2.2. Akademiai Kiado, Budapest.

Erdey-Grúz, T. (1974b). "Transport Phenomena in Aqueous Solutions." §5.2.3. Akademiai Kiado, Budapest.

Erdey-Grúz, T., and Majthenyi, L. (1958). *Acta Chim. Acad. Sci. Hung. 16,* 417.

Graydon, W. F., and Stewart, R. J. (1955). *J. Phys. Chem. 59,* 87.

Greyson, J. (1967a). *J. Phys. Chem. 71,* 259.

Greyson, J. (1967b). *J. Phys. Chem. 71,* 4549.

de Groot, S. R. (1963). "Thermodynamics of Irreversible Processes." North-Holland Publ. Corp., Amsterdam.

Haase, R. (1963). "Thermodynamik der irreversiblen Prozesse." Steinkopf, Darmstadt.

Helfferich, F. (1962). "Ion Exchange." McGraw-Hill, New York.

Hills, G. J., Jacobs, P. W. M., and Lakshminaraynaiah, N. (1961). *Proc. Roy. Soc. A262,* 257.

Illjashik, L. P., Stefanova, O. K., and Shultz, M. M. (1972a). *Elektrokhimiya 8,* 1080.

Illjashik, L. P., Stefanova, O. K., and Shekina, G. I. (1972b). *Elektrokhimiya 8,* 1082.

Ionov, V. I., and Bystrov, G. S. (1970). *Zh. Strukt. Khim. 11,* 207.

Ivanovskaya, I. S., and Shultz, M. M. (1968). *Elektrokhimiya 4,* 1045.

Ivanovskaya, I. S., Gavrilova, V. I., and Shultz, M. M. (1970). *Elektrokhimiya 6,* 1006.

Kamo, N., and Kobatake, Y. (1971). *Kolloid-Z. 249,* 1069.

Kamo, N., Toyoshima, Y., and Kobatake, Y. (1971). *Kolloid-Z. 249,* 1061.

Koryta, J. (1975). "Ion-Selective Electrodes." Cambridge University Press, Cambridge-London-New York-Melbourne.

Lakshminarayanaiah, N. (1966). *J. Phys. Chem. 70,* 1588.

Lakshminarayanaiah, N. (1969). "Transport Phenomena in Membranes." Academic Press, New York.

Lakshminarayanaiah, N. (1976). "Membrane Electrodes." Academic Press, New York.

Lakshminarayanaiah, N., and Subrahmanyan, V. (1964). *J. Polymer Sci. A2,* 4491.

Lorimer, T. W., Botekenbrood, E. Y., and Flermans, T. T. (1956). *Disc. Faraday Soc. 21*, 141.

Lowe, B. M., and Smith, D. G. (1974). *J. Electroanal. Chem. Interfacial Electrochem. 51*, 295.

Markin, V. S., and Chizmadzhev, Yu. A. (1974). "Indutsirovannyi Ionnyi Transport." Nauka, Moscow.

Meares, P., Thain, J. F., and Dawson, D. G. (1972). In "Membranes" (G. Eisenman, ed.), Vol. 1, p. 55. Marcel Dekker, New York.

Mjaghkoy, O. N., Meleshko, V. P., and Rjaguzov, A. I. (1966). In "Ionity i Ionnyi Obmen," p. 3.8. Nauka, Moscow.

Moody, G. J., and Thomas, J. D. R. (1971). "Selective Ion-Sensitive Electrodes." Merrow-Watford, England, Merrow Technical Library, Practical Science.

Ovchinnikov, Yu. A., Ivanov, V. T., and Shkrob, A. M. (1974). "Membrane Active Complexones." Elsevier, Amseterdam.

Prigogine, I. (1961). "Introduction to Thermodynamics of Irreversible Processes." Wiley, New York.

Samoylov, O. Ya., and Yastremsky, P. S. (1971). *Zh. Strukt. Khim. 12*, 39.

Schlögl, R. (1964). "Stofftransport durch Membranen." Steinkopf, Darmstadt.

Shultz, M. M., and Ivanovskaya, I. S. (1967). *Elektrokhimiya 3*, 576.

Shultz, M. M., and Parfenov, A. I. (1958). *Vestn. Leningradsk. Univ.* No. 16, 118.

Shultz, M. M., Stefanova, O. K., and Illjashik, L. P. (1972). In "Termodinamika Neobratimykh Protsessov," p. 157. Chernovtsy.

Spiegler, K. S. (1958). *Trans. Faraday Soc. 54*, 1408.

Stefanova, O. K. (1979). *Elektrokhimiya 15*, 1707.

Stefanova, O. K., and Suglobova, E. D. (1979a). *Elektrokhimiya 15*, 1822.

Stefanova, O. K., and Suglobova, E. D. (1979b). *Elektrokhimiya 15*, 1710.

Stewart, R. J., and Graydon, W. F. (1957). *J. Phys. Chem. 61,* 164.

Szabo, G., Eisenman, G., and Ciani, S. (1970). In "Physical Principles of Biological Membranes" (F. Snell, J. Wolken, G. J. Iverson, and J. Lam, eds.), p. 79. Gordon and Breach, New York.

Thoma, A. P., Viviani-Nauer, A., Arvanitis, S., Morf, W. E., and Simon, W. (1977). *Anal. Chem. 49,* 1567.

Tombalakian, A. S., and Graydon, W. F. (1966). *J. Phys. Chem. 70,* 3711.

Wuhrmann, H.-R., Morf, W. E., and Simon, W. (1973). *Helv. Chim. Acta 56,* 1011.

Yurinskaya, V. E., Stefanova, O. K., Materova, E. A., and Glazunov, V. V. (1979a). *Elektrokhimiya 15,* 419.

Yurinskaya, V. E., Stefanova, O. K., and Materova, E. A. (1979b). *Elektorkhimiya 15,* 723.

ELECTROCATALYTIC PROPERTIES OF METALLOPORPHINS AT THE INTERFACE

M. R. Tarasevich
K. A. Radyushkina

Institute of Electrochemistry
Academy of Sciences of the USSR
Moscow, USSR

I. INTRODUCTION

Organic complexes of metals constitute a new and rather promising class of electrocatalysts, and at present, in a number of countries, extensive fundamental and applied investigations in this field are under way. Among organic complexes of metals, particular attention is paid to compounds that contain, as a ligand, various porphin derivatives (such as metallophthalocyanines, tet-

ISBN 0-12-571814-4

raarylporphins, etc.). The applicability of these compounds as electrode materials is conditioned by three main factors: high activity in various heterogeneous redox reactions, the presence of semiconductor properties, and high chemical and thermal stability. Their practical application is one way of replacing catalysts based on precious metals. Wider use of the new processes of electrochemical technology, for instance, in current sources, in photoelectrochemical conversion of solar energy, etc., will be to a considerable extent dependent on the solution of this problem.

Investigation of the mechanism and kinetics of chemical and electrochemical reactions on metalloorganic complexes is also of considerable theoretical interest, since it allows clarifying the role of collective and local effects in electrocatalysis and establishing the connection between the chemical structure of the catalyst and chemical activity. These results may also be useful in understanding a number of problems of fermentative catalysis. Molecular complexes containing a porphin group serve as active centers of protein macromolecules - ferments and electron carriers in respiratory chains of living organisms, and play a decisive role in photosynthesis.

A most important object of investigations carried out at present in the field of electrocatalysis is to substantiate regularities in selecting catalytically active systems. This problem becomes all the more important in the case of metalloorganic complex catalysts insofar as the number of possible organic compounds is practically unlimited. Most comprehensive fundamental and applied investigations on organic complexes of metals have been carried out for the electroreduction reaction of molecular oxygen. The main trend of the investigations was to study the interconnection between the electrophysical parameters and chemical structure of organic complexes of metals, on the one hand, and the mechanism and kinetics of electroreduction of oxygen, on the other hand. Organic catalysts may have the same electrophysical characteristics, determined by collective intermolecular interaction, e.g., electrical conductivity, but different structure of the organic complex molecule conditions quite different electrocatalytic properties. Concurrently, in the course of engineering developments, a number of versions of catalysts have been created, fit for practical application in porous gas-diffusion electrodes.

In this chapter covering the literature sources published up to January 1977, most detailed analysis is given of the literature data and of the authors' data dealing with the investigation of the mechanism and kinetics of electroreduction of molecular oxygen. Other electrocatalytic and photocatalytic reactions, the investigation of which has been only started, are considered in less detail.

II. ELECTROPHYSICAL AND CATALYTIC PROPERTIES OF METALLOPORPHINS

In electrocatalytic investigations, compounds of the porphin series hold the central place among metal-containing organic complexes possessing semiconductor properties. The results of studying their chemical, electrophysical, and catalytic characteristics are generalized in a number of monographs (Moser and Thomas, 1963; Falk, 1964; Gutman and Lajons, 1970; Kayushin *et al.*, 1973). Therefore, in this chapter only those properties will be rather briefly considered, which are necessary for interpreting the electrocatalytic phenomena.

A. Electronic Structure

Structural formulas of porphin and its derivatives most frequently used in catalysis and electrocatalysis are presented in Fig. 1. Two hydrogen atoms bonded with nitrogen may be substituted by a metal with the formation of metalloorganic complexes. Three main groups of porphin derivatives are

(1) Reduced porphyrins (at I-2, 3-4, etc.).

(2) Porphyrins with substituents at α, β, γ, δ, e.g., tetraphenylporphin shown in Fig. 1b.

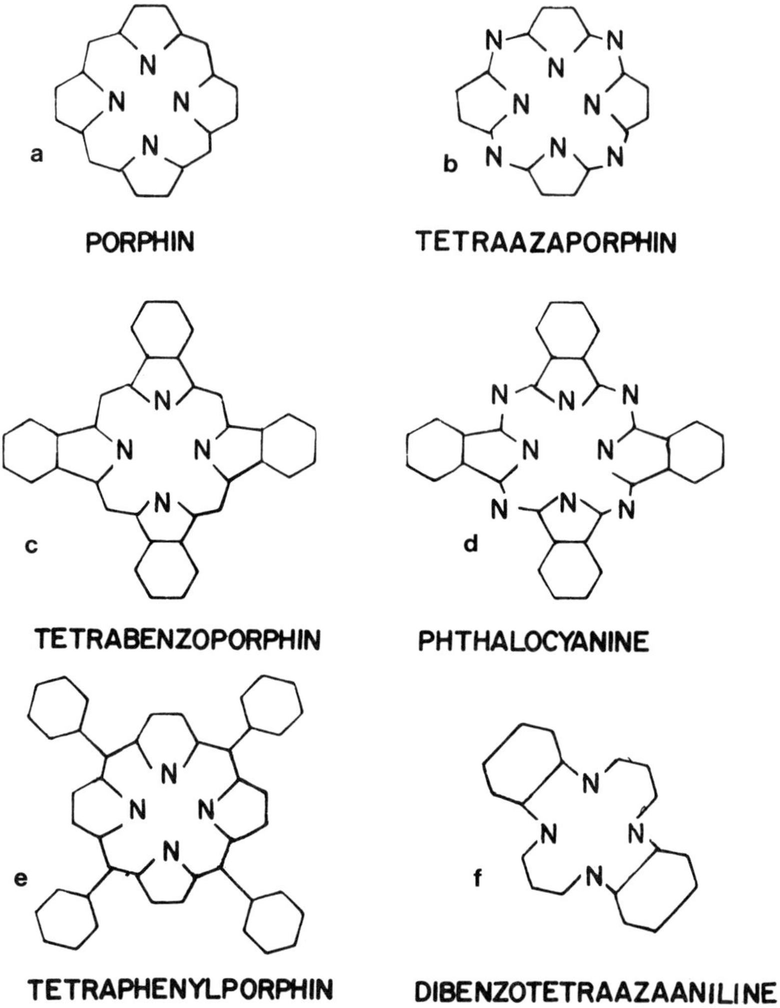

a PORPHIN

b TETRAAZAPORPHIN

c TETRABENZOPORPHIN

d PHTHALOCYANINE

e TETRAPHENYLPORPHIN

f DIBENZOTETRAAZAANILINE

Fig. 1. Structural formulas of organic complexes used in electrocatalysis.

(3) Porphyrins with benzene rings attached via external bonds 1-2, 3-4, etc., e.g., tetrabenzoporphin shown in Fig. 1c.

Phthalocyanines are derivatives of tetraazaporphin (Fig. 1d,e). X-Ray investigations (Moser and Thomas, 1963; Niwa *et al.*, 1974) have shown that phthalocyanines of bivalent Cu, Ni, Fe, Co, and Mn belong to the monoclinic system, symmetry group D_{4h}. Such high-symmetry group is conditioned by the conjugation effect. In the case of a metal-free phthalocyanine (H_2Phc) the symmetry of the molecules lowers to D_{2h}, since hydrogen atoms of imine groups are in transposition to each other. Neutral molecules of porphyrins (e.g., of tetraphenylporphin, H_2TPhP) also have symmetry D_{2h}, since the imine and tertiary nitrogen are not equivalent, which leads to nonequivalence of pyrrole rings. There is a known number of polymeric phthalocyanines of metals. Polyphthalocyanines are high-molecular compounds having a band, network, or three-dimensional structure containing tetraazaporphin macrocycles (C_8N_8) alternating in a certain order (Fig. 2). Linking of monomer units into a polymer may proceed either through benzene rings (without conjugation disturbance), or through bridge groups (CH_2, O, N, etc.) that are insulators or weak conductors of conjugation. A continuous conjugation chain in polyphthalocyanines is not large (4 to 5 units).

Porphin and its derivatives are characterized by the presence of a developed conjugated system of double bonds that embraces the entire molecule. Changes taking place in porphins in chemical reactions are associated with electron transitions in the conjugated system of the ligand and in the electronic system of the central ion of metal. Therefore, to understand the nature of the catalytic and electrocatalytic processes of metal-containing organic complexes, detailed study of their electronic structure is required. Quantum-chemical calculations of porphin complexes of transition metals have been carried out by Gouterman *et al.* (1966, 1972) and Taube (1966). These authors employed for the calculations an extended method of Hückel that covers all valence orbitals of H, C, N atoms and 3d, 4s, 4p orbitals of metal. Conjugated system of the porphin ring contains 26 π-electrons, carbon atoms of the pyrrole ring featuring excess density of π-electrons (~1.03) and carbon atoms of methine bridges featuring deficient density of π-elec-

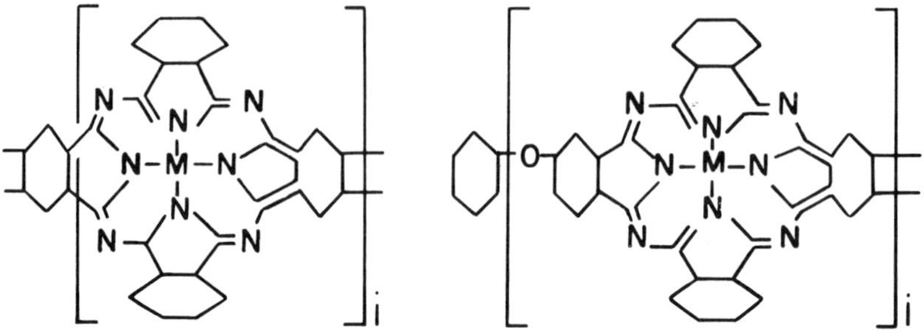

Fig. 2. Polyphthalocyanines.

trons (0.945). In azoporphins, the density of π-electrons on nitrogen atoms (~1.23) is higher than on pyrrole rings (~1.01). A diagram of orbital energies for porphyrins of transition metals is presented in Fig. 3. With the exception of Cu and Zn, orbitals of

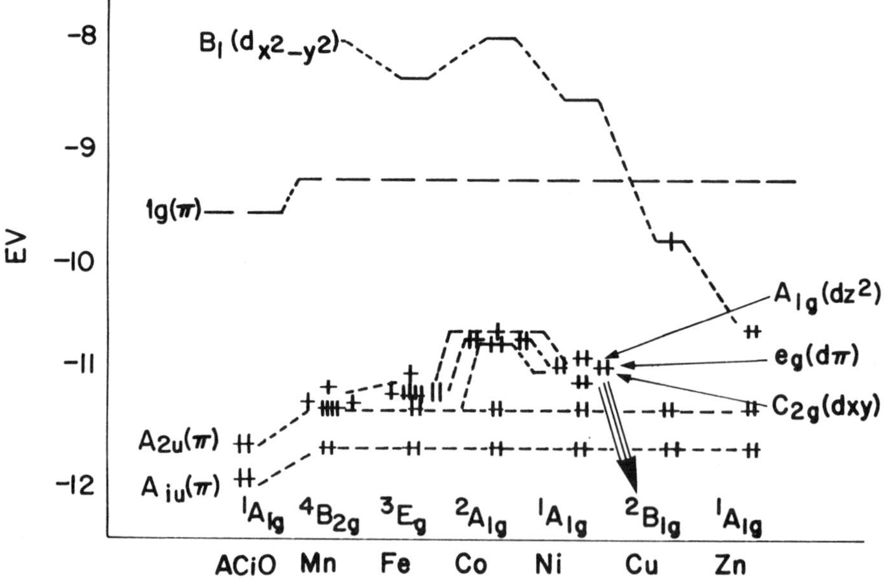

Fig. 3. Calculated energies of upper-filled and lower-vacant orbitals.

metal weakly interact with the porphin π-system. Porphins of Zn and Ni are diamagnetic; those of Cu and Co are paramagnetic; in the case of Fe and Mn, a number of versions of the spin state is possible, since the energies of their orbitals are close to the energy of $a_{2i}(\pi)$ MO. Similar results were obtained (Taube, 1966) for phthalocyanines of bivalent Ni, Co and Fe. A somewhat different electron configuration has been found for MnPhc. The origination of excited states in the visible and ultraviolet region of the spectrum is attributed (Zerner *et al.*, 1966) to orbital $a_{2i}(\pi) \rightarrow e_g^*(\pi)$ and $a_{Ii} \rightarrow e_g^*(\pi)$. Their strong degeneration and mixing leads to the origination of low-energy state *Q* in the visible region and high-energy state in the Soret band region. In the series porphin > tetraazaporphin > tetrabenzoporphin > phthalocyanine *Q*-bands approach one-electron transition (Schaffer and Gouterman, 1972). The origination of N and L bands in the blue region is similarly explained by $a_{2i}(\pi) \rightarrow e_g^*(\pi)$ and $b_{2i}(\pi) \rightarrow e_g^*(\pi)$ transitions. *n* - π transitions in tetraazaporphin and phthalocyanine lie at lower energies than in porphin and tetrabenzoporphin, which must bring about broadening of the Soret band in the first two cases. The results of quantum-chemical calculations satisfactorily correlate with the electron absorption spectra of porphin and its derivatives, though for interpreting all fine particulars of these spectra, it is necessary to take into account the participation of a large number of orbitals of the ligand system and, possibly, of d-orbitals of metal (Stillman and Thomson, 1974).

The absorption spectrum of porphyrins in the visible region is a characteristic "ladder" of bands diminishing in intensity, corresponding to individual electron transitions. Metal-free porphyrins display an intensive band in the region of 400 nm (Soret band), conditioned by the electron transfer, and four other rather weak bands. In metalloporphins, the Soret band narrows and two rather weak bands are observed instead of four in the visible region. Introduction of benzene rings and aza-substitution (when going from porphyrins over to benzoporphins and phthalocyanines) leads to intensification of the visible absorption bands and to their shifting toward long waves. Phthalocyanine and tetrabenzoporphin have two intensive absorption bands in the visible region, which merge into one band when metal is introduced, and a band in the ultraviolet region (~350 nm) on which the introduction of metal exerts no influence. Electron absorption spectra of polyphthalocyanines are, in the main, similar to the spectra of the monomers (Berezin and Shormanova, 1968), but in polymerization there takes place leveling of less intensive bands. Constancy of the long-wave absorption band that characterizes the energy levels of the ground and lowest excited state of phthalocyanine indicates that polymerization has no influence on the character of π-electron interaction of the benzene and tetraazaporphin rings of phthalocyanine.

Absorption spectra of solid amorphous or polycrystalline layers of porphin derivatives substantially coincide with the absorption spectra of these substances in solution. This indicates that optical absorption is effected by separate molecules of organic compounds rather than by the lattice as a whole.

A more detailed analysis of the electronic structure of porphin metal derivatives and, in particular, of the spin state of the central ion of metal, is carried out by comparing theoretical calculations with experimental results obtained by ESR methods (Kayushin *et al.*, 1973; Sidorov and Maslow, 1975; Ingram, 1972; Wolberg and Manassen, 1970; Walker, 1970), by Mössbauer spectroscopy methods (Dale et al., 1968; Appleby *et al.*, 1976), and by magnetic susceptibility measurements (Barraclough *et al.*, 1970). ESR data have shown the unpaired electron in CuPhc to be in $d_{x^2-y^2}$MO (Ingram, 1972), and in CoPhc, in d_{z^2} orbital (Sidorov and Maslov, 1975). This is in agreement with the results of quantum-chemical calculations described above. Works devoted to the investigation of the electron state of iron ions in porphyrin complexes are particularly numerous. Yet this problem is not clear so far. Detailed theoretical investigations have been carried out for porphyrin complexes of Fe(II) and Fe(III) (Zerner *et al.*, 1966). It is shown that, depending on the position of Fe ion in relation to the plane of the ligand, on the introduction of additional ligands (H_2O, O_2, CO, etc.) or side substituents, a low-spin, high-spin, or intermediate state is possible.[1] These calculations have been carried out in porphin field; additional changes may take place as one goes over the phthalocyanine. Thus, according to the data of Dale *et al.* (1968), the electron configuration is $(d_{xy})^2(d_{yz}d_{xz})^3(d_{z^2})^I$, while according to Barraclough *et al.* (1970), it is $(d_{xy}d_{xz})^4(d_{xy})^I(d_{z^2})^I$. According to calculations (Zerner *et al.*, 1966), as a rule, there is preserved the following order of increase in the energy of the orbitals: $d_{xy} < d_{\pi} < d_{z^2} < d_{x^2-y^2}$. However, other authors (Mathur and Singh, 1972), proceeding from the experimental and calculated data concerning the position of the absorption band that corresponds to the electron transfer from metal to ligand, come to a different conclusion as regards the order of the energies: $d_{z^2} < d_{xy} < d_{\pi} < d_{x^2-y^2}$. The results of investigating the magnetic susceptibility of MnPhc crystals (Barraclough *et al.*, 1970) show that most preferable configuration is $(d_{xy})^2(d_{xz})^2(d_{z^2})^I$ with the term $^4\Delta_{2g}$ in the ground state. Evidently, in connection with specific features of the electronic structure of Mn and Fe porphins, the distribution of d-electrons of these central ions is strongly dependent on the method of synthesis and on certain conditions for carrying out the experiment.

The porphin ring is amphoteric: In an acidic medium, due to unshared electron pairs of nitrogen atoms, addition of one or two protons is possible with the formation of mono- and dications, and in an alkaline medium, with the formation of mono- and dianions. In the case of phthalocyanines, protonation of both central and bridging nitrogen atoms is principally possible. Thus, phthalo-

[1] *It should be noted that transition to the high-spin state leads to the appearance of additional bands in the visible region.*

cyanines and porphins noticeably differ in their acidic and basic properties. In the case of porphyrins in protonation dianions are formed, protons add to the central nitrogen atoms already at pH = 2-3 and excess charges are uniformly distributed over the entire π-electronic system of the molecules (Mamaev *et al.*, 1970). In the case of phthalocyanines, addition of the first proton occurs at considerably lower pH (pK $\simeq$ 1.8), probably via bridging atoms of nitrogen (Berezin and Shormanova, 1971). It should be noted that the problem of the number and localization of the added protons for phthalocyanines remains disputable.

In the action of reducers (alkali metals, borohydride, hydrazine, etc.) on compounds of the porphin series or in cathodic polarization, negative molecular ions of tetrapyrrole compounds are formed. Under the action of electron acceptors or in anodic polarization cations are formed. In connection with possible participation of ionic forms of metalloporphins in electrocatalysis, there arise problems of establishing redox potentials of the ring and central ions in nonaqueous and aqueous solutions and of identifying the pathway of electron transitions (ring, metal). The first problem is solved polarographically; the second, by combining the ESR method (Wolberg and Manassen, 1970; Walker, 1970; Fajer *et al.*, 1970), electronic (Clack and Yandle, 1972; Bobrovsky and Sidorov, 1976), and IR absorption spectra (Furhop *et al.*, 1973; Stillman and Thomson, 1974). A wide range of metal-substituted octaethylporphins has been investigated by Furhop *et al.* (1973). It has been shown that irrespective of the nature of metal, the difference $\psi_{1/2}$ for the formation of a monoanion and a monocation is 2.25 ± 0.15 V.[1] The potentials of redox reactions of metals: Co(II) $\rightleftarrows$ Co(I); Mn, Fe, Co, Cr(II $\rightleftarrows$ III); Cr, Fe(III $\rightleftarrows$ IV); MoO(IV $\rightleftarrows$ V) lie within this interval. $\Delta\psi_{1/2}$ between the transi-

[1]*It should be pointed out that reduction was carried out in a dimethyl sulfoxide solution and oxidation, in butyronitrile. In the case of phthalocyanines this difference amounts to 1.40 ± 0.20 V (Rollman and Iwamoto, 1968).*

tion of the first and second electrons to the rings in all cases is equal to ~0.42 V. A close value is reported by Zerner and Gouterman (1966) for the reduction of metal-containing complexes of tetraphenylporphin. The value of $\psi_{1/2}$ for the transition (III $\rightleftarrows$ II) in the ions of metals is linearly dependent on their electronegativity. This fact, as well as constancy of the $\Delta\psi_{1/2}$ value for the oxidation and reduction of the ring confirms the conclusion made by Zerner *et al.* (1966) about the relatively weak interaction of the electronic systems of the metal and ligand. Unlike metalloporphyrins that form mono- and dianions, molecules of the majority of metallophthalocyanines are capable of adding up to four electrons (Taube, 1966). Oxidation of metalloporphin complexes in which the metal ion cannot be oxidized (Zn, Mg, Cd, Cu) proceeds in two one-electron stages with the formation of mono- and dications, e.g., $M\text{-}(Phc)^{+}$ and $M\text{-}(Phc)^{2+}$. In the case of complexes with Co and Fe (Wolberg and Manassen, 1970), evidently, first there takes place elimination of an electron from the metal, and after that the porphin ring is oxidized. It is established that localization of the charge is substantially dependent on the nature of the central ion of metal. In the case of monoanions of metal-free phthalocyanine and tetraphenylporphin, excess charge is localized on the ligand. In monoanions of complexes having cobalt as the central ion (Phc, TPhP), the electron is localized on the central atom (Clack and Yandle, 1972; Rollman and Iwamoto, 1968). In mono- and dianions $(Fe\text{-}Phc)^{-}$ and $(Fe\text{-}Phc)^{2-}$, negative charge is shared between the atom and ligand (Sidorov and Maslov, 1975; Clack and Yandle, 1972).

The electronic structure of the anions of phthalocyanines of transition metals has been considered by Taube (1966) and the order of filling the MO with electrons has been determined. In the case of phthalocyanines of Mn, Fe, and Co the first electron, and in the case of Fe the second electron as well, fill d orbitals of the metal. Subsequent electrons enter the system of π-electrons. These results, in the main, are in agreement with the experimental data described above.

Let us now consider the electrophysical properties of metalloporphins in crystalline state and their activity in heterogeneous catalytic reactions.

B. Electrophysical Properties

Monomeric phthalocyanines can exist in several crystalline modifications, differing in the crystalline structure and electrical properties (Moser and Thomas, 1963). A monoclinic cell accommodates two molecules, the distance between parallel molecules (e.g., in a CuPhc crystal) is 3.38 Å. α-Form has a tetragonal structure and is formed in precipitation from concentrated sulfuric acid or by sublimation under high pressure of vapors. β-Form may be obtained by heating the α-form at a temperature above 300°C. In vacuum the transformation proceeds faster and at a lower temperature than in air. Phthalocyanine molecules in β-form are more densely packed, which leads to stronger interplanar interactions than in the α-form. For instance, the interplanar distances are 11.8 and 9.8 Å, and the size of crystals increases from 100-200 Å to 1 μm for the α and β modifications of CuPhc, respectively. The methods of preparing different polymorphic modifications and their transformations are described in detail by Moser and Thomas (1963).

Important characteristics of the conjugated π-electronic system are electric conductance and photoconductivity. The electric conductance of phthalocyanines was measured on films, prepared by vacuum deposition, precipitated from solutions, on single crystals and pressed samples (Gutman and Lajons, 1970; Usov and Bendersky, 1970; Kanda and Pohl, 1968). For different conditions of measurements, the values of the conductivity (ρ) and of the activation energy of dark conductivity (ε_T) markedly differ: ρ, from 10^{-16} to 10^{-7} $\Omega^{-1}\cdot cm^{-1}$ for monomers and from 10^{-7} to 10^{-4} $\Omega^{-1}\cdot cm^{-1}$ for polymers; ε_T, from 0.6 to 1.9 eV. This is associated with the fact that methods of preparing samples for measurements do not always ensure their required purity (impurities, adsorbed gases are present), so that impurity conductivity rather than intrinsic conductivity is observed. Introducing metal into phthalocyanine causes insignificant changes in the value of the activation energy. On the other hand, electric conductance of metal-containing phthalocyanines increases rather considerably (by 4-5 orders of magnitude as compared to H_2Phc). Evidently, this is conditioned by strong increase in the mobility of charge carriers when hydrogen atoms are replaced by metal atoms. The sign of the main charge carriers depends on the conditions under which a given sample was prepared, on its pretreatment, on the atmosphere in which measurements are carried out, and on the crystalline modification. Obviously, charge carriers in phthalocyanines are electrons, and hole conductivity is the result of adsorption of oxygen that creates levels of acceptor type to which the electrons transfer. According to Usov and Bendersky (1970), in thoroughly degassed samples of H_2Phc the mobility of electrons is 0.43-0.70 $cm^2 V^{-1} sec^{-1}$, and that of holes is 0.24 to 0.56 $cm^2 V^{-1} sec^{-1}$. Electrical characteristics of organic semiconductors can be best explained by the "jump" mechanism of conductivity, consisting in a series of consecutive activation jumps of the carriers from trap to trap. Inclusion of phthalocyanines into a polymer chain brings about an increase in the electric conductivity. The position of the Fermi level in phthalocyanines is also dependent on their pretreatment and the atmosphere: oxygen shifts the Fermi level closer to the valence band, while hydrogen produces an opposite effect (Bendersky *et al.*, 1972; Barbe and Westgate, 1970). In H_2Phc layers, impurity levels of oxygen lie 0.86 eV above the upper edge of the valence bond.

Phthalocyanines and tetraphenylporphins are photoconductors (Usov and Bendersky, 1968, 1970). Photoeffect is associated with processes in the surface layer having a thickness of $\sim 10^{-5}$ cm. The process of formation of photocurrent carriers includes thermal dissociation of excitons, and main carriers of photocurrent are electrons. Photoconductivity, similar to dark conductivity, is essentially dependent on the composition of the ambient atmosphere: presence of adsorbed oxygen increases the photocurrent in H_2Phc and CuPhc films (Usov and Bendersky, 1968).

Obviously, adsorption of various gases exerts substantial influence on the electrical properties of metalloorganic complexes. With the adsorption of oxygen, NO, and NO_2, an increase in the electrical conductivity is observed. Introduction of NH_3 into the system cancels this effect. Phthalocyanines of α-modification adsorb by one order of magnitude more oxygen than their β-form. Adsorption of oxygen is not stable, in the case of H_2Phc oxygen is adsorbed reversibly on the organic groups (Contour *et al.*, 1973). X-Ray measurements (Moser and Thomas, 1963; Niwa *et al.*, 1974) show that oxygen may penetrate between the planes of the phthalocyanine crystal and add to the metal atoms, thus forming either oxygen bridges between the molecules, or compounds of peroxide type. Complexes of oxygen with vitamin B-12 (Bayston *et al.*, 1969) and with CoTPhP (Walker, 1970; Yamamto and Kwan, 1970), detected by the ESR method, were observed in nonaqueous solutions. In alkali aqueous solutions at pH > 12.6, stable oxygen adducts $Co:O_2 = 2:1$ are formed (Wagnerova *et al.*, 1974). At pH < 12.6, the adduct formation reaction does not take place. In a number of works the character of oxygen bonding in its adsorption or in complex formation is discussed. According to the data reported by Blumenfeld *et al.* (1967), oxygen interacts with MgPhc, forming local complexes with charge transfer; the d_{z^2}MO electron of the cobalt ion in CoPhc is transferred to the $2p\pi^*$-orbital of the oxygen molecule (Walker, 1970). In protoporphyrin and methoxytetraphenylporphin (Stynes *et al.*, 1973; Walker, 1973; Kadish and Morrison, 1975), the formation of oxygen adduct $Co^{III}PPhO_2^-$ is accompanied by establishing a feedback between the π^*-orbital of O_2 and the filled d_{xz} orbital of cobalt. The study of the adsorption interaction of the components of catalytic and electrocatalytic reactions with metalloporphins is one of the most important trends in present-day research work.

C. Heterogeneous Catalysis

Considerable experimental material has by now been accumulated on heterogeneous catalysis of a number of reactions on metalloporphyrins and is treated in detail in reviews by Hanke (1969), Roginsky (1973, 1975), and Borisenkova and Rudenko (1976). The range of reactions activated by metalloporphins is extremely wide, but the greatest number of works is devoted to the study of the oxidation reactions of organic substances: mild oxidation of hydrocarbons (cumene, propylene), oxidative dehydrogenation (of alcohols, cyclohexadiene), and decarboxylation. These processes proceed, evidently, through the stage of activated adsorption on a catalyst of molecular oxygen with the formation of O_2^- or O^- ions. In this connection they are of interest for predicting the electrocatalytic activity in the reaction of electroreduction of molecular oxygen and in other electrode reactions that include oxygen adsorption as the intermediate stage.

The data reported in the majority of works on heterogeneous catalysis on metalloporphins testify to the absence of direct connection between the catalytic activity and bulk semiconductor properties (e.g., electric conductance). The main effects influencing the rate and selectivity of redox reactions are attributed to the nature of the central ion of metal, the chemical structure of the metalloorganic complex molecule (influence of substituents in the ligand, polymerization), and to the crystalline structure. Unfortunately, by now there are too few works in which comprehensive investigations of the catalytic and electrophysical properties of a number of metalloporphins would have been carried out under the same conditions. In the works of Manassen and Bar-Ilan (1970, 1973, 1974), the catalytic activity of phthalocyanines and tetraphenylporphins of transition metals (Fe, Co, Mn, Cu, Ni) in the reaction of oxidative dehydrogenation of cyclohexadiene was compared with the first oxidation potentials of these catalysts. In the case of phthalocyanine complexes, the catalytic activity is higher the lower the first oxidation potential (MnPhc > FePhc > CoPhc > NiPhc, CuPhc). Yet, univocality of such comparison is not obvious. As has been shown above, in the case of phthalocyanine and tetraphenylporphin complexes of Fe and Co, on the one hand, and of Ni and Cu, on the other hand, localization of electrons takes place in different regions of the complex, namely, in the electronic system of the central ion and in the π-electronic system of the ligand, respectively. Such correlation is not observed when the catalytic properties of tetraphenylporphin complexes is considered: CoTPhP is more active than FeTPhP, though the first oxidation potential of FeTPhP is -0.32 V, while that of CoTPhP is +0.52 V. Evidently, this is associated with the complicated influence of both the organic ligand and metal on the redox processes and may be indicative of the fact that the catalytically active site is not the central atom alone, but the ligand as well. Recently it was convincingly shown in the work of Borisenkova and Rudenko (1975) for the oxidation of isopropyl alcohol and decomposition of oxalic acid. One should also take into account that the nature of the chemisorption bond may be different in various redox processes; thus, it may be not only π-bond with ions of transition metals of a chelate complex, but also a bond using unpaired electrons. In the latter case catalytic activity can be compared with concentration of paramagnetic centers, as it was found in earlier works (Acres and Eley, 1964) for ortho-para conversion of hydrogen on phthalocyanines of copper. Taube (1966) put forward an interesting idea that, depending on the type of interaction -σ, π, phthalocyanine can behave either as a donor or as an acceptor of electrons. Transfer of electrons occurring in the π-interaction depends on the relative energetic position of the central ion and ligand. The π-system here functions as a kind of "buffer."

In the investigations of the catalytic properties of organic semiconductors, the reaction of decomposition of hydrogen peroxide is very often employed as a model. For electrocatalysis this reaction is of particular interest, since hydrogen peroxide in many cases is an intermediate product of the oxygen electroreduction process. Decomposition of hydrogen peroxide on metallophthalocyanines (Fe, Co, Ni, Cu, Mn, Pt, Pd, Ru, Os, Rh, Ir) was studied in the works of Roginsky and co-workers (1973), Berezin (1968), Levina *et al.* (1966), Waldmeier and Sigel (1970). Most active are monomers of phthalocyanines of Ru, Os, and Fe and the polymer of copper phthalocyanine. From these works it follows that neither the presence of unpaired d electrons in the central atom nor the oxidizability of the central atom are the condition for the catalase activity of metallophthalocyanines, which is, probably, associated with the ability of certain MPhcs to form intermediate complexes with hydrogen peroxide. It has been shown (Levina *et al.*, 1966) that the activity of metal powders (Co, Cu, Ni) coated with a continuous thin film (10^{-5} to 10^{-6} cm) of phthalocyanine of the same metal greatly exceeds the activity of individual components in the reaction of hydrogen peroxide decomposition. In accordance with the electron theory of catalysis (Wolkenstein, 1976), this may be explained by that due to injection of electrons from the metal into the phthalocyanine film, when the thickness of the latter does not exceed the depth of shielding, there takes place a shift of the Fermi level on the surface of phthalocyanine. Evidently, similar phenomena also take place in complexes with charge transfer, obtained by condensation of phthalocyanine vapors on an alkali metal film (Ichikawa *et al.*, 1966; Sudo *et al.*, 1969), which activate even the synthesis of ammonia at room temperature and under pressures below atmospheric pressure.

Explanation for the last two examples can be found within the framework of the electron theory of catalysis on semiconductors, and they may be indicative of the contribution of collective electron properties to the catalytic activity of metalloorganic complexes. However, most of the experimental data now available indicate the decisive role of local electron transitions caused by the local chemical properties of the complex metalloorganic molecule-catalyst and by its electronic structure. In organic semiconductors the interaction between individual molecules is considerably weaker than the interaction between the composite parts of the molecule, and each molecule in the crystal lattice has an opportunity of preserving its individuality. This should be, probably, first of all taken into account in the analysis of results of catalytic and electrocatalytic investigations.

III. KINETICS AND MECHANISM OF OXYGEN ELECTROREDUCTION ON METALLOPORPHINS

As has been pointed out, most detailed study on organic complexes of metals has been made of the electroreduction reaction of molecular oxygen. This reaction takes place on air or oxygen cathodes of chemical current sources. Depending on the nature of the electrocatalyst, reduction of oxygen may proceed along different pathways (Tarasevich, 1973):

$$\begin{array}{ccccccc} & & H_2O & & & & H_2O \\ & & \uparrow k_1 & & & & \uparrow k_3 \\ (O_2)_{vol} & \rightleftarrows & (O_2)_{surf} & \underset{k_5'}{\overset{k_5}{\rightleftarrows}} & (O_2)_{ads} & \underset{k_2'}{\overset{k_2}{\rightleftarrows}} & (H_2O_2)_{ads} \\ & & & & \uparrow & \xrightarrow{\quad k_4' \quad} & \uparrow \end{array}$$

$$\underset{k_6'}{\overset{k_6}{\rightleftarrows}} (H_2O_2)_{surf} \rightleftarrows (H_2O_2)_{vol} \qquad (1)$$

On most active platinum electrode the process proceeds, mainly, through dissociative adsorption of oxygen directly to water (with constant k_1). Hydrogen peroxide may form only in a parallel reaction (k_2). In the case of carbon electrode, it is molecularly adsorbed oxygen that undergoes reduction (k_2). In a two-electron reaction hydrogen peroxide is formed and, since $k_3 << k_2$, its further transformation goes at considerably more negative potentials. On a number of electrode materials hydrogen peroxide may catalytically decompose (k_4), oxidize by a reverse reaction (k_2'). The stages of adsorption and desorption of oxygen (k_5 and k_5') or of hydrogen peroxide (k_6 and k_6') may also be inhibited. In principle, electrocatalysis may reside in acceleration of any of the stages of this process.

In connection with the multielectron character of electroreduction of molecular oxygen, the study of this process must include two main stages:

(1) Elucidation of the mechanism or pathway of the process, i.e., establishing the relationship between partial reactions of the total process.

(2) Analysis of the kinetics of most important partial reactions with a view to elucidating the nature of the inhibited stage.

A. Model Systems

Metalloorganic complexes have low conductivity and, therefore, in studying electrocatalytic reactions metalloorganic compounds are applied as a thin film to an electrically conductive support. Described in the literature is a number of model systems, intended for carrying out investigations under purely kinetic or strictly definite hydrodynamic conditions.

1. Thin films (10^2-10^4 Å) of metalloorganic complexes on a smooth electrically conductive support, which are either applied from solution (Kozawa *et al.*, 1970, 1971; Tarasevich and Bogdanovskaya, 1975) or deposited by vacuum sublimation (Savy *et al.*, 1972, 1975, 1976). As the electrically conductive support, gold or pyrographite are most often used.

2. Extremely thin layers of a catalyst, which is a fine, dispersed (usually carbonaceous) carrier, coated with metalloorganic complexes (Savy *et al.*, 1974; Radyushkina *et al.*, 1975, 1977; Jahnke and Schönborn, 1969; Alt *et al.*, 1971, 1973; Beck *et al.*, 1973; Meier *et al.*, 1973). This system is noted for a higher activity as compared with the first one and is often used in investigations by the disk electrode and disk-ring electrode methods. The disk-ring electrode method that offers a possibility of fixing on the ring electrode the intermediate product (in the given case, hydrogen peroxide) of the reaction taking place on the disk, is most informative from the standpoint of subdividing the process into partial reactions and determining the kinetic parameters of each such reaction (Bagotsky *et al.*, 1972).

3. Suspension electrode (Jahnke and Schönborn, 1969; Alt *et al.*, 1971, 1973; Beck *et al.*, 1973), which allows eliminating diffusion limitations and comparing the electrocatalytic activity of organic complexes of metals, applied to a disperse carrier, under kinetic conditions. A disadvantage of the suspension electrode lies in a complicated character of the charge transfer from the suspension particles to the current lead (Slaidin *et al.*, 1974); under certain conditions this stage may become limiting.

4. Porous gas-diffusion electrodes promoted by organic complexes (Radyushkina *et al.*, 1973, 1977; Meier *et al.*, 1973; Dabrowski *et al.*, 1975). The results in comparing the catalytic activity, obtained by this method, are of qualitative character, since the characteristics of such electrodes are substantially dependent on a number of macrokinetic parameters.

When analyzing the kinetics of electrochemical reactions on metalloorganic complexes, it is necessary to take into account specific features of the structure of the double electrical layer on semiconductor materials, namely, possible distribution of the electrode potential drop between the Helmholtz layer in the electrolyte and the space charge in the semiconductor. Potential distribution depends on the concentration and mobility of free current carriers, on the concentration of surface states, etc. In case the thickness of the semiconductor film is less than the depth of screening (for organic complexes of metals this value is 10^{-6}-10^{-5} cm), additional effects are possible, caused by the injection of current carriers from the conductor (support) into the semiconductor (catalyst). Therefore, investigations are required into the structure of the metalloorganic complex/electrolyte interface for the particular model systems described above.

Measurements of the differential capacity on films of Co, Fe, and metal-free phthalocyanines, sublimated on pyrographite, have shown that the capacity of the film-coated electrode is by an order of magnitude lower than that of the support and amounts to ~0.5 - 0.7 $\mu F/cm^2$. Capacity appears to be substantially dependent on the alternating current frequency. This, evidently, indicates that the capacity measured in such a system is conditioned by rapid surface states rather than determined by the charge of the depleted layer in the semiconductor. This gives grounds to assume that the charge transfer in the film is a rapid process.

When investigating the nature of the catalytic and electrocatalytic activity, the electronic structure of the neutral state of organic complexes is taken into account. However, in the course of electrochemical reactions, with variations in the pH and electrode potential values, there may take place a change of the complexes (origination of ionic forms). This, accordingly, brings about a change in the electronic structure and donor-acceptor properties of the active species. Electrochemical and spectroscopic investigations of the redox processes of metal-containing phthalocyanines in aprotic media (Rollman and Iwamoto, 1968; Clack and Yandle, 1972) have shown that their oxidation (reduction) proceeds in several consecutive one-electron stages up to the formation of triply and quadruply charged ions, and made it possible to determine the place of localization of the electrons (see Section II). It should be pointed out that the majority of investigations of redox processes on metal-containing phthalocyanines, in view of their limited solubility, have been carried out in nonaqueous solutions with the use of various solvents and various electrolytes; this renders the comparison of redox potentials rather difficult. In the case of aqueous solutions (or when traces of water are added to the nonaqueous solutions), the electrochemical oxidation-reduction products chemically interact with water (or with H_3O^+), which leads to merging of the waves in the polarograms. The study of the state of the surface of metal-free phthalocyanine (Fig. 4) and of cobalt phthalocyanine (Fig. 5) by the method of potentiodynamic i - ψ curves in aqueous solutions at pH 0.3-14 allows interesting conclusions. On both phthalocyanines, irrespective of the pH of solution, potentiodynamic i - ψ curves display a number of delays or maxima, among which, in the first approximation, three maxima can be discriminated in the region of potentials (ψ_r, V)[1]: I, 1.2-0.75; II, 0.75-0.4; III, 0.4-0.0. The potentials of these maxima (in terms of the ψ_r scale) are independent of pH. Rollman and Iwamoto (1968) measured the electrode potentials of the processes of electroreduction and electrooxidation of H_2, Co, Ni, and Cu phthalocyanines in dimethyl sulfoxide. Translation of these values for an aqueous solution, taking into account the correction for the interfacial potential jump, gives grounds for suggesting ligand oxidation to take place in acid solutions at ψ_r = 1.2-0.75 V and 0.75-0.4 V (for H_2Phc and CoPhc), accompanied (at ψ_r - 1.2-0.75 V) by a water discharge with the formation of adsorbed oxygen. At ψ_r - 0.4-0.0 V in acid solutions, 0.75-0.4 V in neutral solutions, and 1.2-0.7 V in alkaline solutions the molecules of H_2Phc and CoPhc are neutral. In acid solutions such a neutral molecule discharge of H_3O^+ ions is, evidently, possible. At ψ_r = 0.5-0.4 V in alkaline solutions and at ψ_r = 0.2-0.0 V in

[1]*ψ_r is referred to the hydrogen electrode potential in the same solution; ψ is referred to the normal hydrogen electrode potential.*

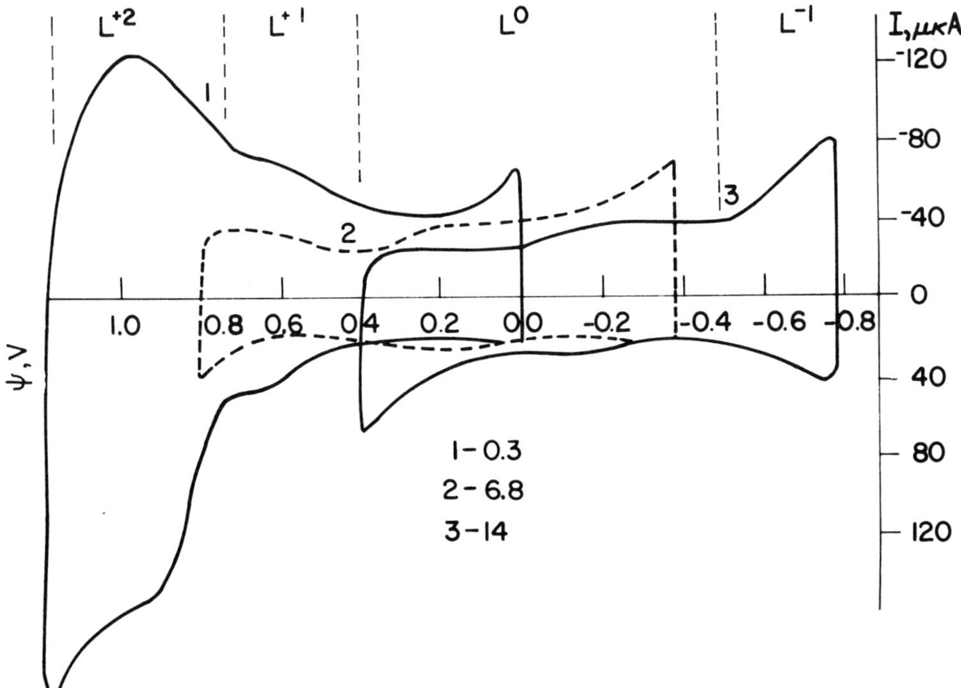

Fig. 4. Potentiodynamic i - ψ curves on H_2Phc at pH 0.3; 6.8, and 14 (v = 50 mV/sec).

neutral solutions, there takes place the process of reduction of CoPhc with localization of the electron on the central atom of cobalt. H_2Phc molecule under these conditions is neutral. With a further negative potential shift (to ψ_r = 0.0 V or ψ = -0.8 V), which may occur only in alkaline aqueous solutions, reduction of the ligand takes place.

Useful quantitative information concerning the state of the surface of films from metalloorganic complexes in an electrolyte can be derived from mirror reflection spectra. Mirror reflection spectra measured on H_2Phc films in an alkaline and neutral solutions at a stationary potential (ψ = -0.05 V) have a band at 650-670 nm, characteristic for the neutral form of the pigment. In case of cathodic polarization ψ_r = 0.0 V (ψ = -0.8 V) in an

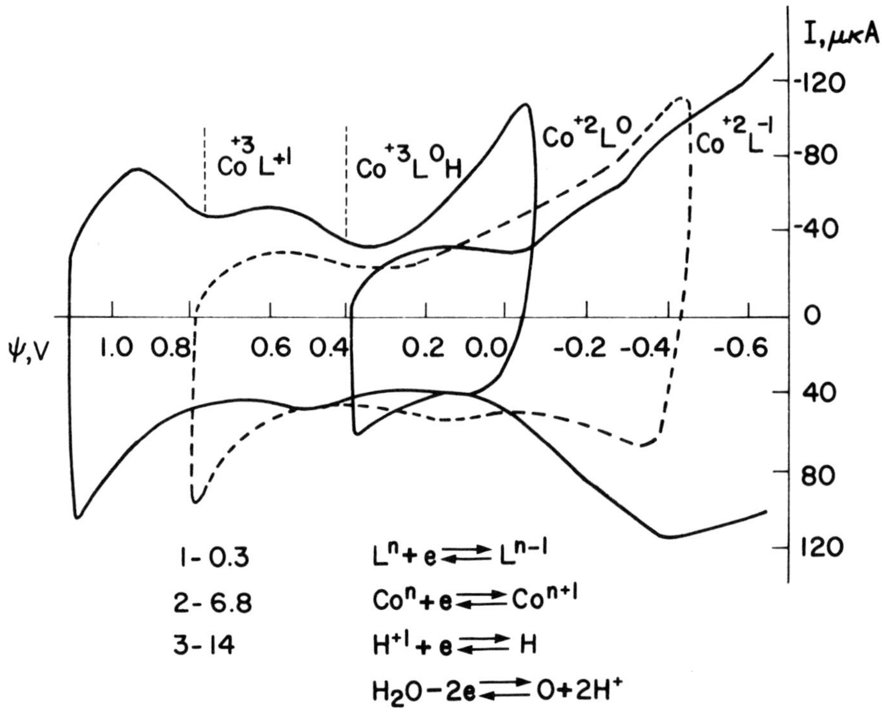

Fig. 5. Potentiodynamic i - ψ curves on CoPhc at pH 0.3; 6.8, and 14 (v = 50 mV/sec).

alkaline medium, the absorption band shifts into the region of shorter waves of 580-540 nm, this being indicative of the presence of a negative form of metal-free phthalocyanine. In neutral solutions, cathodic polarization causes no shift of the absorption band which is characteristic for the neutral molecule. Monomer of cobalt phthalocyanine, like all neutral phthalocyanines of metals, has an absorption band in the region of 600-700 nm. In cathodic polarization to ψ = -0.45 V, the first stage of CoPhc reduction takes place. The neutral molecule absorption band intensity diminishes and a double bond appears at 400-500 nm. At a potential ψ = -0.8 V, a broad band appears in the region of 400-600 nm. Evidently, in this region there occurs superposition of the band due to the intramolecular electron transfer from the metal to the ligand (similar to the CoPhc monoanion) and of the band due to the electron transition between the intrinsic levels of the charged ligand (similar to the H_2Phc monoanion).

Thus, within the range molecular oxygen electroreduction potentials (ψ_r = 1.0-0.0 V), depending on the pH of the solution and on the nature of the central ion, the phthalocyanine molecule can be in different electronic states. Therefore, when considering the dependence of the electrocatalytic activity of metalloporphins on the nature of the ligand and of the central ion, it is necessary to take into account the possibility of the reaction to proceed on charged forms of organic complexes.

At present the information concerning the structure of the metalloorganic complexes/electrolyte interface is rather limited and relates only to certain particular systems used in the study of electroreduction of oxygen. Proceeding from these data, it may be supposed that the potential drop is localized mainly in the electrolyte and that the transfer of charges in the semiconductor film is rapid.

B. Mechanism of Electroreduction of Molecular Oxygen

From the works published by now, it can be inferred that the mechanism and rate of the reaction of electroreduction of molecular oxygen are influenced by the structure of metalloorganic complex catalysts, by their crystalline properties, and by the composition of the solvent. Of greatest importance here is the chemical structure of the catalyst molecule: the type of the ligand and of the substituents, the degree of polymerization, on the one hand, and the nature of the central ion in the molecule, on the other hand.

Macrocyclic complexes containing a system of conjugate π-electrons (Phthalocyanines: Savy *et al.*, 1972, 1974, 1975; Tarasevich and Radyushkina, 1975-1977; Jahnke and Schönborn, 1969; tetraphenylporphins: Alt *et al.*, 1971-1973; Tarasevich and Radyushkina,

1976; dibenzotetraazaannulenes, DTAA: Alt *et al.*, 1973; Beck *et al.*, 1973) prove to be more or less active in the oxygen electroreduction reaction in both acid and alkaline solutions. (It should be noted that the number of π-electrons in the inner conjugate system of these complexes is approximately the same, namely, ~18.) Polarization curves of oxygen reduction on a rotating disk electrode coated with a thin layer of $CoPhc_i$, CoTPhP, and CoTMPhP, in an alkaline solution, unlike pyrographite or carbon black, have one wave (Fig. 6). In the same figure dependences of limiting currents of hydrogen peroxide oxidation on a ring for the compounds cited are given. Quantitative calculations carried out on the

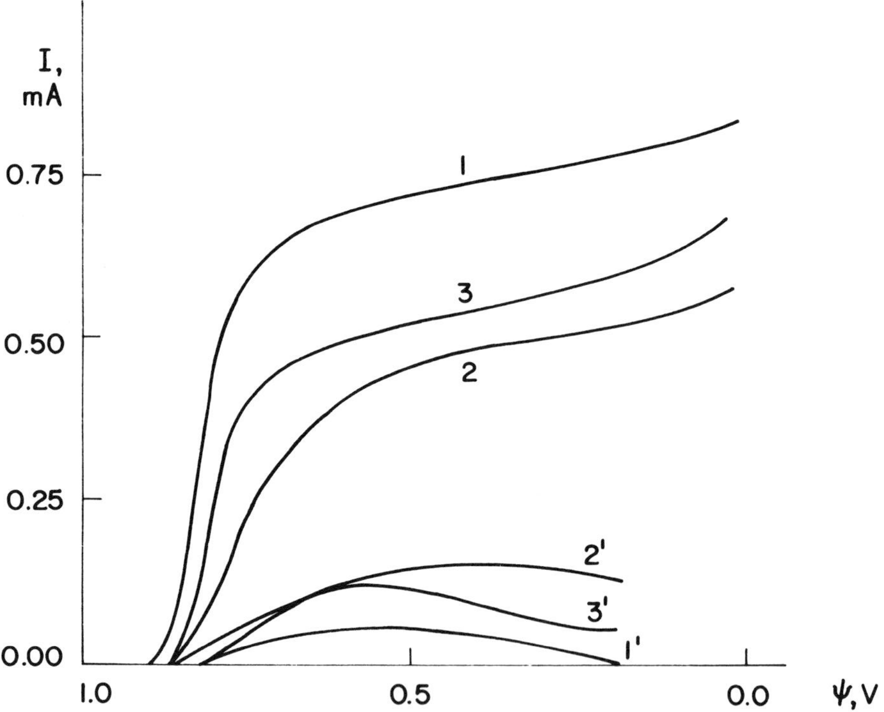

Fig. 6. Polarization curves of oxygen reduction on $CoPhc_i$ (I), CoTPhP (2), and CoTMPhP (3) in 0.1 N KOH and corresponding curves of H_2O_2 oxidation on a platinized platinum ring (m = 680 rpm).

basis of these data with the aid of a method suggested by Gagotsky *et al.* (1972) showed that the molecular oxygen reduction mechanism on these organic complexes is different. On the CoTPhP the reaction proceeds only through the intermediate formation of hydrogen peroxide ($k_1 = 0$) (Tarasevich *et al.*, 1975, 1977). In the case of $CoPhc_i$ and CoTMPhP, there takes place parallel course of the reaction to water. The contribution of this reaction to the total process is 50% for $CoPhc_i$ and 30% for CoTMPhP.

In an acid solution (Fig. 7) on all the investigated complexes, the reaction proceeds only through the intermediate formation of hydrogen peroxide. The value of the constant k_2 decreases in the series CoTMPhP > CoTPhP > $CoPhc_i$ > CoPhc. Phthalocyanines in acid solutions are less active than tetraphenylporphin complexes (Manassen, 1973; Tarasevich and Radyushkina, 1976; Alt *et al.*, 1972, 1973). Introduction of electron donors into the phenyl groups of cobalt tetraphenylporphin increases its activity in the following order (Alt *et al.*, 1973):

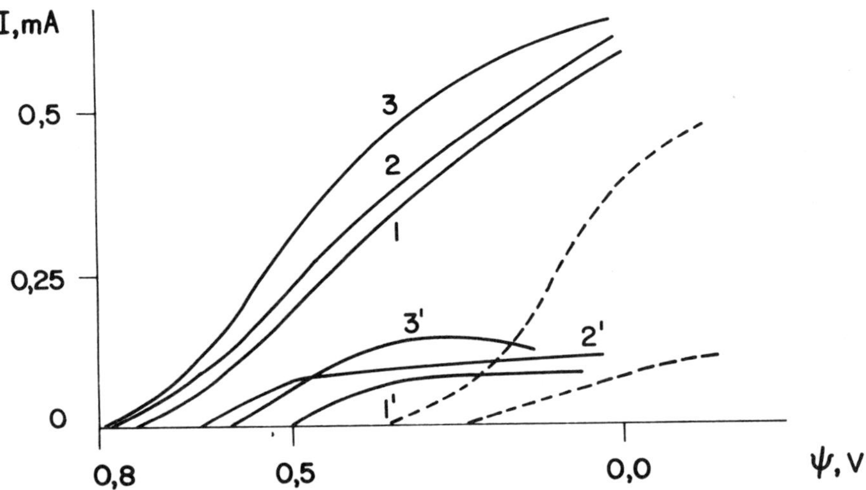

Fig. 7. Polarization curves of oxygen reduction on $CoPhc_i$ (I), CoTPhP (2) and CoTMPhP (3) in 0.1 N H_2SO_4 and corresponding curves of H_2O_2 oxidation on a platinized platinum ring (m = 680 rpm).

$-OCH_3$ > (phenyl) > H

It should be noted, however, that the method of synthesizing the chelate is, evidently, of no small importance. For instance, in the work of Manassen (1973) the catalytic activity of CoTPhP did not change with the introduction of *p*-methoxy or *p*-nitro-substituents. Acceptors (e.g., F^-) exert no influence on the electrocatalytic activity of CoPhc, but reduce it in the case of FePhc. According to the data reported by Beck *et al.* (1973), dibenzotetraazaannulene of Co(CoDTAA) is more active than CoTPhP in an acid solution, neither donor nor acceptor substituents exerting influence on the rate of oxygen electroreduction on this complex.

Thus, as regards the reaction of electroreduction of molecular oxygen, the investigated organic ligands constitute the following series:

alkaline solutions: Phc > TPhP > DTAA

acid solutions: DTAA > TPhP > Phc

A still more substantial influence on the rate and mechanism of the ionization reaction of molecular oxygen on the investigated organic complexes is produced by the nature of the central ion of metal (Savy *et al.*, 1972, 1974, 1975; Tarasevich and Radyushkina, 1976; Alt *et al.*, 1972, 1973). As to the rate of the electroreduction of molecular oxygen, irrespective of the pH value, the activity of metal-containing phthalocyanines changes in the series:[1]

$$Fe(II), Fe(III) > Co(II), Mn(II) > Ni(II) \geq Cu(II) > H_2$$

In the case of tetraphenylporphins at all the pH values and of dibenzotetraazaannulenes in acid and alkaline solutions, complexes with Co(II) are more active than complexes with Fe(II, III):

[1] *The relationship between the electrochemical activity of complexes containing Fe(II) and Fe(III) will be considered in greater detail.*

TPhP: Co(II) > Fe(II,III) > Ni(II), Cu(II)

DTAA: Co(II) > Mn(II) > Fe(II,III) > Ni(II), Cu(II)

Systematic investigations carried out with the aid of the disk-ring electrode method (Tarasevich *et al.*, 1975-1977) have made it possible to establish which reactions in the complicated process of reduction of molecular oxygen are catalyzed by metalloorganic complexes and how the rate of these reactions changes with the change of the chemical structure of the catalyst. In an acid solution in the presence of organic complexes of metals (Phc, TPhP, TMPhP), there takes place acceleration of the reaction that runs to hydrogen peroxide. In alkaline solutions, in the case of FePhc, $FePhc_i$, MnPhc, $MnPhc_i$, $CoPhc_i$, and CoTMPhP, in addition to this, there originates a direct reaction that proceeds with breakage of the O-O bond. Cumulative data for alkaline solutions are given in Table 1. In the same table the data for pure carbonaceous carrier, viz., carbon black, are presented for comparison. Establishing of the fact that oxygen can be reduced to water without the intermediate formation of hydrogen peroxide substantially differs from the view point expressed in the works by Kozawa *et al.* (1970), Savy *et al.* (1972, 1974, 1975), Jahnke and Schönborn (1969), and Alt *et al.* (1972, 1973). In these works the reaction in all cases is assumed to proceed only through the intermediate formation of hydrogen peroxide.

The dependence of the rate of electroreduction of oxygen on organic complexes on the nature of the central ion of metal has been discussed in the literature time and again (Savy *et al.*, 1972, 1974, 1975; Alt *et al.*, 1973; Tarasevich *et al.*, 1976) and finds qualitative explanation when taking into account the electronic structure of metalloorganic complexes. The simplest approach is to establish a correlation between the electrocatalytic activity of phthalocyanines of metals (and, to a smaller extent, of tetraphenylporphins) and their ability to oxidation, expressed by the first oxidation potential (Manassen, 1973; Beck *et al.*, 1973).

TABLE 1. Constants of Partial Reactions of Oxygen Reduction to Water (k_1) and to Hydrogen Peroxide Reduction (k_2) for Various Organic Complexes (0.1 N KOH, $\psi_r = 0.6$ V)

Electrode	Carbon black	H_2Phc	FePhc	CoPhc	$CoPhc_i$	CoTPhP	CoTMPhP	MnPhc	$MnPhc_i$
$k_1 \times 10^3$ cm/sec	-	-	1000	-	55	-	18	10	14
$k_2 \times 10^3$ cm/sec	12	30	30	32	55	15	53	20	24

This leads to a mechanism of the type (L being an organic ligand):

$$(M\text{-}L) + \frac{1}{4} O_2 + H^+ \rightarrow (M\text{-}L)^+ + \frac{1}{2} H_2O; (M\text{-}L)^+ + e \rightarrow (M\text{-}L) \qquad (2)$$

However, at present no kinetic measurements are available, offering evidence in favor of this mechanism. Randin (1974) has found a correlation between the electrocatalytic activity and magnetic properties of MPhcs, determined by the spin state of the central ion.

On the contrary, in the opinion of Dabrowski *et al.* (1975), there is no direct connection between the value of the magnetic moment and the electrocatalytic activity. However, Dabrowski *et al.* (1975) used the data obtained on porous gas-diffusion electrodes, and therefore the plausibility of their conclusions is not immune from probable distortional influence of macrokinetic effects.

Appleby *et al.* (1976), Savy *et al.* (1973-1975), and Alt *et al.* (1972, 1973), for explaining the influence of the nature of the central ion on the rate of the oxygen electroreduction reaction, used the results of quantum-chemical calculations of the electronic structure of metal-substituted porphins and their complexes with oxygen, briefly discussed in Section II. Zerner *et al.* (1966) showed that stable complexes of oxygen with metalloporphins (e.g., iron-substituted ones) are formed when oxygen is adsorbed in the side position. d_{z^2}MO of the metal ion is most favorable for binding the oxygen molecule, since it has the symmetry that ensures the necessary overlapping of the orbitals. It is supposed that in the adsorption of the oxygen molecule on metallophthalocyanines (tetraphenylporphins) there takes place formation of the double σ-, π-bond between the $d_{xz}d_{yz}$ and d_{z^2}MO of the central ion and π^* MO of the oxygen molecule. Partial transfer of the electron from the central ion of the metal to antibonding π^*-orbitals of the oxygen weakens the O-O bond in the molecule. This contributes to acceleration of cathodic reduction of oxygen. Since, for the donor-acceptor bond to form, interaction of filled orbitals of oxygen with vacant d_{z^2} orbitals of metal is necessary, maximum activity must be in the case of complexes with Fe(II,III), Co(II), and Mn(II), while for Ni(II) and Cu(II) only weak catalytic activity or its absence can be expected. An increase in the quantity of shared electrons (polymerization of phthalocyanines or introduction of donor groups into tetraphenylporphin) increases the electron density of the central ion. This facilitates electron transfer to the oxygen molecule and accelerates its reduction. On the other hand, an increase in the electron localization degree on the adsorbed molecule of O_2 brings about weakening of the O-O bond to its breakage, with the direct reaction proceeding to water. To explain changing activity series of the ligands as one goes from acid to alkali solutions, one should take into account a possibility for the reaction to proceed on charged forms of organic complexes (anions, cations) which may exist on the electrode surface within a definite region of pH and potential values.

Consideration of the nature of electrocatalytic effects, carried out on the basis of the data of approximate quantum-chemical calculations, is, naturally, only qualitative and leaves out of account quite a few experimental data, such as the influence of the type of the support-carrier, of the crystalline modification, and of the method for synthesizing the metalloorganic complex. Their explanation requires a more thorough theoretical and experimental study of the structure of the organic catalyst/electrolyte interface in the presence and in the absence of oxygen.

Considerable influence on the electrocatalytic activity of metallochelates is produced by the nature of the carrier, the highest activity being attained with the use of various forms of carbon. Maximum effect takes place in the presence of alkaline groups on the surface of carbon (Jahnke and Schönborn, 1969). At present the mechanism of these phenomena is not yet clear. Investigations carried out with α- and β-modifications of Fe(II)Phc have shown (Savy *et al.*, 1972, 1973) the electrocatalytic activity of the α-form to be higher. Measurements of the electronic spectra in the transmission and reflection mode have shown that as one goes from the β- to the α-form of FePhc, the band at ~15000 cm^{-1} shifts into the long-wave region. In the opinion of Savy *et al.* (1972, 1973) according to Mathur and Singh (1972), this is indicative of a higher concentration of high-spin Fe(III)Phc in the α-modification as compared to β. This is conditioned by oxygen adsorption in the bulk of the α-form and by the transition of the most part of Fe(II)Phc into the high-spin form. Cathodic polarization of an α-FePhc sample leads to the irreversible loss of activity. Elevated electrochemical activity is supposed to be conditioned by the existence of the O_2^--Fe(III)Phc complex, on which there takes place secondary adsorption of the oxygen molecule that takes part in the electrochemical reaction.

The influence of the method of synthesis of monomer and polymer FePhc on their electrocatalytic activity and on the spin state of the central ion has been investigated in detail by Appleby *et al.* (1976). They have established that the polymer synthesized by the gas phase method directly on the carrier is most active, including Fe(III) with the intermediate spin. Least active is the Fe(II) monomer in the triplet state. However, the state of FePhc at the electrolyte interface and the mechanism of adsorption of molecular oxygen still remain unclear.

The analysis carried out shows that local interaction of the oxygen molecule with the catalyst molecule is of prime importance in the electrocatalysis of the oxygen reduction reaction. Collective interaction effects, at least in the case of the considered particular model systems, are only of secondary importance. This conclusion is confirmed by the data reported by Meier *et al.* (1973), where no correlation has been found between the electrocatalytic activity of phthalocyanines of a number of metals and their electric conductivity.

C. Oxygen Reduction Kinetics

For elucidating the nature of the rate-determining stage of partial reactions in the total process of molecular oxygen reduction, it is necessary to study kinetic regularities within a wide range of pH values. Such studies for H_2Phc, CoPhc, $CoPhc_i$, and CoTPhP were carried out by Tarasevich *et al.* (1977). Typical dependences of the kinetic parameters $\sigma\psi/\sigma$ ℓg i and $\sigma\psi/\sigma$pH for the investigated catalysts under various experimental conditions are presented in Figs. 8 and 9 and in Table 2. In the entire investigated range of pH from 0 to 14, the reduction rate of O_2 to H_2O_2 in the presence of organic complexes is higher than in the case of carbon black as the carrier. The kinetic parameters of the reaction of reduction of O_2 to H_2O_2 on various catalysts, with the exception of CoTPhP in acid medium, are close to carbon black. Coincidence of the kinetic parameters of the O_2 reduction on the organic complexes and on carbon black, which is observed in the

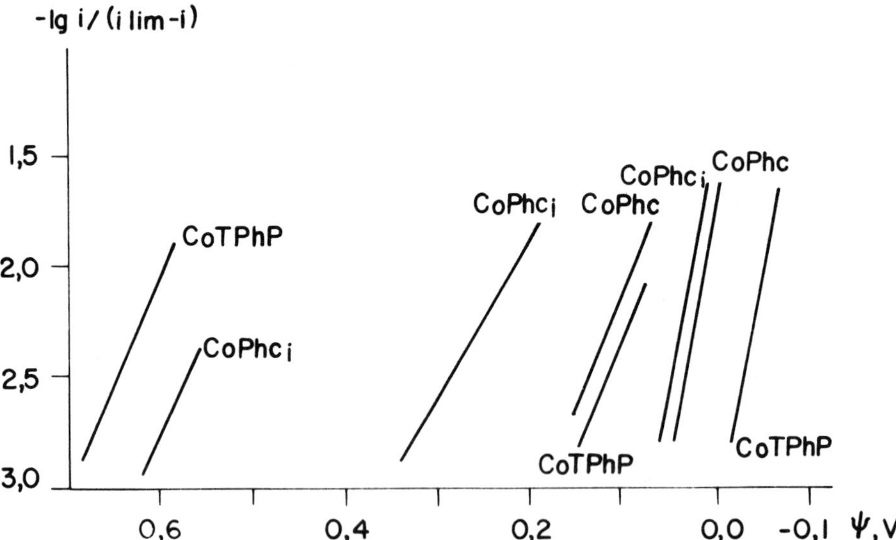

Fig. 8. ℓg i − ψ *dependence for oxygen reduction on organic complexes of metals.*

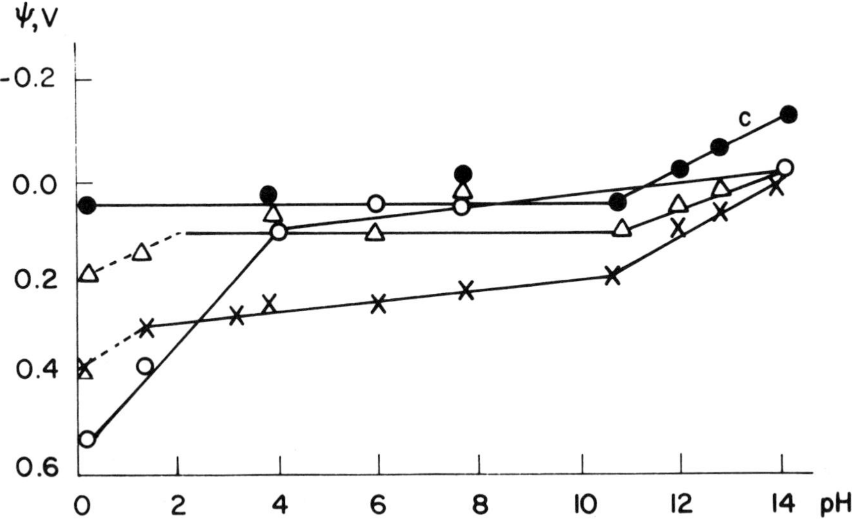

Fig. 9. $\psi_{1/2}$ - pH dependence for the reduction of O_2 on carbon black (o), cobalt phthalocyanine monomer (Δ), cobalt phthalocyanine polymer (×), metal-free phthalocyanine (Δ), and cobalt tetraphenylporphin (o).

TABLE 2. Kinetic Parameters of the Reaction of Oxygen Reduction to Hydrogen Peroxide on Carbon Black and in the Presence of Organic Complexes[a]

Electrode	pH	$\sigma\psi/\sigma pH$ (mV)	$\sigma\psi/\sigma \ell g\ i$ (mV)
Carbon black	2 - 10	0	-90 - -120
	10 - 14	-60 - -30	-40 - -50
CoPhc	2 - 10	0	-100 - -120
	10 - 14	-60 - -30	-40 - -50
CoTPhP	0 - 4	110	-100 - -120
	4 - 10	0 - 10	-100 - -120
	10 - 14	-60	-40 - -60

[a]*Close values of kinetic parameters were observed on phthalocyanine films, 500 - 1000 Å thick, sublimated on pyrographite electrode.*

majority of cases, is indicative of similarity of the rate-determining stage mechanism. Studies carried out on carbon black (Radyushkina *et al.*, 1977) and on other carbonaceous materials (Tarasevich *et al.*, 1971; Taylor and Humffray, 1975) showed that the reaction of cathodic reduction of oxygen is determined by the rate-determining stage of electron transition to the adsorbed molecule of oxygen (in the reverse reaction of anodic oxidation of hydrogen peroxide from the O_2^- ion):

$$\begin{aligned}
&(M\text{-}L) + O_2 \rightarrow (M\text{-}L)O_2 \\
&(M\text{-}L)O_2 + e \rightarrow (M\text{-}L)O_2^- \ \text{(rate-determining stage)} \\
&(M\text{-}L)O_2^- + H_2O \rightarrow (M\text{-}L)HO_2 + OH^- \\
&(M\text{-}L)HO_2 + e \rightarrow (M\text{-}L)HO_2^- \\
&(HO_2^- + H_2O \rightleftarrows H_2O_2 + OH^-)
\end{aligned} \tag{3}$$

Other stages, leading to the formation of H_2O_2 (in acid solution) or to the formation of HO_2^- (in alkaline solution), proceed rapidly. In acid and neutral solutions the reaction is irreversible and $(\sigma\psi/\sigma pH)_{i=const} = 0$, $\sigma\psi/\sigma\ \ell g\ i = 120$ mV with $\alpha = 0.5$. At higher pH the rate of the reverse reaction increases and the system $O_2/H_2O_2(HO_2^-)$ approaches a reversible state. This leads to a negative shift of the potential and to a decrease in the $\sigma\psi/\sigma\ \ell g\ i$ value. It is quite reasonable to suppose that in the case of organic catalysts (except CoTPhP in acid solutions), the reduction of oxygen also proceeds by this mechanism.

The behavior of CoTPhP in acid solutions cannot be interpreted within the framework of the scheme described above. Yet, this phenomenon is quite interesting and worthy of particular attention. Formal explanation may be offered on the assumption that the mechanism of the rate-determining stage is as follows:

$$(M\text{-}L)O_2 + H^+ + e \rightarrow (M\text{-}L)HO_2 \tag{4}$$

With $\alpha = 0.5$ $\sigma\psi/\sigma\ \ell g\ i = 120$ mV and $\sigma\psi/\sigma pH = 120$ mV. A change in the character of the O_2 reduction mechanism in acid solutions on CoTPhP and CoTMPhP may be associated with other character of their protonation as compared to phthalocyanines.

The kinetics of oxygen reduction at various pH was also investigated on monomer and polymer phthalocyanines of cobalt and iron in the works of Savy *et al.* (1973) and Appleby *et al.* (1976). Detailed comparison of the data reported by Savy *et al.* (1973), Appleby *et al.* (1976), and Tarasevich *et al.* (1975-1977) shows a certain parallelism to exist between them. In both cases, the reaction is of the first order as regards molecular oxygen: the value of $\sigma\psi/\sigma\ \ell g\ i$ diminishes with the growth of pH and at pH > 13, it is ~40 mV; in alkali solutions the potential undergoes a negative shift with an increase of pH; inclinations of polarization curves on the promoted and nonpromoted carbonaceous carriers coincide. Yet, unlike Tarasevich *et al.* (1975-1977), Savy *et al.* (1973) and Appleby *et al.* (1976), come to the conclusion that in alkali solutions the rate-determining stage is the chemical stage that includes breakage of the O-O bond:

$$HO^-_{2\ ads} + H_2O \rightarrow 2OH + OH^- \tag{5}$$

It is clear that this mechanism is of limited importance, since it offers no explanation for the entire corpus of the experimental data, especially in nonalkali media.

The above discussion shows that at present the pathway of the reaction of cathodic reduction of oxygen on organic complexes of metals is, generally, clear. At all pH on all the investigated metalloporphins there takes place reduction of molecular oxygen by a consecutive mechanism that includes intermediate formation of hydrogen peroxide. In alkali solutions (pH = 14-10), in addition, on a number of organic complexes there takes place parallel reaction of oxygen reduction to water without intermediate formation of hydrogen peroxide, i.e., including breakage of the O-O bond either in the oxygen molecule, or, more likely, in the molecular ion O_2^-. In the case of iron phthalocyanine direct reaction to water is the main one.

Less clear is the problem concerning the nature of the rate-determining stage of the reaction. The opinions of different authors are rather contradictory. First of all this is associated with the absence of data on the mechanism of oxygen adsorption on the organic catalyst/electrolyte interface and on the electronic structure of the adsorbed state.

IV. THE USE OF ORGANIC COMPLEXES OF METALS FOR CREATING GAS-DIFFUSION ELECTRODES

High activity of a number of organic complexes of metals in the reaction of electroreduction of molecular oxygen has made possible their use for the activation of oxygen (air) electrodes of electrochemical current sources (Jahnke and Schonborn, 1969; Mrha *et al.*, 1973; Radyushkina *et al.*, 1973, 1975), detectors (Yao *et al.*, 1974), and other electrochemical devices. Creation of active and stable gas-diffusion electrodes requires the solution of a number of macrokinetic problems concerning optimal distribution of the catalyst, electrolyte, and gas in a porous, electrically conducting matrix with a view to maximum surface development for the electrochemical reaction to proceed. Detailed consideration of these problems has been given by Chizmadjev *et al.* (1971). Here we shall consider only those data that are directly pertinent to the use of metalloorganic complexes. Because of their low electric conductivity, organic complexes of metals are used in combination with a high dispersed electrically conductive carrier (carbonaceous materials, nickel powder). The catalyst is introduced into the carrier either by their mechanical mixing (Jasinski, 1968; Tanaka *et al.*, 1971), or by impregnating the carrier with an organic complex solution in concentrated sulfuric acid or in an appropriate solvent [pyridine, dimethyl formamide, etc. (Radyushkina *et al.*, 1973, 1975; Mrha *et al.*, 1973; Meier *et al.*, 1972; Appleby and Savy, 1976; Alt *et al.*, 1973; Hiller *et al.*, 1972; Jahnke *et al.*, 1976)].

This method is most widespread, since it allows uniformly distributing the catalyst over the surface of the carrier. Also used is a method, including the synthesis of the metalloorganic complex directly on the surface of the fine dispersed carrier (Dabrowski *et al.*, 1975; Appleby and Savy, 1976; Pobedinsky *et al.*, 1974). The basic structural type of the electrode in the works dealing with metalloorganic complexes are hydrophobized electrodes in which stable gas/electrolyte interface is created by introducing into the active mass a hydrophobing agent (polyethylene, fluorinated plastic), which simultaneously acts as a binder. Active mass is applied to a gauze or porous metallic base that functions as a current lead. Other particulars concerning the design and manufacturing technology of the electrodes can be found in Radyushkina *et al.* (1973, 1975), Mrha *et al.* (1973), and Appleby and Savy (1976).

In an alkali electrolyte in porous gas-diffusion electrodes quite a number of monomer phthalocyanines display a rather high electrochemical activity (Table 3). The rate of oxygen electroreduction on phthalocyanines of iron, cobalt, manganese, platinum, and palladium are several times higher than on pure carbon black. It is interesting to point out that introducing metal-free phthalocyanine into carbon black increases the reaction rate by approximately 1.5 times. In the case of cobalt, molybdenum, and copper phthalocyanines, the use of the polymeric complex instead of the monomeric one (Fig. 10) leads to a two- or threefold increase in the current density. A similar effect was observed in the case of polymeric iron phthalocyanine by Appleby and Savy (1976). For the polymeric phthalocyanine of cobalt, the influence of the method used for its preparation on the activity of gas-diffusion electrodes (Radyushkina *et al.*, 1975) was investigated. Highest characteristics were displayed by the polymeric cobalt phthalocyanine (polymerization degree from 2 to 6), synthesized by baking from pyromellitic dianhydride, phthalic anhydride, urea, and cobalt salt, with ammonium molybdate used as the catalyst. Polymeric

TABLE 3. Comparison of the Electrochemical Activity of Oxygen Electrodes Prepared from Carbon Black Promoted with 20% of Monomeric Phthalocyanines of Metals.[a]

Catalyst	Without catalyst	H_2Phc	NiPhc	CuPhc	MoO_2Phc	CoPhc	FePhc	MnPhc	PtPhc	PdPhc
i (mA/cm^2) at $\psi_r = 0.80$ V	40	60	8	60	70	130	120	140	110	140

[a] 12N KOH, 90°C.

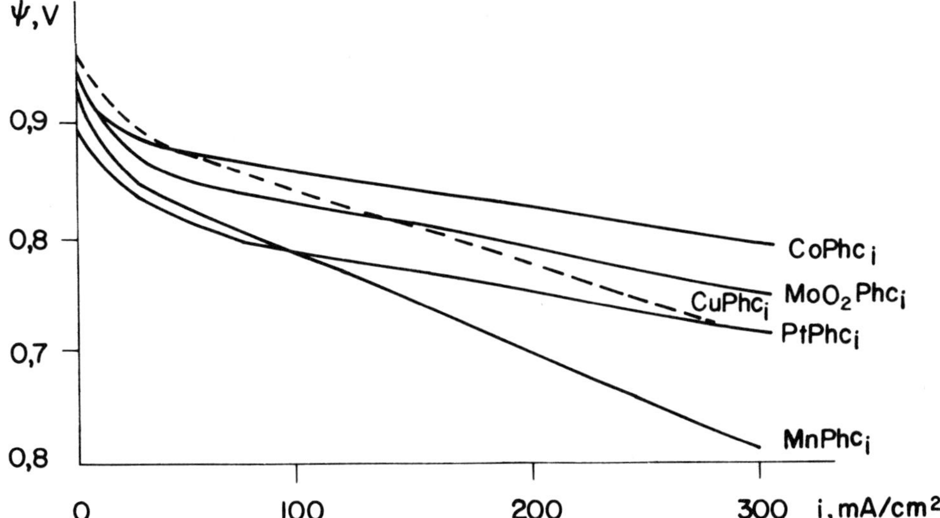

Fig. 10. Polarization curves of oxygen reduction on porous carbon electrodes promoted with polyphthalocyanines in 12N KOH, 90°C.

preparations synthesized by other methods had markedly lower electrochemical activity.

The activity and stability of gas-diffusion electrodes are substantially dependent also on their macrokinetic parameters that are determined by the porous structure of the hydrophobized active layer and by the distribution of the gas and electrolyte in it. In most detail these problems have been investigated by Dribinsky *et al.* (1978) for hydrophobized electrodes based on carbon black promoted with polymeric cobalt phthalocyanine. Judging from the mercury porometry data, the content of $CoPhc_i$ to 10 wt.% has no influence on the electrode structure. An increase in the $CoPhc_i$ content to 20% leads to an increase in the total porosity; with an increase of the $CoPhc_i$ content to 40%, blocking of a portion of narrow pores is observed. Investigations carried out by the method of mercury porometry with frozen electrolyte (Dribinsky *et al.*, 1971) showed that, with the fluorinated plastic content being the same, introduction of $CoPhc_i$ leads to a considerable increase of liquid porosity. In Fig. 11 gas-diffusion electrode current density-fluorinated plastic content and active layer liquid porosity-fluorinated plastic content dependence curves are compared. Correlation of these values indicates the existence of direct connection between the structure and electrochemical activity of the electrodes. Optimization of the structure of the electrode active layer has made it possible to effect current generation in the mode close to the optimum, i.e., intrakinetic one. The value of the apparent activation energy for porous electrodes promoted with $CoPhc_i$ and FePhc at ψ_r = 0.83 V is 4 to 5 kcal/mole, which confirms the predominantly kinetic character of the process. Highest electrochemical characteristics in an alkali electrolyte (0.3 $Å/cm^2$

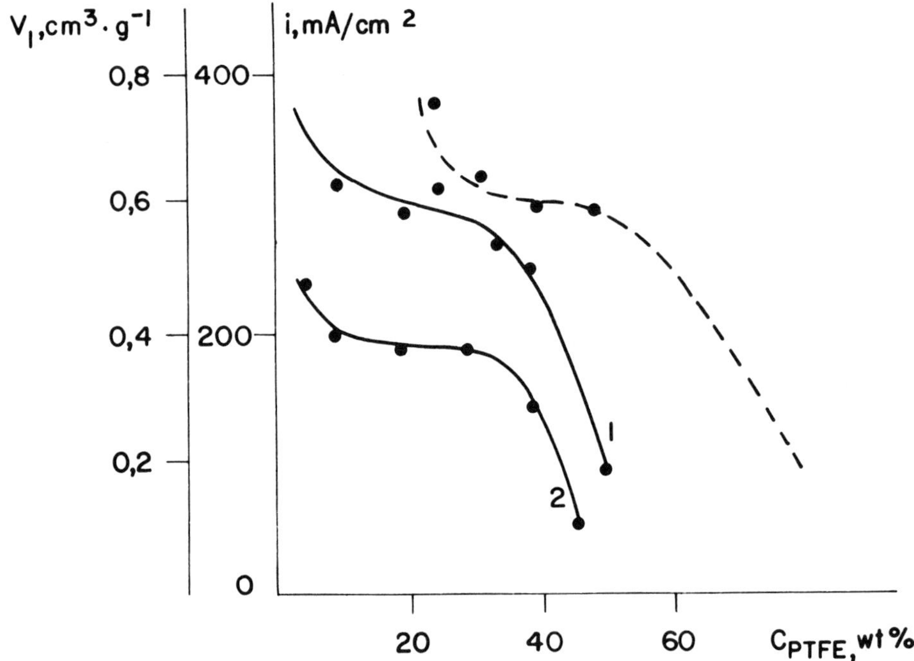

Fig. 11. Electrochemical activity dependence at ψ_r(V): 1 - 0.80, 2 - 0.83 of electrodes promoted with $CoPhc_i$ and liquid porosity dependence (dashed curve) on the polytetrafluoroethylene content.

with ψ_r = 0.80 V) are displayed by carbon electrodes promoted with polymeric cobalt phthalocyanine. These data are substantially higher than the characteristics of the electrodes with metalloorganic catalysts, described in the works of Jasinski (1965), Appleby and Savy (1976), Richter and Henkel (1971), and Pobedinsky *et al.* (1974).

An urgent task in the field of electrochemical current sources is the development of an oxygen (air) cathode for acid media, which would not contain critical metals. Solving this problem will make it possible to overcome the difficulties associated with carbonization of alkali electrolytes when organic fuels are used or contaminated hydrogen and air oxygen are employed as the oxidant. Yet, so far no description can be found in the literature of any inorganic catalysts sufficiently stable in acid media, which contain no noble metals. Certain organic complexes of metals are rather promising in this respect. Given in Table 4 are the electrochemical characteristics of carbon oxygen electrodes, promoted with organic complexes of metals, for acid solutions, that are described in the literature (Jahnke and Schönborn, 1969; Alt *et al.*, 1973; Mrha *et al.*, 1973; Kretzschmar and Wiesener, 1976; Hiller *et al.*, 1972; Jahnke, 1973; Reber *et al.*, 1973; Tarasevich *et al.*, 1977). According to Jahnke and Schönborn (1969), the potential of the carbon electrode promoted with $FePhc_i$, at current density of 10 mA/cm^2 is 0.85 V. The electrodes, however, are unstable. Meier *et al.* (1972, 1973) and Kretzschmar and Wiesener (1976), by modifying the method of the iron polyphthalocyanine synthesis, have succeeded in attaining certain stabilization of the work of the electrodes. Instability of the electrochemical characteristics of phthalocyanine-containing oxygen carbon electrodes in acid solution is associated, probably, with partial dissociation of tetrabenzotetraazaporphin complexes in acid solutions. Negative effect may also be caused by hydrogen peroxide that forms in acid solutions as an intermediate product of the reduction of oxygen (Tarasevich and Radyushkina, 1976; Tarasevich *et al.*, 1977).

TABLE 4. Electrochemical Characteristics of Carbon Electrodes Promoted with Organic Complexes of Metals[a]

Composition of active layer	ψ_r, V at $i_{(mA/sm^2)}$				Working hours	Authors
	10	20	40	50		
1 Carbon black + $CoPhc_i$ + polyethylene	0.70	0.60	0.50	0.43	-	Jahnke and Schonborn (1969)
2 Carbon black + $FePhc_i$ + polyethylene	0.85	0.78	0.72	0.70	-	Jahnke and Schonborn (1969)
3 Norit BRX + 50% $FePhc_i$ + Teflon		0.80	0.75		400	Meier et al. (1972, 1973)
4 Carbon + $FePhc_i$ + Teflon	0.78	0.71	0.60	0.56	1000	Musilova et al. (1973)
5 Norit BRX + $CoPhc_i$ + Teflon	0.72	0.69	0.64		150	Kretzschmar et al. (1976)
6 Norit FNX + CoTMPhP + Teflon	0.75	0.70	0.65	0.63	300	Alt et al. (1973)
7 Carbon AG-3 + Co pyropolymer + Teflon	0.82	0.79	0.76	0.75	10,000	Tarasevich et al. (1977)

[a] 3-6 N H_2SO_4, 25-30°C.

Hydrogen peroxide oxidizes phthalocyanines of metals with scission of the macrocycle (Moser and Thomas, 1963). This, evidently, imposes limitations to the direct use of metal-containing porphin complexes in acid electrolytes.

As is known, heat treatment of organic complexes of semiconductors at temperatures up to 1200°C leads to the formation of pyropolymers that are characterized by a narrow forbidden band and a substantially higher electric conductivity. Such heat treatment, evidently, results in the formation of polymers from highly unsaturated molecules having an aromatic structure (Berlin, 1975), these polymers displaying high chemical stability. This group of polymeric compounds is rather promising for the electrocatalysis.

Richter and Luft (1972) used polyacrylonitrile pyrolized at 600-1000°C for activating an oxygen electrode in acid electrolyte. With current density of 50-80 mA/cm^2, the potential of such electrode is 0.7 V. In durability tests, however, the current density noticeably diminishes.

Levina and Radyushkina (1975) proposed a method for manufacturing a catalyst based on carbon promoted with organic complexes of cobalt (TPhP, TMPhP, Phc), which ensures durable (over 10,000 hr) and stable operation of oxygen (air) electrodes in an acid (4.5 N H_2SO_4) electrolyte. A mixture of carbon and an organic complex (e.g., CoTMPhP) was subjected to heat treatment in the atmosphere of an inert gas (helium) at a temperature of 800-900°C. The heat treatment (Fig. 12) does not change the initial activity of the investigated catalysts, but exerts substantial influence on the stability of their operation (Table 5). The x-ray and elementary analysis data show that the heat treatment of cobalt chelates gives novel polymeric compounds of cobalt, whose chemical stability in sulfuric acid considerably exceeds that of the initial compounds.

Investigations have shown that the heat treatment does not change the main parameters of the porous active layer: total porosity and the ratio of liquid and gas pores. From the data obtained by the disk-ring electrode method, it follows that microkinetic

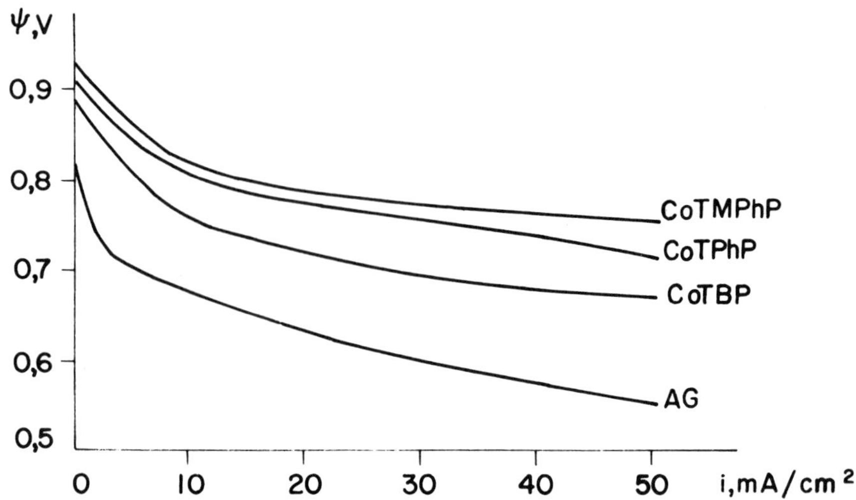

Fig. 12. Polarization curves of oxygen reduction on porous carbon electrodes promoted with organic complexes of cobalt in 4.5 N H_2SO_4, 20°C.

TABLE 5. Influence of Heat-Treatment Temperature on Duration of Operation of Carbon Oxygen Electrodes[a]

Heat-treatment temperature, °C	400	600	700	800	900	1000
Stable operation hours	150	350	500	10.000	10.000	1000

[a]With 5 wt.% of CoTMPhP in 4.5 N H_2SO_4 at i = 10 mA/cm^2.

factors are decisive in stabilization of the electrochemical characteristics of electrodes with the heat-treated catalyst. In the case of thermally stabilized catalysts, the rate of hydrogen peroxide conversion increases by an order of magnitude as compared to the initial organic complexes, which leads to lower concentration of the hydrogen peroxide in the pores of the active layer and to stabilization of the operation of oxygen electrodes in the acid electrolyte.

V. ELECTROCHEMICAL REACTIONS OF HYDROGEN PEROXIDE AND OTHER ELECTROCATALYTIC REACTIONS

Other electrocatalytic processes on organic complexes of metals have been studied less in detail than the reaction of electroreduction of molecular oxygen.

A. Reactions of Hydrogen Peroxide

Investigation of the electrode behavior of hydrogen peroxide is of interest in various aspects. As was shown above, in a number of cases hydrogen peroxide is the intermediate product of oxygen reduction, and its further transformation determines the performance efficiency of the oxygen (air) cathode. On the other hand, the reaction of hydrogen peroxide decomposition is used as model reaction in the investigations of heterogeneous catalysis on organic complexes. In such cases, however, the possibility for the electrochemical reactions of hydrogen peroxide to proceed on semiconductors is, as a rule,[1] not taken into account.

Qualitative investigations of cathodic reduction of hydrogen peroxide on phthalocyanines of certain metals have been carried out by Kozawa *et al.* (1970, 1971) and Meier *et al.* (1972, 1973). In the works of Meier *et al.* (1972, 1973), the catalytic activity of $CoPhc_i$ and $FePhc_i$ specimens in the reaction of hydrogen peroxide decomposition is compared with their specific conductivity. Absence of correlation between these parameters induced the authors to attribute the catalase activity of phthalocyanines to acceptor sites, such sites, in the authors' opinion, being defects, perturbations, etc.

Detailed studies of chemical and electrochemical reactions of hydrogen peroxide on monomeric phthalocyanines of iron and cobalt

[1] *With the exception of works carried out by S. Z. Roginsky and co-workers (1975).*

and on polymeric phthalocyanine of cobalt have been carried out by Zakharkin and Tarasevich (1975, 1977). Combination of the methods of polarization measurements, gasometry, and isotopic analysis of oxygen evolving in the decomposition of $H_2O_2^{18}$ has allowed separate investigation of the chemical and electrochemical reactions of hydrogen peroxide. In Fig. 13 typical polarization and gasometric curves for CoPhc are presented. In cathodic polarization of the electrode (curve 1) there takes place reduction of H_2O_2, and in anodic polarization, its oxidation with evolution of molecular oxygen (curve 2). In addition, there takes place decomposition of H_2O_2 vis the chemical mechanism. This is indicated by the maximum on the gas evolution versus potential dependence curve, within the region of which the reaction of H_2O_2 decomposition is bimolecular. At more anodic or cathodic potentials, the reaction order diminishes to 1. The results of the isotopic analysis of the oxygen evolved in the decomposition of H_2O_2 with the label O^{18} indicate absence of the

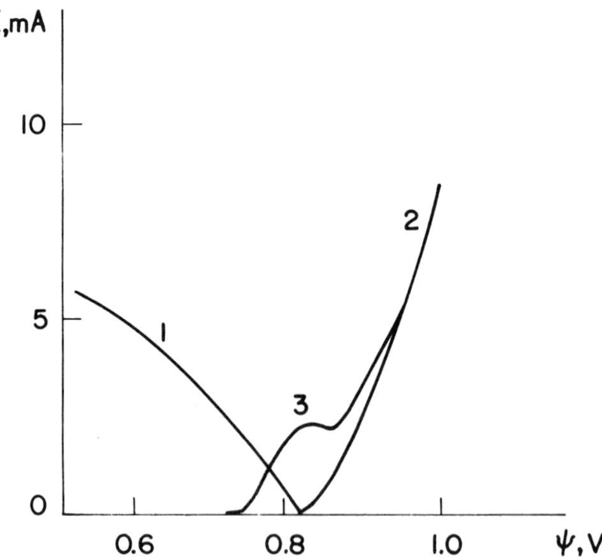

Fig. 13. Polarization (1, 2) and gasometric (3) curves in 1 N KOH with 5×10^{-3} N H_2O_2 on pyrographite electrode coated with CoPhc.

O-O bond cleavage in the oxygen molecule. These data allow suggesting the following mechanism (e.g., alkali solutions) of hydrogen peroxide decomposition at a stationary potential:

$$\boxed{M}\!-\!\bar{O}^{16}H + HO_2^{-\,18-18} \xrightarrow[-O^{16}H]{} \boxed{M}\begin{matrix}\leftarrow O^{18}H \\ | \\ \leftarrow O^{18}\end{matrix}$$

$$+ HO_2^{-\,18-18} \xrightarrow[-O^{18}H]{} \boxed{M}\!-\!O^{18}H + O_2^{18-18} \qquad (6)$$

This is followed by rapid isotopic exchange:

$$\boxed{M}\!-\!O^{18}H + O^{16}H^{-} \rightleftarrows \boxed{M}\!-\!O^{16}H + O^{18}H^{-} \qquad (7)$$

Therefore, the evolving oxygen has isotopic composition of the initial hydrogen peroxide. Functioning of the phthalocyanine molecule is possible only after the completion of the whole chain. Intermediate compounds, complexes of phthalocyanine with one adsorbed molecule or with two molecules (ions) of H_2O_2, are, actually, Chance and Oguri complexes that are formed in the process of hydrogen peroxide decomposition by the catalase.

With the superposition of electrochemical polarization, the central ion of metal becomes able to donate or accept electrons, which provides for the proceeding of the stationary electrochemical reaction and leads to the change of the reaction order to unity. On the basis of the results of kinetic measurements carried out by Zakharkin and Tarasevich (1975, 1977), a mechanism of anodic oxidation and cathodic reduction of hydrogen peroxide on phthalocyanines of iron and cobalt is suggested. According to the pH dependence of the cathodic reaction, two regions may be distinguished: at pH > 12 $\sigma\psi_k/\sigma\text{pH} \simeq 120$ mV; at pH > 12 $\sigma\psi_k/\sigma\text{pH} \simeq 0$. This indicates that the cathodic reaction is controlled by the rate-determining stage

$$H_2O_2 + e \rightarrow OH + OH^- \tag{8}$$

As regards both chemical (catalase) and electrochemical activity, the investigated phthalocyanines constitute the series: $CoPhc_i$ > FePhc > CoPhc.

B. Photocatalytic Reactions

As has been mentioned above (see Section II), phthalocyanines are photoconductors. On exposure to radiation in the region of 0.5-0.8 μm wavelengths, free charge carriers are formed in thin layers of phthalocyanines and internal photoeffect takes place. The concentration of free charge carriers may grow by more than two orders of magnitude. In several recent works (Alferov and Sevastyanov, 1975; Schumov and Heyrovsky, 1975; Schreiber and Savy, 1976) the influence of illumination on the process of O_2 electroreduction on H_2, Cu, Co, and Fe phthalocyanines was investigated. These phthalocyanines were either deposited by sublimation on a support (pyrographite, Au, Pt, etc.) (Alferov and Sevastyanov, 1975; Schreiber and Savy, 1976), or deposited from solution on the surface of a mercury drop (Schumov and Heyrovsky, 1975). It is noted that photoactivity, like dark conductivity, is substantially influenced by the nature of the central ion, by the support material, and by the solution pH. Light exerts maximum influence on the rate of oxygen electroreduction on H_2Phc, photoelectrochemical activity diminishes in the sequence H_2Phc > CuPhc > CoPhc, FePhc. Thus, when the electrode is exposed to light, the electrocatalytic activity order of the phthalocyanines observed in the oxygen reduction reaction is opposite that under dark conditions. In the works of Alferov *et al.* (1975) and Schreiber and Savy (1976), it is supposed that under the action of light there takes place an increase in the concentration of O_2^- ions without changes in the mechanism of the oxygen electroreduction reaction. Schumov and Heyrovsky (1975) are of the opinion that under the action of light there takes place an increase in the rate of cathodic reduction of hydrogen peroxide.

Shown in Fig. 14 are polarization curves of oxygen reduction on a thin film of H_2Phc, sublimating on pyrographite, in 0.1 N KOH. The photocurrent (curve 3) exceeds or is equal to the dark current, this, evidently, being indicative of a change in the reaction mechanism when H_2Phc is exposed to light, namely, of the origination of a direct reaction to water.

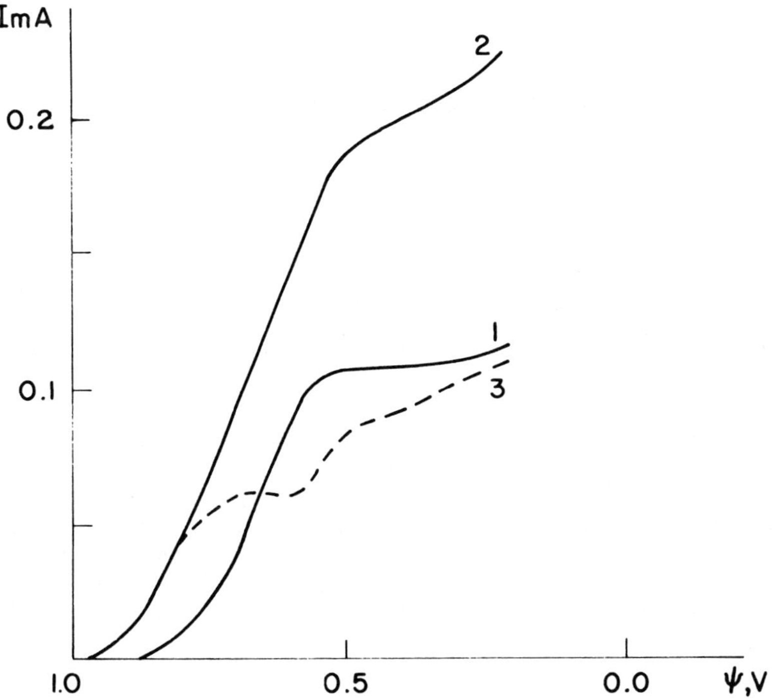

Fig. 14. Polarization curves of O_2 reduction on H_2Phc film (1000 Å) in 0.1 N KOH: 1, without exposure to light; 2, with exposure to white light; 3, photocurrent.

C. Other Reactions

As can be inferred from patent publications, organic complexes of metals are starting to be used for activating the reactions of electrooxidation of carbon oxide (Ziener *et al.*, 1974); hydrogen (Richter and Henkel, 1971; Matsuda *et al.*, 1971); hydrazine (Ziener *et al.*, 1974; Matsuda *et al.*, 1971); or organic compounds: methanol (Matsuda *et al.*, 1971), formic acid (Ziener *et al.*, 1974; Zimmerman *et al.*, 1976), oxalic acid (Ziener *et al.*, 1974); as well as electroreduction of carbon dioxide gas (Meshitsuka *et al.*, 1974) (Table 6). As in the reaction of electroreduction of oxygen, phthalocyanines, tetraphenylporphins, dibenzotetraazaannulenes of a number of metals are employed. Active compounds in the reaction of oxidation of organic and inorganic compounds are metal-free phthalocyanines, phthalocyanines of Cu, Pd, Cr, VO, tetraphenylporphins of Ni, Zn, and Pt, i.e., compounds that have low activity in the reaction of oxygen electroreduction. For the reaction of electroreduction of CO_2 monomers of Co, Ni, Mn, Pd, Cu, and Fe, phthalocyanines were investigated. High rate of electroreduction of CO_2 to oxalic and gluconic acids was observed only in the presence of CoPhc and NiPhc. The mechanism of these reactions was

TABLE 6. Comparison of Electrochemical Activity of Organic Complexes in Oxidation Reactions

Compound	*Catalyst*	*Oxidation rate*
HCOOH	*CoTAA*	*11.0 mA/mg* (ψ_r = 0.5 V) *in 2 N* H_2SO_4
	CoTAA, heat-treated	*6.8 mA/mg* (ψ_r = 0.35 V) *in 2 N* H_2SO_4
N_2H_4	*CoTAA*	*7.5 mA/mg* (ψ_r = 0.5 V) *in 2 N* H_2SO_4
	Co-bis-1-hydroxyanthraquinone	*10 mA/mg* ($\Delta\psi$ = 0.1 V) *in KOH, 90°C*
H_2	*PdPhc*	*500* mA/cm^2 (ψ_r = 0.05 V) *in 2 M* H_2SO_4, *95°C*

not investigated, but this trend of research seems to be promising and will be, evidently, developed further both toward extending the range of electrochemical reactions and catalytically active organic complexes, and in the theoretical aspect.

Extensive studies in the electrocatalysis on organic complexes of metals habc bccn going on during thc last 5 yr. During this short period of time, significant advances have been made both in understanding the nature of the electrocatalytic activity and in practical application of organic catalysts for activating electrochemical reactions. It has been shown that the main role is played by the interaction of the reaction components with the catalyst molecule. In the case of oxygen electroreduction, the direction and rate of the process are determined by the nature of the central ion of metal of the organic complex and, hence, by the character of the adsorption interaction of the oxygen molecule with it. Detailed kinetic studies have made it possible to suggest phenomenological schemes of electrode reactions. A number of metalloporphins have been found to be promising electrocatalysts for alkali electrolytes. The application of thermally stabilized organic polymers offers new possibilities, particularly in acid media.

The object of further research is to extend the range of electrocatalytically active organic systems and electrocatalytic reactions. The use of organic compounds having both electron-donor and electron-accepting properties will allow activating novel electrochemical and photocatalytic reactions. In the field of fundamental research, thorough studies are needed in the adsorption interaction of the reaction components with the catalyst, responsible for the main effects of the electrocatalysis. This requires developing of combined physical and electrochemical methods for investigating the structure of the organic catalyst/electrolyte interface.

REFERENCES

Abkowitz, M. A., and Lakatos, A. J. (1972). *J. Chem. Phys. 57,* 5033.

Acres, G. J. K., and Eley, D. D. (1964). *Trans. Far. Soc. 60,* 1157.

Assour, J. M. (1965). *J. Am. Chem. Soc. 87,* 4701.

Alt, H., Binder, H., Lindner, W., and Sandstede, G. (1971). *J. Electroanal. Chem. 31,* App. 19.

Alt, H., Binder, H., and Sandstede, G. (1973). *J. Catal. 28,* 8.

Alt, H., Binder, H., Lindner, W., and Sandstede, G. (1973). DOS 2.049.008.

Alferov, G. A., and Sevastyanov, V. I. (1975). *Elektrokhimiya 11,* 827.

Andrianova, T. I., Sherle, A. I., and Berlin, A. A. (1973). *Izv. ANVSSSR, Div. Chem. Sci. 3,* 531.

Appleby, A. J., and Savy, M. (1976). *Electrochim. Acta 21,* 567.

Appleby, A. J., Fleisch, J., and Savy, M. (1976). *J. Catal. 44,* 281.

Bagotsky, V. S., Tarasevich, M. R., and Filinovsky, V. Yu. (1972). *Elektrokhimiya 8,* 84.

Bagotsky, V. S., Tarasevich, M. R., Radyushkina, K. A., Levina, O. A., and Andruseva, S. I. (1977, 1978). *Power Sources 2,* 233.

Bar-Ilan, A., and Manassen, J. (1974). *J. Catal. 33,* 68.

Barraclough, C. G., Martin, R. L., Mitra, S., and Sherwood, R. C. (1970). *J. Chem. Phys. 53,* 1638, 1643.

Barbe, D. F., and Westgate, C. (1970). *J. Chem. Phys. 52,* 4046.

Bayston, J. H., King, N. K., Looney, F. D., and Winfield, (1969). *J. Am. Chem. Soc. 91,* 2775.

Beck, F., Dammert, W., Heiss, J., Hiller, H., and Polster, R. (1973). *Z. Naturforsch. A-28,* 1009.

Bendersky, V. A., Belkind, A. I., Fedorov, M. I., and Aleksandrov, S. V. (1972). *FTT 14,* 790.

Berezin, B. D., and Shlyapova, L. N. (1968). *Izv. VUZ, Khimiya i Khimich Tekhologiya 12,* 1641.

Berezin, B. D., and Shlyapova, L. N. (1968). *Izv. VUZ, Khimiya i Khimich Tekhnologiya 12,* 1641.

Berezin, B. D., and Shormanova, L. P. (1968). *Vysokomolekul. Sojedineniya* (A)X, 384.

Berezin, B. D., and Shlyapova, L. N. (1971). *Izv. VUZ, Khimiya i Khimich. Tekhnologiya 14,* 1665.

Berlin, A. A. (1975). *Uspekhi Khimii 44,* 502.

Bett, J. A. S., and Ludquist, J. I. (1969). *Am. Chem. Soc. Div. Fuel Cell Progr. 13,* 44.

Beyer, W., and Sturm, F. (1972). *Angew. Chemie 84,* 154.

Blumenfeld, L. A., Bendersky, V. V., Lyubchenko, L. S., and Stupchak, P. A. (1967). *Zh. Strukt. Khimii 8,* 829.

Bobrovsky, A. P., and Sidorov, A. N. (1976). *Zh. Strukt. Khimii 17,* 63.

Borisenkova, S. A., and Rudenko, A. P. (1976). *Vestn. MGU 17,* 3.

Bosch, R. (1969). Pat. Fr. No. 1.582.905.

Brown, C. J. (1968). *J. Chem. Soc.* (A) 2488.

Caughey, W. S., Deal, R. M., Weiss, C., and Gouterman, M. (1965). *J. Mol. Spectros. 16,* 451.

Chizmadzkev, Yu. A., Markin, V. S., Tarasevich, M. R., and Chirkov, Yu. G. (1971). "Makrokinetika processov poristych sredach," Moscow, "Nauka."

Clack, D. W., and Yandle, I. R. (1972). *Inorg. Chem. 11,* (1965).

Cohen, J. A., Ostfeld, D., and Lichtenstein, B. (1972). *J. Am. Chem. Soc. 94,* 4522.

Contour, J. P., Lenfant, P., and Vijh, A. K. (1973). *J. Catal. 29,* 8.

Dabrowski, R., Matysiak, M., and Witkievioz, Z. (1975). *Electrochem. Curr. Sources,* p. 178.

Dale, B. M. (1969). *Trans. Faraday Soc. 65,* 331.

Dale, B. M., Williams, R. J. P., Edwards, P. R., and Johanson, C. E. (1968). *J. Chem. Phys. 49,* 3445.

Day, P., and Price, M. (1969). *J. Chem. Soc.* (A) 236.

Dribinsky, A. V., Tarasevich, M. R., and Burstein, R. Kh. (1971). *Elektrokhimiya 7,* 1144.

Dribinsky, A. V., Tarasevich, M. R., Burstein, R. Kh., Radyushkina, K. A., and Levina, O. A. (1978). *Elektrokhimiya 14,* 22.

Dribinsky, A. V., Tarasevich, M. R., Burstein, R. Kh., Lezhnev, N. N., Radyushkina, K. A., Levina, O. A., and Zusman, R. I. (1977). *Elektrokhimiya 13,* 1631.

Falk, I. E. (1964). "Porphyrins and Metalloporphyrins." Elsevier, New York.

Fajer, J., Borg, D. C., Forman, A., Dolphin, D., and Felton, R. H. (1970). *J. Am. Chem. Soc. 92,* 3451.

Furhop, I. H., Kadish, K. M., and Davis, D. G. (1973). *J. Am. Chem. Soc. 95,* 5140.

Grenoble, D. C., and Drickamer, H. D. (1971). *J. Chem. Phys. 55,* 1624.

Gutman, F., and Lajons, L. (1970). "Organic Semiconductors." Nauka Publishers, Moscow.

Guzy, C. M., Raynor, J. B., Stodulski, L. P., and Symons, M. C. R. (1969). *J. Chem. Soc.* (A) 997.

Hanke, W. (1969). *Ztschr. Chemie 9,* 1.

Hanson, L. K. (1973). *J. Am. Chem. Soc. 95,* 4822.

Harrison, S. E. (1969). *J. Chem. Phys. 50,* 4739.

Hiller, H., Polster, R., Beck, F., and Guthke, H. (1972). DOS 2.046.354.

Ichikawa, M., Soma, M., Onishi, T., and Tamaru, K. (1966). *J. Phys. Chem. 70,* 2069.

Ingram, D. (1972). "Electron Spin Resonance in Biology.:

Jahnke, H., Schönborn, M., and Zimmermann, G. (1976). Topics in Current Chemistry 61, 133.

Jahnke, H. (1973). US Pat. No. 3.773.878.

Jahnke, H., and Schönborn, M. (1969). *C. R. Troisiéme Journee Internat. d;Etude des Piles à Combustible,* Bruxelles, p. 60.

Jasinski, R. (1965). *J. Electrochem. Soc. 112,* 526.

Johanson, L. T., Mrha, J., and Larrsson, R. (1973). *Electrochim. Acta 18,* 255.

Jonescu, M. N., Schuster, R. H., Trestianu, D., and Mihailescu, A. (1975). *Z. Chemie 15,* 496.

Kadish, K. M., and Morrison, M. (1975). *Third Internat. Symp. Bioelectrochem. Juelich,* p. 100.

Kanda, S., and Pohl, H. A. (1968). "Organic Semiconducting Polymers" (J. E. Kanton, ed.), p. 87. Marcel Dekker, New York.

Kaufhold, J. (1965). *Ber. Bunsenges 69,* 168.

Kayushin, L. P., Gribova, Z. P., and Azizova, O. A. (1973). "Elektronnyi paramagnitnyi rezonans fotoprotsessov biologicheskiph sojedinenij."

Kozawa, A., Zilionis, V. E., and Brodd, R. J. (1970). *J. Electrochem. Soc. 117,* 1470, 1474.

Kozawa, A., Zilionis, V. E., and Brodd, R. J. (1971). *J. Electrochem. Soc. 118,* 1705.

Kretzschmar, Chr., and Wiesener, K. (1976). *Z. Phys. Chemie 257,* 39.

Kropf, H., and Witt, O. J. (1971). *Z. Phys. Chemie. N.F. 76,* 331.

Kwaskowska-Chec, E., Fried, K., Przywalska-Boniecka, H., and Ziolkowski, J. J. (1975). Soobsccheniya po kinetike i katalizy *2,* 245.

Lanese, J. G., and Wilson, G. S. (1972). *J. Electrochem. Soc. 119,* 1039.

Levina, S. D., Andrianova, T. I., Sakharov, M. M., Golovina, O. A., Lobanova, K., and Rotenberg, Z. (1966). *ZhPhKh 40,* 1229.

Levina, O. A., and Radyushkina, K. A. (1975). Invertor's Certificate of the USSR No. 542416.

Mamaev, V. M., Ponomarev, G. V., Zenin, S. Z., and Yevstigneeva, R. P. (1970). *Teoret. Eksper. Khimiya 6,* 40.

Manassen, J. (1973). *J. Catal. 33,* 133.

Manassen, J., and Bar-Ilan, A. (1970). *J. Catal. 17,* 86.

Martin, L., and Mitra, S. (1969). *Chem. Phys. Letters 3,* 183.

Mathur, S. C., and Singh, J. (1972). *Int. J. Quant. Chem. 6,* 56.

Matsuda, S., O'Connel, J., and Freerks, M. S. (1971). US Pat. No. 3,617,388.

Meier, H., Albrecht, W., Tschirwitz, U., and Zimmerhackl, E. (1973). *Ber. Buns. Gesell. 77,* 843.

Meier, H., Zimmerhackl, E., Albrecht, W., and Tschirwitz, V. (1972). *Ber. Buns. Gesell. 76,* 1104.

Meshitsuka, S., Ichikawa, M., and Tamaru, K. (1972). *Ber. Buns. Gesell. 76,* 1104.

Meshitsuka, S., Ichikawa, M., and Tamaru, K. (1974). *J. Chem. Soc. Chem. Comm. 5,* 158.

Moser, F. H., and Thomas, A. L. (1963). "Phthalocyanine Compounds." Reinhold, New York.

Mrha, J., Musilova, M., and Jindra (1973). *J. Appl. Electrochem. 3,* 213.

Myl'nikov, V. S. (1968). *Uspekhi Khimii 37,* 78.

Niwa, Y., Kobayashi, H., and Tsuchiya T. (1974). *J. Chem. Phys. 60,* 799.

Peychal-Heiling, G., and Wilson, G. S. (1971). *Anal. Chem. 43,* 550.

Pleskov, Yu. V. (1972). *Itogi nauki i Tekniki. Elektrokhimiya 8,* 5.

Pobedinsky, S. N., Trofimenko, A. A., Bazanov, M. I., Aleksandrova, A. N., Belonogov, K. N., and Al'yanov, M. I. (1974). III Vsesojuzny simposium po kinetike i mekhanizmu reaktsij s uchastiem kompleksnykh sojedinenij, p. 102.

Pülman, B., and Pülman, A. (1965). "Quantum Biochemistry." Mir Publishers, Moscow.

Radyushkina, K. A., Burstein, R. Kh., Berezin, B. D., Tarasevich, M. R., and Levina, O. A. (1973). *Elektrokhimiya 9,* 410.

Radyushkina, K. A., Levina, O. A., Tarasevich, M. R., Burstein, R. Kh., Berezin, B. D., Koifman, O. A., and Shormanova, L. P. (1975). *Elektrokhimiya 11,* 989.

Radyushkina, K. A., Tarasevich, M. R., and Andruseva, S. I. (1975). *Elektrokhimiya 11,* 1079.

Radyushkina, K. A., Tarasevich, M. R., and Andruseva, S. I. (1977). *Elektrokhimiya 13,* 483.

Randin, J. P. (1974). *Electrochim. Acta 19,* 83.

Reber, H., Jahnke, H., and Steiner, W. (1973). US Pat. No. 3,778,313.

Richter, G., and Henkel, H. J. (1971). US Pat. No. 3,585,079.

Richter, G., and Luft, G. (1972). *Ber. Buns. Gesell. 76,* 1105.

Roginsky, S. Z. (1973). *Problemy Kinetiki i Kataliza 15,* 12.

Roginsky, S. Z. (1975). "Elektronnyje Yavleniya v Geterogennom Katalize." Nauka Publishers, Moscow.

Rollman, L. D., and Iwamoto, R. T. (1968). *J. Am. Chem. Soc. 90,* 1455.

Savy, M., Magner, G., and Perslerbe, M. (1972). *C. R. Acad. Sci. 275,* 163.

Savy, M., Andro, P., Bernard, C., and Magner, G. (1973). *Electrochim. Acta 18,* 191.

Savy, M., Andro, P., and Bernard, C. (1974). *Electrochim. Acta 19,* 403.

Savy, M., Bernard, C., and Magner, G. (1975). *Electrochim. Acta 20,* 383.

Savy, M., Bernard, C., and Magner, G. (1976). *Electrochim. Acta 20,* 383.

Schaffer, A. M., and Gouterman, M. (1972). *Theoret. Chim. Acta 25,* 63.

Schreiber, B., and Savy, M. (1976). *C. R. Acad. Sci. 282,* 787.

Schumov, Yu. S., and Heyrovsky, M. (1975). *J. Electroanal. Ch. 65,* 469.

Sevastyanov, V. I., Alferov, G. A., Asanov, A. N., and Komissarov, G. G. (1975). *Biofizika 20,* 1004.

Sidorov, A. N., and Maslov, V. G. (1975). *Usp. Khimii 44,* 577.

Simonov, A. D., Keier, N. P., Kundo, N. N., Mamaev, E. K., and Glazneva, G. V. (1973). *Kinetika I Kataliz 14,* 988.

Slaidin', G. Y., Bagotsky, V. S., Tarasevich, M. R., Balode, L. Yu, Shimsheleviya, J. B., Steinberg, G. V., and Chakste, Ya. E. (1974). Materialy Vsesojuznoj konferentsii po kataliticheskim reaktsiyam v zhidkoj faze, Alma-Ata, p. 346.

Steinbach, F., and Hiltner, K. (1973). *Ztschr. Phys. Chem. N.F. 83,* 126.

Steinbach, F. (1975). *Kinetika I Kataliz 16,* 1094.

Steinbach, F., and Schmitt, H. H. (1975). *J. Catal. 39,* 190.

Stillman, M. J., and Thomson, A. (1974). *J. Chem. Soc. Faraday II 70*, 805.

Stynes, D. V., Stynes, H. C., James, B. J., and Iberis, J. A. (1973). *J. Am. Chem. Soc. 96*, 1796.

Stillman, M. J., and Thomson, A. J. (1974). *J. Chem. Soc. Faraday II 70*, 790.

Sudo, M., Ichikawa, M., Soma, M., Onishi, T., and Tamaru, K. (1969). *J. Phys. Chem. 73*, 1174.

Tanaka, H., Negishi, A., and Ono, J. (1971). *Denki Kagaku 38*, 596.

Tarasevich, M. R. (1973). *Elektrokhimiya 9*, 599.

Tarasevich, M. R., and Bogdanovskaya, V. A. (1975). *Bioelectrochem. Bioenerg. 2*, 69.

Tarasevich, M. R., and Radyushkina, K. A. (1976). *Izv. VUZ. Khim. i Khim. Tekhn. 19*, 1639.

Tarasevich, M. R., Radyushkina, K. A., and Andruseva, S. I. (1977). *Bioelectrochem. Bioenerg. 4*, 18.

Taube, R. (1966). *Ztschr. Chemie 6*, 6.

Taylor, R. J., and Hummfray, A. A. (1975). *J. Electroanal. Chem. 64*, 63.

Usov, N. N., and Bendersky, V. A. (1968). FTP 2, 699.

Usov, N. N., and Bendersky, V. A. (1970). *Phys. Stat. Solidi 37*, 535.

Wagnerova, D. W., Schwertnerova, E., and Veprek, J. (1974). *Collect. Czechoslov. Chem. Commun. 39*, 1980.

Waldmeier, P., and Sigel, H. (1970). *Chimia 24*, 196.

Walker, F. (1970). *J. Am. Chem. Soc. 92*, 4235.

Walker, F. (1973). *J. Am. Chem. Soc. 95*, 1154.

Wolberg, A., and Manassen, J. (1970). *J. Am. Chem. Soc. 92*, 2982.

Wolkenstein, F. F. (1976). "Fiziko-khimiya poverkhnosti polyprovodnikov." Nauka Publishers, Moscow.

Yamamto, Y., and Kwan, T. (1970). *J. Catal. 18*, 354.

Yao, S. J., Michuda, M., Markby, F., and Wollson, S. E. (1974). Inf. Bull. PPTE & TE, Issue 4/141.

Zakharkin, G. I., and Tarasevich, M. R. (1975). *Elektrokhimiya 11*, 1019.

Zakharkin, G. I., and Tarasevich, M. R. (1977). *Reaction Kinet. Catalysis Lett.* 6, 77.

Zerner, M., and Gouterman, M. (1966). *Theoret. Chim. Acta 4,* 44.

Zerner, M., Gouterman, M., and Kobayashi, H. (1966). *Theoret. Chim. Acta 6,* 363.

Ziener, H., Weber, L., Jahnke, H., Magenau, H., and Zimmerchakl, E. (1974). US Pat. No. 3,821,028.

Zimmermann, G., Schönborn, M., Magenau, H., Jahnke, H., and Becker, B. (1976). US Pat. No. 3,930,884.

ADSORPTION FROM BINARY GAS AND LIQUID PHASES

K. V. Chmutov
O. G. Larionov

Institute of Physical Chemistry
Academy of Sciences of the USSR
Moscow, USSR

1. INTRODUCTION

The adsorption of liquid solutions may be considered as being the first adsorption process used by man for practical purposes. So far, however, its theoretical basis and relation to other adsorption phenomena, e.g., to more thoroughly studied adsorption of

ISBN 0-12-571814-4

individual substances and adsorption of binary gas mixtures, have not been sufficiently developed. This is due to a number of reasons, one of these being undoubtedly the difficulty of obtaining by conventional methods the data on the adsorption of liquid solutions, and the much lesser accuracy and reproducibility of these data. Therefore, up to the present time, practically no data have been available in literature that could be used for verification of any particular theoretical concepts. However, it is clear to everybody that there is an intimate connection between the adsorption of individual substances and their mixtures, and simultaneous investigation of these phenomena should give a new impetus to the study of adsorption as a whole.

The aim of the present paper is to acquaint the reader with the main directions of works on the adsorption of mixtures. In this chapter, main attention will be given to the works dealing with theoretical problems and treating the adsorption from various phases as one process.

The authors are not concerned with the adsorption of gas mixtures. Therefore, the studies in this field will be considered only as related to the main aim of this review. The studies on the adsorption of gas mixtures are treated in more detail in the work of Schirmer and his associates (Bülow *et al.*, 1972).

II. CONCEPTS OF EXCESS AND ABSOLUTE ADSORPTION

The adsorption, i.e., the concentrating of substance near the interface, is due to decrease in the free energy of the system. As the result of adsorption, a certain region of change in the properties and concentration of substance is observed near the interface, a region of inhomogeneity. We have no definite knowledge of the extent of this region and the distribution of substance in it. Therefore, the only experimentally possible method of determining the extent of adsorption and the properties of the system is the excess functions method, the Gibbs Method, in which the properties of a real system including the interface are compared with a hypothetical system with the same values of intensive properties but in the absence of the interface. However, in spite of its rigorousness, the method has a very important draw-back, viz., that it ignores the thickness of the surface layer, its structure, etc. This means that it gives no information on the properties of the surface phase, which are of considerable interest for investigation of some problems. Therefore, along with the Gibbs method, the layer of finite thickness method is coming into use, in which the surface layer is ascribed a finite thickness (Rusanov, 1967a).

In accordance with the two methods outlined, two extents of adsorption are recognized: excess and total, absolute adsorptions.

As has been pointed out above, it is possible to determine experimentally only the extent of excess adsorption, which is defined as the difference in the amount of substance determined in the system in the absence and presence of adsorbent

$$n^{e} = n^{0} - n \tag{1}$$

In treating adsorption as the total content of substance in the surface layer (absolute adsorption), it is assumed that when adsorbent is introduced, the substance present in the system can be represented as the adsorption layer containing n' moles of substance and the remaining bulk solution containing $n^{0} - n'$ moles.

A. Individual Substances

If we express the quantities n^{0}, n, and n' in terms of the volumes and densities of substance in respective phases, we can obtain the relation between excess and absolute adsorption of pure substance

$$n' = n^{e} + v'\rho \tag{2}$$

where ρ is the molar density of substance in bulk, v' is the volume of the surface layer.

At small pressures in the system, as is usual in most cases of investigation of vapors adsorption, $v'\rho \ll n^e$ and the extents of adsorption determined experimentally can be equated to the values of n'. At large pressures, these corrections should be taken into account (Payne *et al.*, 1968).

B. Mixtures

For mixtures, Eq. (1) is written

$$n_i^e = x_i^0 \sum n_i^0 - x_i \sum n_i^0 \quad (3)$$

The relation between the extents of absolute and excess adsorptions is given by the expressions

$$n_i^e = n_i' - x_i \sum n_i' = n_i' - n'x_i \quad (4)$$

whence it follows that

$$n_i^e = n'(x_i' - x_i) \quad (5)$$

Only the values of n_i^e can be determined experimentally. Therefore, in dealing with concentrations in the surface layer, use should be made of Eqs. (4) or (5). In this case, the value of n' is usually unknown and is calculated from the given adsorption layer model. It should be emphasized that all subsequent conclusions based on the analysis of the total content values are defined by the chosen layer model.

C. Surface Layer Models and Determination of the Limiting Extents of Adsorption

The most widely used models are those of a monolayer or of several layers, and the model of the filling of pore volumes. In calculating the total content values by means of the monolayer model, an additional assumption of the additivity of the molar area of the adsorption layer is usually made

$$a = \sum a_i^0 x_i' ; \qquad \frac{1}{n'} = \sum x_i'/n_{mi}' \tag{6}$$

The molar areas of pure components a_{0i} or the equivalent quantities, the extents of the limiting adsorption n_{mi}', are usually determined from the data on the adsorption of individual vapors (Kipling, 1965; Kurbanbekov *et al.*, 1973a-c; Berezkina *et al.*, 1972a-c, 1973), or from the data on the adsorption of solutions (Scherbakova and Kiselev, 1947; Schay *et al.*, 1960; Schay and Nagy, 1972a,b).

Sometimes (Rusanov, 1967b) a correction for the nonadditivity of molar areas is suggested, but this is hardly expedient. The nonadditivity is associated with the interactions between the solution molecules. In the surface layer these interactions and the mutual packing of molecules can differ significantly from those in the bulk solutions. Therefore, we do not know beforehand whether such corrections will bring us nearer to the real picture of the surface solution or, on the contrary, take us further away.

The determination of the limiting extents of adsorption from the data on the adsorption of solutions is of considerable interest inasmuch as our conclusions about the surface layer structure and its properties depend on the correctness of this determination. Moreover, reliable values of n_{mi}' make it possible to determine the specific surface areas from the data on the adsorption of solutions. For this reason we shall dwell on this problem in more detail.

1. Scherbakova - Kiselev's and Schay's Methods

Scherbakova and Kiselev proposed to determine the limiting extent of adsorption (more precisely, the adsorption volume) by means of Eq. (4). Assuming that in the linear section of the excess adsorption isotherm $n_i' =$ constant and hence $n' =$ constant, they obtained after differentiating Eq. (4) with respect to x_i, the expression

$$dn_i/dx_i^e = -n' \tag{7}$$

Later, a method for determination of the limiting extent of adsorption was proposed, which is in many respects similar to that described above (Schay *et al.*, 1960). Writing Eq. (4) for a binary solution as

$$n_1^e = n_1' - \left(n_1' + n_2'\right)x_1 = a - bx \tag{8}$$

Schay *et al.* (1960) came to the conclusion that the linear section on the adsorption isotherm corresponds to the condition $n_1' = a$, $n_1' + n_2' = b$, and thus from the intercept on the ordinate axis at $x_1 = 0$, we can obtain n_1' and from the slope of the linear section of the isotherm $n_1' + n_2'$.

However, Larionov *et al.* (1965, 1967a) showed that because Eq. (8) is a functional equation, a large number of functional dependences $n_i'(x_i)$ can be found which satisfy the condition of linearity. One of these dependences was noted by Cornford *et al.* (1962). The linear dependence (8) does not imply the constancy of the coefficients a and b and can be observed in any case, when

$$n_1' = \frac{b + \left(\beta n_{m1}' - b\right)x_1}{1 + (\beta - 1)x_1} \tag{9}$$

where $\beta = n_{m2}'/n_{m1}'$. Larionov *et al.* (1967a) also proved experimentally the erroneousness of Schay's method for a number of cases.

Elovich and Larionov (1962a) showed that the quantities b, n'_{mi}, and K [see Eq. (37)] are related by the expression

$$-b = -\left(dn_1^e/dx_1\right)_{x_1 \to 1}$$
$$= n'_{m1}[1 - \beta/K^{1/\beta}] \quad (10)$$

Therefore, Scherbakova - Kiselov's and Schay's methods can be used for determination of the values of n'_{mi} only if $\beta/K^{1/\beta} >> 1$.

Kiselev - Schay's methods were criticized by Larionov (1972), Serpinskii (1972), and Bering and Serpinskii (1970).

Analyzing Kiselev - Schay's method with reference to the thermodynamic condition

$$\left(dx'_i/dx_i\right) > 0 \quad (11)$$

Bering and Serpinskii showed that Scherbakova - Kiselev's and Schay's methods are based on the condition $dx'_i/dx_i = 0$ and so are at variance with thermodynamics.

Bering and Serpinskii came to the conclusion that it is impossible in principle to find the values of n'_{mi} from the isotherm of adsorption from solution, and, in order to calculate the total content, it is necessary to use the molar areas values determined from the vapors adsorption data. In this case, however, as was pointed out by Larionov *et al.* (1967), there is no reason to believe that the orientation of molecules during adsorption from vapor phase is always the same. Gas adsorption arises as the result of interaction of a molecule with the adsorbent alone, whereas adsorption from solution is due to interaction of a molecule both with the adsorbent surface and with the bulk solution.

2. *Rusanov's Method*

Of interest is the thermodynamically consistent method of determining the minimum thickness of the adsorption film, suggested by Rusanov (1972). According to this method, the minimum thickness of the adsorption film is expressed by the equation

$$A\tau > \left(v_1^0 - v_2^0\right)n_1^e - v\left(dn_1^e/dx_1\right) \tag{12}$$

in which v_1^0 and v_2^0 are the molar volumes, τ-the thickness of the absorption film, $v = v_1^0 x_1 + v_2^0 x_2$. At $v_1^0 = v_2^0$, condition (12) turns into

$$A\tau > -v\left(dn_1^e/dx_1\right) \tag{13}$$

from which it follows that Kiselev - Schay's method permit to find only the minimum value of the adsorption film thickness. At $v_1^0 > v_2^0$, Kiselev - Schay's method gives a too high value of the minimum adsorption layer thickness and at $v_1^0 > v_2^0$, a too low one. Rusanov emphasized that this estimate cannot be used for evaluation of the true thickness of the adsorption film and thus cannot serve as an indication, let us say, of the monomolecular kind of adsorption.

3. Methods Based on the Equation of the Adsorption Isotherm

Other methods for the determination of the limiting extent of adsorption are based on analysis of the equations of the adsorption isotherm. Thus, for the case of an ideal adsorption system, Klinkenberg (1959) obtained an expression relating the extent of adsorption and the concentration at maximum to the adsorption layer capacity

$$n'_{m1} = n_1^e/(1-2x_1) \tag{14}$$

A similar relation, but a more complex one, can be obtained also for the case when $n'_{m1} \neq n'_{m2}$ (Elovich and Larionov, 1962b)

$$A(n'_{m1})^2 + Bn_{m1} + C = 0$$

$$A = x_1x_2\beta[(x_2+\beta x_1)^2 - (x_2+\beta^2x_1)]$$

$$B = n_1^e(x_2-\beta x_1)[2(x_2+\beta x_1)^2 - (x_2+\beta^2x_1)]$$

$$C = -(n_1^e)^2[2(x_2+\beta x_1)^2 - (x_2+\beta^2x_1)] \tag{15}$$

x_1 and n_1^e are the molar fraction and the extent of excess adsorption at maximum, respectively.

From the equation of the adsorption isotherm for an ideal adsorption system, Everett (1964) obtained an expression relating the excess adsorption and the adsorption phase capacity to the equilibrium constant [see Eq. (29)]

$$\frac{x_1x_2}{n_1^e} = \frac{1}{n'}\left(x_1 + \frac{1}{K-1}\right) \tag{16}$$

A similar expression for volume fractions was obtained by Klinkenberg (1959).

Equation (16) allows to determine from the plot x_1x_2/n_1^e versus x_1 the values of n' and K. Some authors (Davis *et al.*, 1973; Wright, 1966, 1967; Kagiya *et al.*, 1971) used Eq. (16) for this purpose.

It should be emphasized that the uncritical use of this method can lead to significant errors. Equation (16) was deduced for ideal adsorption systems, i.e., for the systems in which $\gamma'_i = \gamma_i = 1$ and $a_1^0 = a_2^0$, and can give correct results only for such systems.

Larionov and Kurbanbekov calculated three theoretical adsorption isotherms by means of Eqs. (98) and (99), taking $a_1(\sigma_1^0 - \sigma_2^0)$ equal for the three cases and using different β (Fig. 1); then they applied Eq. (16) to the adsorption isotherms obtained. As is clear from Fig. 2, sufficiently good straight lines in the coordi-

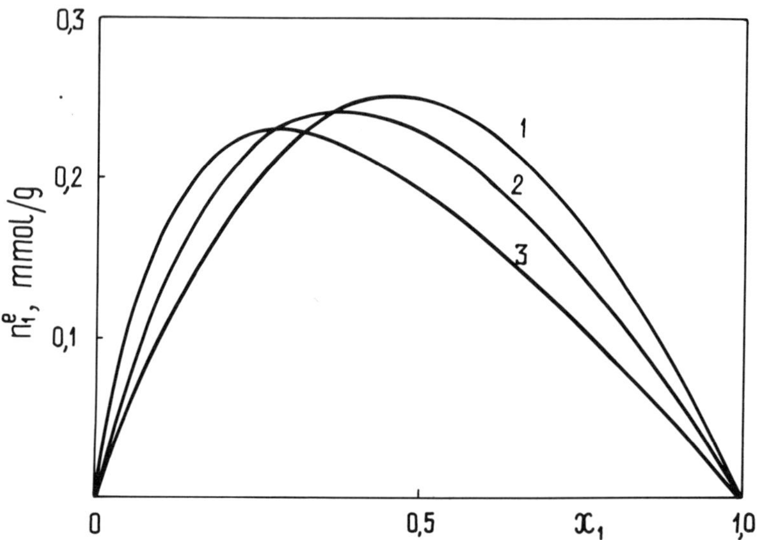

Fig. 1. The excess adsorption isotherms calculated by means of Eqs. (98)-(99). $\sigma_2^0 - \sigma_1^0 = 10$ erg/cm^2, $_1 = 240 \times 10^{10}$ cm^2/mole. 1 - $\beta = 0.5$; 2 - $\beta = 1$; 3 - $\beta = 1.5$.

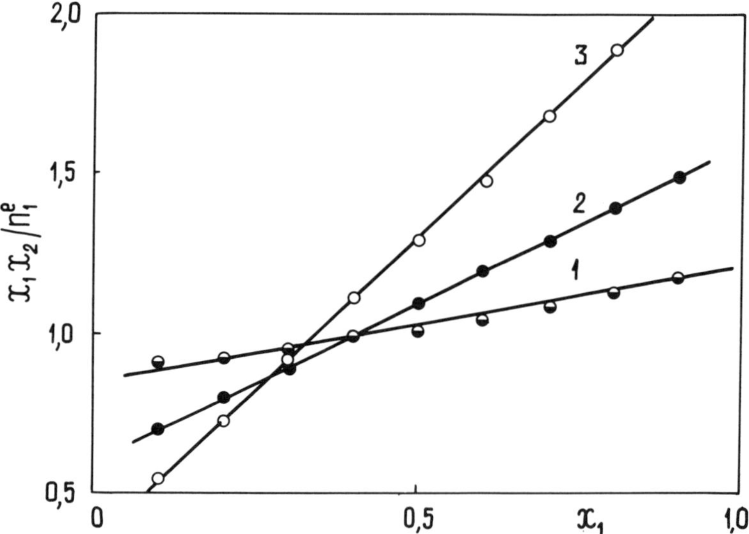

Fig. 2. The excess adsorption isotherms of Fig. 1 plotted in the coordinates of Eq. (16). The numbers correspond to Fig. 1.

nates x_1x_2/n_1^e versus x_1 were obtained for all isotherms. However, both the values of n' (slopes of the curves) and the equilibrium constants (intercepts on the ordinate axis at $x_1 = 0$) differ sharply, which points to the formal validity of Eq. (16).

Schay and Nagy (1972a) suggested a correction for Eq. (16), so that it could be used for the case $n'_{m1} \neq n'_{m2}$, i.e., at $\beta \neq 1$.

It seems to us, however, that the equation obtained by the above authors

$$\frac{x_1x_2}{n_1^e} = \frac{1}{n_{m1}^s}\left[\frac{1}{\beta(S-1)} + \frac{(S-1/\beta)}{S-1}x_1\right] \quad (17)$$

should not in general give (theoretically) a straight line in the coordinates x_1x_2/n_1^e versus x_1, since in this case the separation factor $S = x'_1x_2/x'_2x_1$ is itself a function of composition. Therefore, the quantities obtained from Eq. (17) have no physical sense. Only in the cases when $S \gg 1$, we have a straight line issuing from the origin with the slope $1/n'_{m1}$.

We have treated the theoretical isotherms shown in Fig. 1 by means of Eqs. (14), (16), and (17) and obtained the values listed in Table 1.

TABLE 1.

β	K			n'_{mi}			
	Theor.	*Eq. (17)*	*Eq. (16)*	*Taken*	*Eq. (17)*	*Eq. (16)*	*Eq. (14)*
0.5	*1.624*	*2.847*	*2.176*	*1.00*	*1.27*	*2.777*	*3.125*
1.0	*2.680*	*2.680*	*2.680*	*1.00*	*1.00*	*1.00*	*1.000*
1.5	*4.415*	*4.305*	*3.857*	*1.00*	*0.524*	*0.524*	*0.494*

As can be seen from Table 1, in spite of a good linear dependence obtained in Fig. 2, the found values of K and n'_m at $\beta \neq 1$ are not their true values, the discrepancies between the values obtained in calculating the isotherms and those found by means of Eqs. (14), (16), and (17) being quite significant.

The picture becomes even more complicated if account is taken of the activity coefficients in the bulk and surface solutions. Equation (15) gives a correct value of n'_{m1} for all theoretical isotherms, but the results of calculations with the use of this equation are very sensitive to the accurate determination of the position of the maximum on the adsorption isotherm. Thus, e.g., at $\beta = 1.5$, the change in the position of the maximum from 0.26 to 0.29 molar fractions alters the value of n'_{m1} from 0.73 to 1.06.

The extent of limiting adsorption, more precisely that of the minimum limiting adsorption, that is consistent with the thermodynamic condition

$$dx'_i/dx_i > 0$$

can be found by means of Eq. (10), and the calculation can be carried out by the method of successive approximations. Since [see Eq. (34)]

$$K = \exp\left[\frac{a_1\left(\sigma_2^0 - \sigma_1^0\right)}{RT}\right] = \exp\left[\frac{\sigma_2^0 - \sigma_1^0}{n'_{m1}RT}\right]$$

then finding the value of $(\sigma_2^0 - \sigma_1^0\ A/RT$ by integration of Eq. (60), taking $n'_{m1} = -(dn_1^e/dx_1)_{x_1 \to 1}$, we can determine K, and by means of Eq. (10) find the approximate value of n'_{m1}. Recalculating with the use of the value of n'_{m1} found, we can refine the value of n'_{m1}, etc.

It should be stressed, however, that this method, just as that of Klinkenberg - Everett, is based on the application of the equation of the adsorption isotherm. Therefore, although the method in question is not limited by the condition $\beta = 1$, the conditions of the ideality of the surface solution impose their restrictions.

It can be shown that for the case of nonideal surface and bulk solutions, Eq. (11) can be written as

$$-\left(\frac{dn_1^e}{dx_1}\right)_{x_1 \to 1} = n'_{m1}\left[1 - \frac{\beta\gamma_2}{\gamma'_2 K^{1/\beta}}\right] \tag{18}$$

It is clear from Eq. (18) that the nonideality of the bulk and surface solutions can distort greatly the value of n'_{m1}, which we find by Kiselev - Schay's method. Since the activity coefficients of diluted surface solutions usually show considerable negative deviations with positive deviations of the activity coefficients of bulk solutions, $\gamma_2/\gamma'_2 > 1$. This means that the value of n'_{mi} found by means of Eq. (10) without taking account of the activity coefficient will be too low. It is interesting to note that by means of Eq. (18) it is possible in principle to determine n'_{m1} from the *S*-shaped isotherms. In this case the value of γ_2/γ'_2 is so large that the expression in brackets becomes less than zero, and since $n'_{m1} > 0$, then dn^e_1/dx changes sign.

4. *The Case of the Filling of Pore Volumes*

In the case of filling of pore volumes, the total volume of solid adsorbent pores is assumed to be the inhomogeneity region. In this case, Eq. (6) is written as

$$V_p = \sum \bar{v}_i x'_i \tag{19}$$

where V_p is the pore volume, and $\bar{v}_i$ the partial molar volume of the component i.

The quantity $\bar{v}_i$ is usually taken to be approximately equal to the molar volume of the liquid component v^0_i at the temperature of experiment.

However, this method of determination of the inhomogeneity region seems to be valid only for the case of adsorption in micropores, when there are sufficient physical reasons to suppose the whole solution contained in the pore space of the adsorbent to be acted upon by the adsorption forces. In the case of wide-pored adsorbents, as was shown in the adsorption studies of vapor mixtures by Bering (1957) and Serpinskii, the inhomogeneity region can be confined to the monolayer.

Thus, the analysis of the methods for determination of the values of n'_m from the data on the adsorption of solutions shows that none of the existing methods gives reliable values of n'_m for nonideal systems.

5. Bering's Method

According to Bering (1972), very useful information on the properties of the inhomogeneity region and its extent can be obtained by studying the adsorption of a binary vapor mixture. In this case, as was shown above, it is possible to equate the experimentally determined extents of adsorption to the values of n', and to calculate the values of x'_i, γ'_i, and n' directly from experimental data without making any assumptions about the surface layer structure.

It was shown for a number of experimentally studied systems that in the case of adsorption on nonporous and wide-pored adsorbents, the composition of the *i*th and subsequent adsorption layers does not depend on the layer number and adsorbent nature and is determined only by the solution nature. This regularity was observed both for polymolecular adsorption and for capillary condensation. For instance, it was found for the systems chloroform - acetone on silica gel, glass spheres (Bering, 1957) and carbon black (Pavlutchenko, 1969) and also for the system O_2 - N_2 - TiO_2 (Arnold, 1949) that $i = 1$, and for the system isooctane - benzene - graphitized carbon black that $i \sim 3$ (Bering *et al.*, 1973a).

Although the list of works on this problem is practically covered by the given references, this direction of investigations should be recognized as being one of the most interesting for elucidation of the adsorption mechanism of mixtures and the structure of the surface phase in the case of adsorption from liquid solutions.

III. THERMODYNAMICS OF THE ADSORPTION OF SOLUTIONS

In recent years, the interest of the investigators in the thermodynamics of adsorption has significantly increased. This is due to the necessity of a more exhaustive interpretation of experimental data, of obtaining from them information on the surface layer structure, of establishing the relationship between the adsorption phenomena at various interafaces and from different phases, and also due to the need of developing a theory that would make it possible to predict the adsorption equilibrium of mixtures on the basis of the properties of adsorbates and adsorbent.

In the studies on the adsorption of solutions, the following parameters can be measured: the extent of adsorption n_i^e, the dependence of n_i^e on temperature, the heats of immersion.

The aim of the adsorption thermodynamics is to find the relationship between these quantities and the main parameters characterizing the adsorption system: the amounts of components, the energies of their interaction with one another and with adsorbent, etc. At present there are two approaches to the solution of this problem.

A. The Gibbs Method. Thermodynamics of Excess Properties

Myers and associates (Sircar *et al.*, 1972) and Schay (1973) formulated the thermodynamic of adsorption of solutions based on the excess properties determined by the general expression

$$M^e = n'(m'-m) \tag{20}$$

where M^e is the excess property and m' and m are the molar extensive quantities (enthalpy, free energy, molar fraction, etc.) in the adsorption layer and in the bulk phase. Then the excess adsorption, enthalpy, free energy, and entropy are determined by the expressions:

$$n_i^e = n^0\left(x_i' - x_i\right) \tag{21}$$

$$H^e = n^0(h^0 - h) + \Delta H \tag{22}$$

$$G^e = A\sigma + \sum n_i^e \mu_i \tag{23}$$

$$S^e = -(\partial G^e/\partial T)_{n_i^e} \tag{24}$$

The excess enthalpy of adsorption is related to the extent of excess adsorption by the expression

$$H^e = n^0\left[h^m\left(x_i^0,\ T\right) - h^m\left(x_i,\ T\right)\right] + \sum n_i^e h_i^0 - Q \tag{25}$$

where h_i^0 is the molar enthalpy of pure liquid, Q the heat of immersion, and h^m the molar enthalpy of mixing. For pure substance, $H^e = -Q$.

The value of $\Delta H = -Q$ can be found from the derivative $A\sigma$ with respect to temperature by means of the equation

$$\frac{\partial}{\partial T}\left(\frac{A\sigma}{RT}\right)_{x_i} = \frac{\Delta H'}{RT^2} + \sum n_i^e \left(\frac{\partial \ln a_i}{\partial T}\right)_{x_i} \tag{26}$$

where

$$\Delta H' \equiv \Delta H - n^0\left[h^m\left(x_i\right) - h^m\left(x_i^0\right)\right]$$

Since the values of $A\sigma/RT$ can be found by means of Eq. (60), then from the excess adsorption data at several temperatures it is possible to find the values of Q and to compare them with those determined by calorimetric measurements.

The temperature dependence of adsorption is related to the heat of immersion by the expression

$$\left[\frac{\partial n_1^e}{\partial T}\right]_{x_i} = \frac{\frac{1}{RT^2}\left[\frac{\partial(\Delta H')}{\partial x_1}\right]_T + \left[\frac{\partial n_1^e}{\partial x_1}\right]_T \left[\left(\frac{\partial \ln a_1}{\partial T}\right)_{x_1} - \left(\frac{\partial \ln a_2}{\partial T}\right)_{x_1}\right]}{\left[\left(\frac{\partial \ln a_1}{\partial x_1}\right)_T - \left(\frac{\partial \ln a_2}{\partial x_1}\right)_T\right]} \tag{27}$$

Thus, this theory makes it possible to relate the main experimentally determined quantities to the thermodynamic characteristics of the system.

A similar treatment of the thermodynamics of adsorption of solutions is given by Schay in a series of papers (1971a,b, 1973).

Believing that for colloidally dispersed solids the notions surface tension, interfacial area no longer have the physical sense ascribed to them in the usual treatment of surface phenomena by the Gibbs method, Schay proposed to replace them by a certain specific quantity $\varepsilon = \sigma A$, which cannot be divided into two factors σ and A, as is usually done in dealing with surface phenomena. As the result, he obtained for a binary solution an analog of the Gibbs adsorption equation in the form

$$d(\Delta\varepsilon) = \frac{n_1^e}{m^\alpha x_2}\, d\mu_1 \tag{28}$$

in which $\Delta\varepsilon = \varepsilon^\alpha - \varepsilon$ and the specific adsorption n_1^e/m^α are substituted for σ and the surface concentration, respectively, in the Gibbs equation. These substitutions permit to treat more freely the phenomena at the solid/liquid interface, but the main thermodynamic expressions in Schay's work coincide with those obtained by Myers if σA is substituted by $\Delta\varepsilon$

Schay's treatment is very similar to that proposed by Bering *et al.* (1970) for the solid/gas interface, if in the work of Bering *et al.* we put $\mu_a = \mu^\alpha + \varepsilon$.

The advantage of the thermodynamic approach, based on the use of excess functions, is the unambiguous relation of the calculated thermodynamic parameters to the quantities determined experimentally. However, as was pointed out by some authors (Kiselev and Pavlova, 1962), the derivatives at constant n_i^e have two values for one n_i^e, as is the case, e.g., in Eq. (24). For this reason, the attempts to obtain the adsorption isostere $(\partial \ln a_i/\partial T^{-1})_{n_i^e}$ by analogy with the adsorption isostere in the thermodynamics of adsorption of one substance $(\partial \ln p/\partial T^{-1})_{n_i'}$ have failed. Moreover, as has been pointed out above, in dealing with excess quantities the surface layer properties are not considered.

Therefore, some authors developed the thermodynamics of adsorption of solutions based on the concept of the layer of finite thickness.

B. *Thermodynamics of the Layer of Finite Thickness*

In this case the surface layer is treated as a separate phase with parameters different from those of the bulk solution. In this approach, the thermodynamic quantities characterizing the surface layer depend on our assumptions about the extent of the inhomogeneity region, which we take to be the surface layer.

1. *Equations of the Adsorption Isotherm*

Choosing the Langmuir monolayer model as a perfect surface solution and considering the equilibrium constant of the exchange reaction between the phases for such solution, Everett (1964) obtained the expression

$$\ln \frac{x_1' x_2}{x_1 x_2'} = -\left[\frac{\Delta U_1 - \Delta U_2}{RT} - \frac{\Delta S_1^* - \Delta S_2^*}{R}\right]$$

$$= \ln K \tag{29}$$

in which ΔU and ΔS^* are the changes in the energy and in the thermal part of entropy of the solution components during adsorption.

The chemical potential of such perfect surface solution is

$$\mu_i' = \mu_i^0 + RT \ln x_i' - \left(\sigma - \sigma_i^0\right) a_i \tag{30}$$

This expression for the chemical potential coincides with the expression obtained by Rusanov (1967a) with the use of a strictly thermodynamic method for an ideal surface solution. It is interesting to note that it does not follow from the determination of an ideal surface solution by means of Eq. (30), that this solution must obey Eq. (29). Therefore, Eqs. (29) and (30) seem to determine different ideal solutions. According to Eq. (29), a perfect surface solution is a solution characterized not only by the absence of interaction between solution molecules, but also by the equality of the molar areas of components. Equation (30) requires only the absence of interactions.

In his next work, Everett (1965) considering the case of adsorption of molecules of different size, writes that a perfect system must obey the equation

$$(x_1'/x_1)(x_2/x_2')^{1/r} = K \tag{31}$$

where

$$r = \frac{a_2}{a_1} = \frac{1}{\beta}$$

$$K = \exp\left\{\left[\mu_1^0 - \mu_1^{10}\right] - \frac{1}{r}\left[\mu_2^0 - \mu_2^{10}\right]\right\}/RT \tag{32}$$

The determination of a perfect system by means of Eq. (31) is equivalent to that of the chemical potential of an ideal surface solution.

The expressions similar to Eqs. (31) and (32) were also obtained by Rusanov (1960), and Elovich and Larionov (1962a), the latter two authors showing that in the general case $1/r$ is to be considered as being the mutual displacement factor, and $1/r = -dn_2'/dn_1'$. In the case of volume displacement, $1/r = v_1/v_2$ (Schuchowitzky, 1938) and in the case of surface displacement, $1/r = a_1/a_2$. Larionov *et al.* (1967b) demonstrated

$$\mu_i^{10} - \mu_i^0 = \sigma_i^0 a_i \tag{33}$$

Therefore

$$K = \exp\left[a_1\left(\sigma_2^0 - \sigma_1^0\right)/RT\right] \tag{34}$$

The statistical theory of solutions and the experimental data on bulk solutions show that for description of the properties of solutions whose components differ in molar volumes, it is expedient to use volume fractions rather than molar ones. This means that writing the chemical potential of an ideal solution as Eq. (30), we assume at the same time that $a_1 \simeq a_2$ and therefore the determinations of an ideal surface solution by means of Eqs. (29) and (30) coincide.

The expressions for the chemical potential of surface solution in the terms of the surface fractions (coverages) were obtained for the liquid/vapor interface by Prigogine and Marechal (1952)

$$\mu_1' = \mu_1^{10} - \sigma a + RT \ln \theta + RT(1 - 1/r)(1-\theta) \tag{35}$$

$$\mu_2' = \mu_2^{10} - r\sigma a + RT \ln(1-\theta) + RT(1-r)\theta \tag{36}$$

where $\theta = n_1'/(n_1' + rn_2')$. Equating the chemical potentials in bulk and surface solutions, we have instead of Eq. (31)

$$\frac{\theta_1}{x_1\gamma_1}\left(\frac{x_2\gamma_2}{1-\theta_1}\right)^{1/r} = K \tag{37}$$

In the general case, $1/r = \beta$ and the constant in Eq. (37) is given by the expression (Larionov, 1966)

$$K = \exp\left[\frac{a_1\left(\gamma_2^0 - \gamma_1^0\right)}{RT} + (\beta-1)_T\right] \tag{38}$$

The value of r in the surface layer depends on the orientation of the adsorption solution molecules and may differ significantly from that in the bulk solution, in which it is equal to the ratio of molar volumes. Therefore, unlike (Eq. (31), Eq. (38) allows the presence of the adsorption azeotrope $n_1^e = 0$ at $0 < x_1 < 1$ even for an ideal bulk solution (Elovich and Larionov, 1962a).

Analysis of Eq. (37) showed (Everett, 1965) that already relatively small changes in r significantly influence $x_i' - x_i$. Just for this reason, even small deviations of β from unity can affect considerably the calculated values of n_{mi}', as has been shown above.

Although the ideal adsorption systems are a rare exception, serving as a reference standard, as it were, they are of considerable theoretical interest for development of the adsorption theory of real solutions. In developing this theory, it is important first of all to find out how the nonideality of bulk solution affects the extents of excess adsorption.

Using the statistical thermodynamics method and assuming the surface solution to be an ideal one and the molecules to be of equal size, Sircar and Myers (1970) obtained an expression relating the surface tension of the surface solution at the solid/liquid interface and the surface tensions of pure solution components at the same interface to the activities of bulk solution

$$\exp(-\sigma a/RT) = x_1\gamma_1 \exp\left(-\sigma_1^0 a/RT\right) + x_2\gamma_2 \exp\left(-\sigma_2^0 a/RT\right) \tag{39}$$

Solving simultaneously Eqs. (39), (5), the Gibbs - Duhem equation,

$$x_1 d\ \ell n\ \gamma_1 x_1 + x_2 d\ \ell n\ \gamma_2 x_2 = 0 \tag{40}$$

and the Gibbs equation

$$A d\sigma / RT d(x_1\gamma_1) = - n_1^e / x_1 x_2 \gamma_1 \tag{41}$$

the above authors obtained an equation relating the excess adsorption to the activities of bulk solution and the interfacial tensions of pure components

$$\frac{n_1^e}{n_m'} = \frac{x_1\gamma_1 - x_1(\gamma_1 x_1 + K\gamma_2 x_2)}{\gamma_1 x_1 + K\gamma_2 x_2} \tag{42}$$

where

$$K = \exp\left[\left(\sigma_1^0 - \sigma_2^0\right)a/RT\right] \quad , \quad a = \frac{A}{n_m'} \tag{43}$$

at $\gamma_1 = \gamma_2 = 1$, Eq. (42) turns into (44):

$$\frac{n_1^e}{n_m'} = \frac{x_1 x_2 (1-K)}{x_1 + K x_2} \tag{44}$$

i.e., into Eq. (16), obtained by Everett (1964) for a regular surface solution. Larionov and Myers (1971) obtained an equation relating the excess adsorption to the composition of the bulk solution for the case of nonideal bulk and adsorption solutions and unequal molar areas of the components being adsorbed

$$n_1^e = \frac{x_1 x_2 n_{m_1}' (S-1)}{S x_1 + n_{m_1}' x_2 / n_{m_2}'} \tag{45}$$

where

$$S = \frac{x_1' x_2}{x_2' x_1} = \frac{\gamma_2' \gamma_1}{\gamma_1' \gamma_2} \exp\left[\frac{\sigma A}{RT}\left(\frac{1}{n'_{m_2}} - \frac{1}{n'_{m_2}}\right) + \frac{A}{RT}\left(\frac{\sigma_2^0}{n'_{m_1}} - \frac{\sigma_2^0}{n'_{m_2}}\right)\right] \tag{46}$$

in which γ_i' is the activity coefficient in the surface phase determined by Eq. (52).

Kagiya *et al.* (1971) and Davis *et al.* (1973) studied the dependence of the adsorption equilibrium constant K (29) on the electronic structure of adsorbates. They found for substituted aromatic compounds and phenols that in the case of adsorption on silica gel, $\ln K$ (29) decreases linearly with increasing Hammett substituent constant and the ionization potential.

2. Activity Coefficients of the Surface Layer

The nonideality of the bulk solution resulting from the difference in the energies of intermolecular interaction and in the size of molecules manifests itself to a greater or lesser degree in the surface layer as well. Therefore, by analogy with the bulk solution in the theories treating the surface layer as a separate phase, the departures from ideality in the behavior of the bulk phase are described by means of the activity coefficients or excess thermodynamic functions.

If we choose the ideal bulk solution as an ideal solution, since

$$RT \ln \gamma_i = \mu_i - \mu_i^{id} \tag{47}$$

then

$$RT \ln \gamma_i' = \mu_i' - \mu_i^{id} \tag{48}$$

whence it follows (Blackburn *et al.*, 1957)

$$\gamma_i' = \frac{x_i \gamma_i}{x_i'} \tag{49}$$

Everett (1965) determines the activity coefficient of the surface solution with reference to the perfect surface solution determined by Eq. (29). In this case,

$$RT \ln \gamma_i' = \left(\mu_i' + \sigma a_i\right) - \left(\mu_i'^{\text{perf.}} + \sigma^{\text{perf.}} a_i\right) \tag{50}$$

and the chemical potential of the nonideal surface solution is written as

$$\mu_i' = \mu_i^{10} + RT \ln x_1' \gamma' - \left(\sigma - \sigma_i^0\right) a_i \tag{51}$$

It follows from the equality of the chemical potentials in the bulk and surface solutions that

$$\gamma_i' = \frac{x_i \gamma_i}{x_i'} \exp\left[a_i\left(\sigma - \sigma_i^0\right)/RT\right] \tag{52}$$

i.e., this determination of γ_i' coincides with that of the activity coefficient introduced by Butler (1932) for the surface solution at the liquid/vapor interface.

This determination is most widely used (Schay *et al.*, 1962; Rusanov, 1967a, 1972; Khopina, 1972; Eltekov *et al.*, 1972; Lu and Lama, 1967).

If as an "ideal" solution we choose an athermal solution whose chemical potential is determined by Eqs. (35) and (36), then the activity coefficient of this solution (Larionov, 1966) is

$$\gamma_1' = \frac{x_1 \gamma_1}{\theta} \exp\left[(\beta-1)(1-\theta)\right] \exp\left[a_1\left(\sigma - \sigma_1^0\right)/RT\right]$$

$$\gamma_2' = \frac{x_2 \gamma_2}{1-\theta} \exp\left[\theta\left(\frac{1}{\beta} - 1\right)\right] \exp\left[a_2\left(\sigma - \sigma_2^0\right)/RT\right] \tag{53}$$

In this case γ_i' is the thermal part of the activity coefficient and at small h^m, $\gamma_i' \simeq 1$ (Larionov and Kurbanbekov, 1972).

In considering the thermodynamics of a vapor mixture, Myers and Prausnitz (1965) showed that at constant surface pressure (or interfacial tension) the thermodynamics of surface solutions can be described in terms of the thermodynamics of solutions. In the case of liquid solutions, this means that the standard state of pure components should be chosen in such a way that in this state pure component should have the surface tension equal to that of the solution.

The effect of surface tension on the chemical potential of pure component is expressed by the Gibbs equation

$$-Ad\sigma = n_i' d\mu_i = n_i' RT\, d\, \ln f_i' \tag{54}$$

where f_i' is the fugacity. Integrating Eq. (54) at constant $n_i' = n_{mi}'$ and taking into consideration the fact that if the fugacity of pure adsorbate $f_i^0 = p_{si}$, $\sigma_i = \sigma_i^0$, we find the fugacity of pure substance in the standard state

$$f_i^* = f_i^0 \exp\left[-\frac{A\left(\sigma - \sigma_i^0\right)}{n_{mi}' RT}\right] \tag{55}$$

Equating the fugacities of bulk and surface solutions, we can find the expression for the activity coefficient of surface solution, coinciding exactly with Eq. (52).

Larionov and Myers (1971) demonstrated that the activity coefficients determined by Eq. (52) fit the Gibbs - Duhem equation. This can be also proved for the activity coefficients determined by Eq. (53).

The activity coefficients thus calculated take into account all the effects of the surface phase nonideality, including its inhomogeneity, caused by the surface inhomogeneity.

The departures from ideal behavior can be described also by means of the energy distribution function for heterogeneous surface (Rudzinski, 1973). However, since the energy distribution function is found from the condition of the best description of the experimental adsorption isotherm $n^e(x_i)$ by means of a set of theoretical adsorption isotherms for homogeneous areas, it is necessary first of all to make sure that for homogeneous surface the system under investigation can be described by the equation of the adsorption isotherm chosen as an equation for energetically homogeneous areas. Hence, it becomes clear that it is important to investigate the properties of adsorption solutions on energetically homogeneous surfaces. Apparently information on the energetic inhomogeneity can be obtained from the adsorption data on individual substances. The influence of surface heterogeneity on adsorption of solutions was considered by Coltharp and Hackerman (1973a,b).

Several methods for calculation of the activity coefficients in the surface layer are available. It should be stressed once more that irrespective of the method used for calculation of the activity coefficients of surface solution, they include the assumptions made regarding the surface layer model (Rusanov, 1972).

Everett (1965) proposed to calculate the activity coefficients by means of the formula

$$\ln \gamma_2' = x_1' \ln \frac{x_1'\gamma_2 x_2}{x_2' x_1 \gamma_1} - \int_0^{x_1} \ln \frac{x_1' x_2 \gamma_2}{x_2' x_1 \gamma_1}\, dx_1 \tag{56}$$

Larionov and Myers used for this purpose the expression

$$\ln \gamma_1' = \left[\frac{\Delta g^E}{RT}\right]' + x_2 \frac{\partial}{\partial x_1'} \left[\frac{\Delta g^E}{RT}\right]'_T \tag{57}$$

where $[\Delta g^{E}]'$ is the excess molar free energy of adsorbed phase, calculated by means of the expression

$$\left[\frac{\Delta g^{E}}{RT}\right]' = \int_{x_1'=0}^{x_1'} \ln \frac{\gamma_1'}{\gamma_2'}\, dx_1' \tag{58}$$

in which $\ln \gamma_1'/\gamma_2'$ can be found from Eq. (46).

$$n \frac{\gamma_2'}{\gamma_1'} = \ln \frac{\gamma_2}{\gamma_1} + \ln S - \frac{A\left(\sigma - \sigma_1^0\right)}{n_{m1}'\, RT} + \frac{A\left(\sigma - \sigma_2^0\right)}{n_{m2}'\, RT} \tag{59}$$

where the value of $A(\sigma - \sigma_i^0)/n_{mi}'\, RT$ can be found by integration of the Gibbs equation (41) (Schay *et al.* 1962)

$$\frac{A\left(\sigma - \sigma_1^0\right)}{RT} = - \int_{x=1}^{x_1} \frac{n_1^e}{x_1 x_2 \gamma_1}\, d(\gamma_1 x_1)$$

$$\frac{A\left(\sigma - \sigma_2\right)}{RT} = - \int_{x_1=0}^{x_1} \frac{n_1^e}{x_1 x_2 \gamma_1}\, d(\gamma_1 x_1) \tag{60}$$

The activity coefficients can also be calculated directly from Eqs. (52) and (53).

3. Excess Thermodynamic Functions of Surface Solution

It is known that apart from the activity coefficients, the properties of solutions can be characterized by the excess thermodynamic functions

$$z_i^{E} = z_i - z_i^{id} \tag{61}$$

where Z is the chemical potential, the Gibbs excess free energy, entropy, enthalpy, etc.

The excess chemical potential of surface solution is defined as follows:

$$\mu_i^{'E} = \mu_i^{'} - \mu_i^{'id} = RT \ln \gamma_i^{'} \tag{62}$$

In choosing the surface solution as standard state of pure substances at P_i, T_i, σ_i equal to P, T, σ, we can consider it as an analog of the bulk solution subjected to such external action, e.g., compression, that the change of its chemical potential is equal to that due to surface forces.

The Gibbs excess free energy is

$$g^{'E} = \sum \mu_i^{'E} x_i^{'} \tag{63}$$

The excess entropy is

$$S^{'E} = -\left[\frac{\partial g^{'E}}{\partial T}\right]_{P,n_i^{'}} \tag{64}$$

The excess enthalpy is

$$h^{'E} = g^{'E} - T\left[\frac{\partial g^{'E}}{\partial T}\right]_{P,n'} \tag{65}$$

The value of $g^{'E}$ can be calculated from the activity coefficients of surface solution. Having at one's disposal the data for several temperatures, one can find the remaining excess functions of surface solution. The excess functions of surface solutions were obtained by Kazarjan *et al.* (1973a,b) and Kurbanbekov *et al.* (1973b, c). Everett (1965) expressed the heat of immersion in real solution in terms of the heat of immersion in perfect solution and the heats of mixing of bulk and surface solutions

$$\Delta H = \Delta H^{\text{perf.}} + nh^{\text{E}}(x_1) + n'h'^{\text{E}}\left(x_1'\right) - n^0h^{\text{E}}\left(x_1^0\right) \tag{66}$$

where $\Delta H^{\text{perf.}}$ is found by means of Eq. (67). This method also gives the heat of mixing of surface solution (Lu and Lama, 1967).

4. Heat of Immersion. Determination of Thermodynamic Functions from the Temperature Dependence of Adsorption

The changes of the properties of an adsorption system as a whole can be characterized by the change in the thermodynamic functions as the result of adsorption. As has been pointed out, these changes can be calculated by means of Eqs. (20)-(27), in which excess quantities are used.

Some authors (Kiselev and Pavlova, 1962; Bering and Pokrovskii, 1969) considered the thermodynamics of adsorption on the basis of the layer of finite thickness model.

Everett (1964, 1965) obtained the expressions relating the change of enthalpy upon immersion in a perfect system to the enthalpies of immersion in pure components

$$\Delta H^{\text{perf.}} = x_1'\Delta H_1^0 + x_2'\Delta H_2^0 \tag{67}$$

Some authors (Wright, 1966, 1967; Wright and Powell, 1972; Lu and Lama, 1967; Matayo and Wightman, 1973) measured the heats of immersion in pure substances and solutions of various concentrations. In systems that are sufficiently close to the perfect system, the measured heats of immersion obey Eq. (67).

In finding the main thermodynamic functions, Bering and Pokrovskii proceeded from the Gibbs - Helmholtz equation written for the change of the differential molar value of the Gibbs free energy ΔG upon transition from the standard state (pure liquid component) into the adsorption layer at n_i' = constant.

$$\Delta G_i = \Delta H_i - T\Delta S_i \tag{68}$$

Since

$$\Delta G_i = -RT \ln \left(\frac{P_{si}}{P_i}\right) = -RT \ln \gamma_i x_i \tag{69}$$

$$\Delta H_i = -q_i = H_i^0 - H_i' \tag{70}$$

where q_i is the differential heat released upon transition of one mole i from the standard state into the surface layer

$$\Delta S_i = S_i^0 - S_i' \tag{71}$$

then

$$\ln x_i \gamma_i = -\frac{q_i}{RT} + \frac{S_i' - S_i^0}{R} \tag{72}$$

if $\ln x_i \gamma_i$ is a linear function of T^{-1}, then

$$q_i = -R\left(\frac{d \ln x_i \gamma_i}{dT^{-1}}\right)_{n_i'} \tag{73}$$

The values of q_i cannot be obtained from measurement of the heats of immersion because when one substance is introduced into the surface layer, the other substance is desorbed according to Eqs. (6) or (19). Therefore, the calorimetrically measured differential heat of immersion should be compared with the heat-taking account of the desorption of the other component. In the case of a binary solution

$$q_{12} = q_1 - \beta q_2 \tag{74}$$

the values of ΔG_i and ΔS_i are found by means of Eqs. (68) and (69).

Parfitt and Thompson (1971) found the differences in free energies, enthalpies, and entropies of immersion in pure substances from the temperature dependence of the adsorption equilibrium constant (34). Kurbanbekov *et al.* (1973c) and Kazarjan *et al.* (1973a,b) showed that the values thus found correspond to those obtained by integration of the differential values of q_{12}, ΔG_{12}, and ΔS_{12} found by Bering and Pokrovskii's method (1969).

IV. ADSORPTION AT DIFFERENT INTERFACES

A. *Analogy between the Liquid/Vapor, Liquid/Liquid and Liquid/Solid Interfaces*

It has been pointed out that the thermodynamic treatment of surface phenomena at the liquid/vapor and liquid/solid interfaces is in many respects similar. According to Schay, who seemed to be the first to give attention to this analogy and to carry out systematic studies in this direction (Schay and Nagy, 1972b; Schay *et al.*, 1962), this is explained by the fact that in the case of the liquid/vapor interface, the components concentrations in the gas phase are usually negligible, as compared to those in the liquid phase (in the case of the solid/liquid interface, the solutes concentrations are equal to zero). Therefore, the gas phase can be considered as being the surface of a solid with zero energy of interaction with the solution components. The same treatment is applicable to the liquid/liquid interface, provided the mutual solubility of liquid phases can be ignored.

B. *Analogy between the Liquid/Solid and Vapor/Solid Interfaces*

In our opinion, it is of greatest interest in adsorption studies on solid adsorbents to investigate simultaneously the adsorption from gas and liquid phases.

In adsorption studies on individual substances, we can obtain independent information on the quantities characterizing individual adsorbents, which is as necessary in the adsorption studies on solutions as the knowledge of the properties of pure substances in the studies on bulk solutions. The adsorption studies on vapor mixtures offer a possibility of experimental investigation of the properties of surface solutions, for in this case we can take $n_i^e = n_i'$, and thus study them independent of the adopted surface layer model, which is impossible in the case of adsorption of liquid solutions.

1. Adsorption of Pure Substances

The relationships between the integral characteristics of adsorption of pure substance and the quantities characterizing the immersion in this substance are given by the expressions (Kiselev, 1945, 1946; Schay, 1971b)

$$F_{SLi} = F_{SVi} + \sigma_{LVi} A' \tag{75}$$

$$H_{SLi} = H_{SVi} + H_{LVi} A' \tag{76}$$

In Eqs. (75) and (76), $F_{SVi} \simeq G_{SVi}$ is the integral free energy of saturation of adsorbent with vapor

$$-F_{SVi} = \int_0^{P_{Si}} \ln \frac{P_S}{P} \, dn_i'$$

$$= \int_0^{P_S} n_i' d \ln P_i = \left(\frac{\pi A}{RT}\right)_{P=P_S} \tag{77}$$

H_{SVi} is the integral enthalpy of saturation of adsorbent with vapor

$$H_{SVi} = - \int_0^{P_S} Q_i dn'$$

$$H_{LVi} = \sigma_{LVi} + T \frac{d\sigma_{LVi}}{dT} \tag{78}$$

where Q_i is the differential heat of adsorption, A' the surface area of the liquid/vapor interface disappearing during immersion in liquid substance of adsorbent, which is in equilibrium with saturated vapor. Equation (75) is actually an analog of the Dupre - Young equation

$$\sigma_{SL} = \sigma_{SV} - \sigma_{LV} \cos \theta \tag{79}$$

which establishes the relation between the interfacial tensions at the solid/vapor (SV), solid/liquid (SL) and liquid/vapor (LV) interfaces.

Let us consider a system consisting of an ideal flat adsorbent and vapor. During vapor compression in this system, the free surface energy decreases and at $P = P_S$ it is equal to σ_{SV} (state 1). As vapor compression proceeds, condensation occurs and two new interfaces SL and LV are formed. If their surface areas are equal to that of adsorbent, then according to Eq. (79), at $\cos \theta = 1$, we obtain state 2, equivalent to state 1, which is in equilibrium with saturated vapor. If afterwards the system is flooded with liquid so that the liquid/vapor interface area should be equal to zero (state 3), the energy equal to $A\sigma_{LVi}$ is released and state 3 of the system will differ from states 1 and 2. In principle, the first two states can be reached during the investigation of vapor adsorption on nonporous adsorbent, state 3 is reached during the investigation of vapors adsorption on porous adsorbents and corresponds to immersion experiments. In the general case, during investigation of vapors adsorption, an intermediate state between 2 and 3 is realized. Therefore, in comparing the data on adsorption of vapors and liquids, we are faced with the problem of determining σ_{SL} from the data on vapors adsorption.

If during vapors adsorption state 1 is realized, the properties of the interface SL can be found from the data on adsorption of individual vapor and the surface tension of liquid by means of the equation

$$\frac{A(\sigma_0-\sigma_{SL})}{RT} = \frac{\pi_{SLi}A}{RT}$$

$$= \int_0^{P_S} n_i' d \ln P_i + \frac{A\sigma_{LVi}}{RT} \qquad (80)$$

In reality, when the adsorbent surface is not flat and there are contacts between separate adsorbent particles, vapors adsorption is accompanied by capillary condensation. As the result, in the general case of P_S the three interfaces can all be present on the adsorbent surface. When such adsorbent is immersed in liquid substance, the interface SV is replaced by the interface SL, and the surface area of the interface LV becomes equal to that of liquid in the vessel used for immersion, i.e., it can be neglected. In this case, according to Eq. (79), the free energy of immersion is equal to

$$A_{SV}'\sigma_{LVi} \cos \theta$$

where A' is the adsorbent surface area, which at P_S has the surface energy σ_{SV}.

Berezkina *et al.* (1972a-c; 1973) showed that on nonporous adsorbents in the region of medium relative pressures, when the capillary condensation in the gaps between particles can still be considered to be of no great importance, πa is a linear function of πA (Fig. 3). Extrapolating this straight line to $\pi a = 0$, i.e., to the region of polymolecular layers of infinite thickness, we

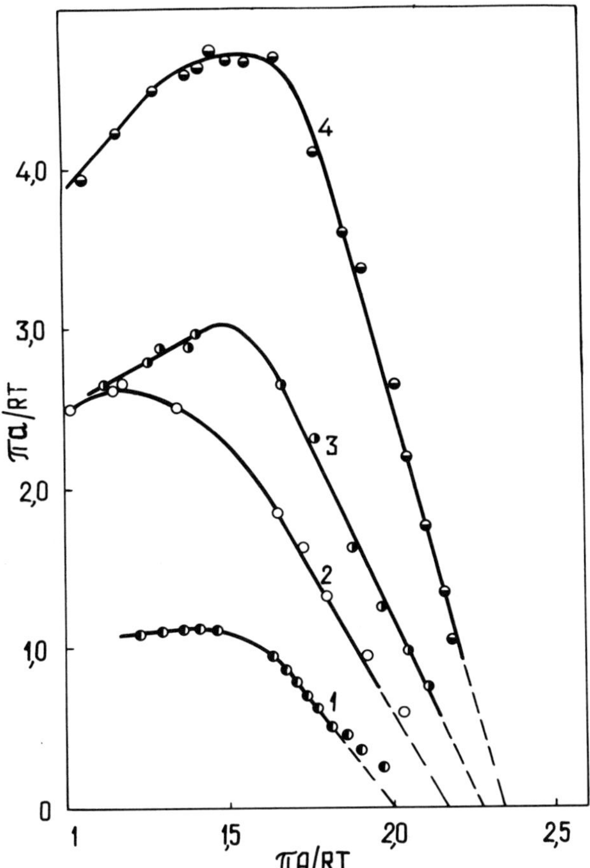

Fig. 3. The dependence of πa on πA for individual vapors on graphitized carbon black: 1-$C_2H_5OH_2$ 2-C_6H_6, 3-CCl_4, 4-iC_8H_{18}.

can find the value of πA which we would have in the absence of capillary condensation at P_S. Using the found value of πA, by means of Eq. (80), we can calculate the free energy of immersion for the case when at P_S, $A' \neq A$ and $A' \neq 0$.

In the case of porous adsorbents, when as the result of capillary condensation the total volume of pores at P_S is filled with liquid adsorbate, the whole interface SV is replaced by SL. In this case, immersion of adsorbent in liquid substance is not accompanied by any effects and since $A_{LV} = 0$, then $\sigma_{SV} = \sigma_{SL}$. Integration of Eq. (77) gives the value of the free energy of immersion. As was shown by Bering *et al.* (1970), in the general case, the integral in Eq. (77) is the function ϕ, determined by the expression

$$\phi \equiv \left(\frac{\pi_A}{n^{\alpha}}\right) - \left(\mu_a - \mu_a^0\right) \tag{81}$$

Thus, in order to obtain the immersion characteristics from the adsorption data on idnvidual vapors, in the general case, we must know the value of A'. The problem is simplified if capillary condensation is absent (in this case $A' = A$) and also when as the result of capillary condensation (or filling of pore volumes), $A' = 0$.

Myers and Sircar (1972a) obtained the condition for the thermodynamic consistency test, which they assume to be valid for adsorption of any pair of vapors or their liquid solutions

$$\int_{P=0}^{P_{S2}} n_2' d \ln P - \int_{P=0}^{P_{S1}} n_1' d \ln P$$

$$+ \int_{x_1=0}^{x_1=1} \frac{n_1^e}{\gamma_1 x_1 x_2} d(\gamma_1 x_1) = 0 \tag{82}$$

It should be noted in connection with what has been said that this test is undoubtedly correct, but it does not take into account the conditions under which adsorptions of vapors and liquids are usually compared. It is valid only in the case if we compare adsorptions of saturated vapor and liquid at constant and equal interface areas, i.e., when $A_{SV} = A_{SL} = A_{LV} = A$.

However, in reality, adsorption of liquids is investigated at $A_{LV} \cong 0$ and, therefore, it is determined in the absence of the interface LV. Such system in the general case is not equivalent to a system in equilibrium with saturated vapor, in which $A_{LV} \neq 0$.

Solving simultaneously Eqs. (60), (75), and (77) and taking into account all that has been said above, we can show that Eq. (82) is valid also in the cases when $\sigma_{LV1} - \sigma_{LV2}$ or $A_1' = A_2' = 0$. The first condition is encountered very seldom, the second condition seems to be always fulfilled for porous adsorbents.

It is known, however, that during adsorption on activated carbons, macropores are not filled with condensate even at P_S.

Myers and Sircar (1972a) checked their condition by studying adsorption on silica gel and obtained a satisfactory validity of Eq. (82). The authors of this review showed that Eq. (82) is valid in the case of adsorption on silica gel, pressed aerosil, but not valid for adsorption on graphitized carbon black.

2. *Adsorption of Vapor Mixtures*

Adsorption studies on vapor mixtures can solve many obscure problems of adsorption of liquid solutions, but the works in this field are not numerous. Among them are those by Bering (1957), Bering *et al.* (1973b), and Pavlutchenko (1969).

Myers and Sircar (1972b) considered the analogy between adsorption from liquids and vapors. In developing their theory, they proceeded from Eq. (82), which can be written as

$$\int_{x_1=0}^{x_1} \frac{n_1^e}{\gamma_1 x_1 x_2}\, d(\gamma_1 x_1) = -\frac{\phi S_1 - \phi S_2}{RT} \tag{83}$$

where ϕ_{Si} is the value of ϕ for saturated vapor. Therefore, the excess adsorption from saturated vapor and from solution must be the same.

Assuming that adsorption from saturated vapors can be expressed in terms of unsaturated vapors adsorption, we have

$$n_i^e = \lim_{P\to P_S} \left[n_i^{ev}\right] = \lim_{P\to P_S} \left[n'\left(x_i' - x_i\right)\right] \tag{84}$$

where n' and x are the functions of pressure, and the molar fraction x_i' can be determined as the composition of liquid that would be in equilibrium with vapor of composition y upon saturation.

Sircar and Myers (1973) considered two cases. In the case of adsorption on microporous adsorbents, n' and $x_i' - x_i$ tend to a certain limit, in the case of adsorption on smooth surfaces, $n' \to \infty$ and $(x'-x) \to 0$.

It should be pointed out, however, that under actual conditions of comparison of experimental data, Eq. (82), as has been shown, is valid only for the case of adsorption on porous adsorbents, when in the course of capillary condensation the interface LV disappears. Therefore, in considering the analogy between adsorption from vapors and liquids on nonporous adsorbents, a correction associated with the presence of the interface LV should be introduced.

V. PREDICTION OF THE ADSORPTION OF MIXTURES FROM THE ADSORPTION PROPERTIES OF PURE COMPONENTS

Just as in the theory of bulk solutions one of the most important problems is the determination of the properties of solutions from those of pure substances, one of the most interesting problems of the theory of adsorption of solutions is the calculation of the adsorption equilibrium of solutions from the characteristics of individual substances.

Because at the present time the adsorption of individual substance can be most readily characterized by its adsorption isotherm, it is of greatest interest to compare adsorption of solutions with that of vapors of individual substances.

A. Adsorption of Vapor Mixtures

A critical review of the methods for calculation of the adsorption equilibrium from the adsorption isotherms of individual substnces was given by Bülow *et al.* (1972). The methods of calculation of microporous adsorbents were considered by Bering and Serpinskii (1971, 1972b). It should be noted that now the most widely used method is that of the ideal adsorption solution. Myers and Prausnitz (Myers and Prausnitz, 1965; Myers, 1968), and Bering and Serpinskii (1972c) developed the theoretical basis of this method. Myers and Prausnitz showed that along the line of constant ϕ for surface solution the same thermodynamic relations are valid as for bulk solutions.

Then Raoult's law for ideal surface solution is written

$$P_{yi} = P_i^* x_i' \tag{85}$$

where P_i^* is the vapor pressure of pure i at the same temperature and ϕ as in the adsorption mixture.

Using for ideal surface solution the relations (6) and $n_i' = n' x_i'$, we can calculate the extents of adsorption for vapor mixtures.

As was shown by many investigators, in the region of not very large coverages, the assumption of the ideality of surface solutions is justified. In the region of high coverages, surface solutions deviate from ideality and the values of the activity coefficients of these solutions approach those for bulk solutions (Bering *et al.*, 1973a,b; Pavlutchenko, 1969; Bering and Serpinskii, 1972a). In view of this fact, Bering *et al.* (1973b) suggest to use in the region of low coverages $\gamma_i' = 1$, and in the region of high coverages (thick adsorption layers) to assume $\gamma_i' = \gamma_i$ at $x_i' = x_i$. The adsorption in the intermediate region can be found by interpolation.

Payne *et al.* (1968) proposed to use for description of adsorption of individual substances and vapor mixtures Eyring's two-dimensional equation of state

$$\pi = \frac{RT}{\alpha - c_2 b_2^{1/2} \alpha^{1/2}} - \frac{a_2}{\alpha^2} \tag{86}$$

where π is the surface pressure, α the molar area, c a constant depending on the adopted layer model, and a and b are the two-dimensional analogs of the van der Waal's constants. For mixtures, the constants of Eq. (86) can be expressed in terms of the combination of the constants of pure substances and the composition of the adsorption phase can be determined without the assumptions of the ideality of surface solution.

B. *Adsorption of Liquid Solutions*

According to Myers and Sircar (1972b), adsorption from liquid mixtures can be explained completely on the basis of the extents of adsorption of unsaturated vapors of the same substances.

In their works, Myers and Sircar (1972a,b) developed a theory allowing to predict the adsorption at the solid/liquid interface from that of unsaturated vapors of pure substances. The general scheme of prediction is that on the basis of the adsorption isotherms of individual vapors and the assumption of ideality of adsorption phase, the adsorption of vapor mixture is calculated by Myers and Prausnitz's method, which in the limit (84) is equal to that from liquid solution. However, in the region of high coverages, the values of ϕ for two substances can differ so much that it becomes impossible to choose a common value of ϕ. Therefore, for this case Sircar and Myers (1973) proposed to use the reduced surface potential ϕ_r, determined by the expression

$$\phi_r = \frac{\phi}{\phi_S} \tag{87}$$

which at $P \to P_S$ tends to 1.

Then the extent of excess adsorption can be expressed by the equation

$$n_1^e = \lim_{\phi_r \to 1} \left[\frac{x_1 x_2 (1-K)}{\left(x_1/n_1'^*\right) + K\left(x_2/n_2'^*\right)} \right] \tag{88}$$

where

$$K = \frac{\gamma_2}{\gamma_1} \exp\left[\frac{\phi_S - \phi_{S2}}{n_2^{'*} RT} - \frac{\phi_S - \phi_{S1}}{n_1^{'*} RT}\right] \tag{89}$$

if at $P \to P_S$, $n_i^{'*}$ tends to the finite value of n'_{mi}, by integrating the Gibbs equation with account taken of the Gibbs - Duhem equation and Eqs. (88), (89), we obtain Eq. (90), similar to the equation obtained for the interface LV by Šprow and Prausnitz (1966)

$$\gamma_1 x_1 \exp\left[\frac{\phi_S - \phi_{S1}}{n'_{m1} RT}\right] + \gamma_2 x_2 \exp\left[\frac{\phi_S - \phi_{S2}}{n'_{m2} RT}\right] = 1 \tag{90}$$

Given $\phi_{S1} - \phi_{S2}$, n'_{mi}, and γ_i, we can solve Eq. (90) simultaneously with Eqs. (88) and (89) and find n_1^e.

In the case when $\lim n_i^{'*} = \infty$ and the adsorption isotherm of individual vapors can be expressed by the equation

$$\ln P/P_S = -\delta/\left(n_i^{'*}\right)^r \tag{91}$$

for ideal bulk and surface solutions, the extent of excess adsorption can be found by means of the equation

$$n_1^e = cx_1x_2 \left[\frac{\phi_{Sr} - \phi_{S1}}{RT}\right] / (x_1 + cx_2)^2 \tag{92}$$

where

$$c = [\delta_1\phi_{S2}/\delta_2\phi_{S1}]^{\frac{1}{s-1}} \tag{93}$$

It is interesting to note that Eq. (92) does not contain the quantities characterizing the surface layer. The above authors showed that using Eqs. (88) - (90), it is possible to predict semi-quantitatively the adsorption of solutions from the adsorption isotherms of individual vapors.

It should be noted that, as has been pointed out above, the starting point of the theory, viz., that ϕ_S is equal to the potential of immersion, allows to compare only the results obtained in investigation of adsorption on porous adsorbents.

Bering *et al.* (1973b) proposed a method for predicting adsorption of liquid solutions from the adsorption isotherms of indivdual substances on nonporous adsorbents. They showed that in the case of polymolecular adsorption, the lines y = const in the coordinates $n_2' = f(n_1')$ are straight lines, and the activity coefficients of such surface solution are equal to those of bulk solution. Therefore, the partial pressures over the surface solution with the composition x_i' are

$$P_i = P_i^* x_i \gamma_i \tag{94}$$

and the composition of the vapor phase y_i is

$$y_i = P_i/(P_1+P_2) \tag{95}$$

Figure 4 explains the scheme of calculation. The line $n_1' - n_2'$ lies in the region of polymolecular adsorption. The line $n_{m1}' - n_{m2}'$ corresponds to the saturation pressure.

It was established experimentally (Bering, 1957; Pavlutchenko, 1969; Bering *et al.*, 1973b) that starting from a certain value of n' in the region of polymolecular adsorption or capillary condensation, the line of constant composition of gas phase y = const goes parallel to the line drawn from the origin with the slopes x_2/x_1, where x_2 is the molar fraction in the bulk solution over which the vapor composition is y_2.

It is clear from the figure that the intercept $DD_1 = MM_1 = n_2' - n_1' x_2/x_1 = n_2^e/x_1$ (see Eq. (5)). Thus, the extent of adsorption of liquid solution can be found in the region of polymolecular adsorption of vapors. To determine n_2^e, we preset the value of ϕ in the region of large n_i', find the values of n_1' and n_2', and draw a straight line through the points corresponding to n_1' and n_2'. With

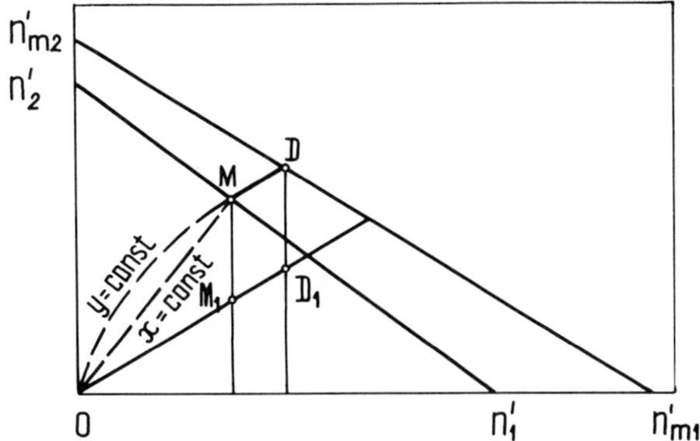

Fig. 4. Illustration of the method of prediction of excess adsorption according to Bering et al. (1973a).

preset molar fraction of surface solution, we find from the phase diagram of bulk solution the values of γ_2' and calculate y_2. Then we find x_2 by means of the phase diagram, corresponding to y_2, and draw through the origin a straight line with the slope x_2/x_1, whereupon we find MM_1 and, multiplying it by x_1, calculate n_2^e.

Bering *et al.* (1973b) obtained quantitative agreement between the experimental and calculated extents of adsorption. It should be pointed out, however, that the method is very sensitive to the accuracy of the data on the liquid/vapor equilibrium and can be used only in cases when in the region of polymolecular adsorption of both components it is possible to choose a common value of ϕ.

The above authors calculated the adsorption equilibrium on the basis of the adsorption characteristics of individual substances using the adsorption isotherms of unsaturated vapors. For calculation in this case, it is necessary to know the activity coefficients in the surface layer.

In the method proposed by Sircar and Myers (1973), it is assumed that $\gamma_i' = 1$, in Bering's method $\gamma_i' = \gamma_i$.

However, in principle, the adsorption characteristic of individual substance can be given by other quantities as well: the free energy of immersion, heat of immersion, or adhesion tension. In this case, calculation can be carried out by means of the equation of the adsorption isotherm.

If the difference of the free energies of immersion is known, then for the case of ideal surface solution it is possible to use Eqs. (88)-(90), and by their simultaneous solution to calculate the value of n_1^e. The solution for the case of nonideal solutions was proposed by Myers (Larionov and Myers, 1971), who obtained the expressions relating the selectivity (46) and the excess adsorption n_1^e (45) to the activity coefficients in surface and bulk phases, the change in the free energy of immersion, and the free energies of immersion in pure components.

Since the change of the free energy of immersion is related to the extent of adsorption by the Gibbs equation (60), then substituting Eq. (42) into Eq. (60) and differentiating, we obtain

$$\frac{d(\sigma A/RT)}{d(\gamma_1 x_1)} = \frac{-m(S-1)}{\gamma_1\left(Sx_1 + n'_{m1}x_2/n'_{m2}\right)} \tag{96}$$

Solving Eq. (96) and taking into consideration Eq. (46) and the equation

$$x'_1 = \frac{Sx_1}{Sx_1 + x_2} \tag{97}$$

it is possible by numerical integration to determine the values of n_i^e.

Chmutov, Larionov, and others used Eq. (37) to relate the adsorption characteristics of solutions and individual substances (Larionov *et al.*, 1967b)

$$\frac{\theta\gamma'_1}{x_1\gamma_1}\left[\frac{x_2\gamma_2}{(1-\theta)\gamma'_2}\right]^{\beta} = \exp\left[\frac{\sigma_2^0 - \sigma_1^0}{n'_{m1}RT} + \beta - 1\right] \tag{98}$$

in which the coverage θ is related to the excess adsorption n_1^e by the expression

$$n_1^e/n'_{m1} = \theta(x_2+\beta x_1) - \beta x_1 \tag{99}$$

The right-hand side of Eq. (98) is a constant and can be found from the properties of pure components.

In order to calculate the extents of excess adsorption by this method, it is necessary to know the values of $\sigma_2^0 - \sigma_1^0$, n'_{m1}, β, and γ'_i.

As has been pointed out, if the values of σ_i^0 are not given, they can be found from the adsorption of individual substances or from the data on adsorption in other systems including the components of given system in different combinations, just as it is done in thermochemistry in calculating the heats of reactions. For example, if we have the data on adsorption in the systems including components 1-2 and 2-3, then the constants of Eq. (98) for the system 1-3 can be found (Sircar and Myers, 1971; Larionov and Kurbanbekov, 1972) from the condition

$$\sigma_1^0 - \sigma_3^0 = \left(\sigma_1^0 - \sigma_2^0\right) + \left(\sigma_2^0 - \sigma_3^0\right) \tag{100}$$

and

$$K_{13} = K_{12}\, K_{23}^{1/\beta_{12}}, \quad \beta_{13} = \beta_{12}/\beta_{23} \tag{101}$$

The determination of n'_{mi}, and hence of β is considered in Section II.C.

The determination of the values of γ'_i presents the greatest difficulty. Since there exist no methods of independent determination of the activity coefficients in the surface layer, these can be found only from the adsorption experiments, or calculated theoretically from some surface layer model.

Two simplest assumptions can be made regarding the concentration dependence of the activity coefficients in the surface layer: the adsorption solution is ideal $\gamma'_i = 1$ and $\gamma'_i = \gamma_i$ at $x'_i = x_i$. Both the assumptions require a thorough experimental verification.

The present authors and collaborators carried out systematic investigations on the properties of surface solutions on nonporous and large-pored adsorbents: carbon black graphitized at 3200°C with specific surface area with respect to nitrogen equal to 85 m^2/gm (ω_{N2} = 0.162 nM^2), aerosil, and large-pored silica gel (Kazarjan *et al.*, 1973a,b; Kurbanbekov *et al.*, 1973a-c).

It was shown that during adsorption from bulk solutions with insignificant positive deviations from Raoult's law, the activity coefficients in the surface layer calculated by means of Eq. (53) differs negligibly from 1 and therefore the assumption $\gamma_i = 1$ is justified for such systems. It was found for systems with larger deviations from Raoult's law in the bulk phase that on graphitized carbon black the adsorption from solutions is polymolecular. In the systems benzene-ethanol and isooctane-ethanol, the minimum thermodynamically possible thickness of the surface layer is equal to 3-4 monolayers. At these thicknesses of the adsorption layer, the activity coefficients of surface solutions, found by means of Eq. (52), almost exactly coincide with those for bulk solutions at $x_i' = x_i$. However, usually the concentration dependences of the activity coefficients are of complex character. The activity coefficients of surface solutions exhibit a tendency for negative deviations, particularly in the region of diluted surface solutions.

VI. DETERMINATION OF THE SPECIFIC SURFACE AREA BY THE METHOD BASED ON ADSORPTION FROM SOLUTIONS

These methods were considered by Schay (1970), Schay and Nagy (1972a), Schay *et al.* (1960), Larionov (1973), and Eltekov (1970).

Usually the problem of finding the specific surface area reduces to that of determining the capacity of monolayer and the molar area of the preferentially adsorbed component, which are related by the formula

$$A = \omega n_m^{'} N \tag{102}$$

The problems involved in the determination of the values of $n_m^{'}$ were considered in Section II.C.

The determination of ω is a common problem for adsorption of vapors and gases, and is considered in any monograph on adsorption of gases and vapors. The summarized data on molecular areas on solid adsorbents can be found in the work of McClellan and Harnsberger (1967).

For nonporous and macroporous adsorbents, it is possible by parallel investigation of adsorption from gas and liquid phases to determine the absolute value of the specific surface area of adsorbent (Larionov, 1970).

The method is based on the application of Eqs. (80) and (60). With the adsorption isotherm of solutions known, it is possible by means of Eq. (60) to determine the value of $(\sigma_2^0 - \sigma_1^0)A/RT$.

From the adsorption isotherm of vapors in the absence of capillary condensation, it is possible by means of Eq. (77) to determine the value of $\pi A/RT$. It follows from Eq. (79) that

$$\frac{\pi_{SLi}A}{RT} = \frac{\pi_{SVi}A}{RT} + \frac{\sigma_{LVi}A}{RT} \tag{103}$$

Therefore, the value of $(\sigma_2^0 - \sigma_1^0)A/RT$ found from the adsorption of solutions can be equated with the difference of the values of $\pi_{SLi}A/RT$ found from the adsorption of vapors

$$\frac{A\left(\sigma_2^0 - \sigma_1^0\right)}{RT} = \frac{A\pi_{SV1}}{RT} - \frac{A\pi_{SV2}}{RT} + \frac{A(\sigma_{LV1} - \sigma_{LV2})}{RT} \tag{104}$$

Therefore,

$$A = RT\,\frac{\dfrac{A\left(\sigma_2^0 - \sigma_1^0\right)}{RT} - \dfrac{A\pi_{SV1}}{RT} + \dfrac{A\pi_{SV2}}{RT}}{\sigma_{LV1} - \sigma_{LV2}} \tag{105}$$

In Eq. (105), the first term in the numerator is the difference between the specific free energies of immersion in pure components, and the two other terms are the free energies of saturation of adsorbent with vapor. All terms can be found by integration of the specific adsorption isotherms. The denominator is the difference between the surface tensions of pure liquids in equilibrium with their vapors. Therefore, by means of Eq. (105), it is possible to determine the specific surface area of adsorbent irrespective of the assumed values of the molecular area and the monolayer capacity.

Larionov (1973) gives the specific surface areas of graphitized carbon black for adsorption of five binary systems.

TABLE 2

System	A (m^2/gm)	($\Delta A/A$)
Benzene-isooctane	89.5	-6.4
Benzene-carbon tetrachloride	93.3	-2.4
Benzene-ethanol	105.4	+10.2
Carbon-tetrachloride-isooctane	88.7	-7.2
Isooctane-ethanol	101.0	+5.6
Mean value	95.6	
Nitrogen, $\omega_{N_2} = 0.162\ nM^2$	85.0	

As can be seen from Table 2, within the experimental errors and within the accuracy of the data on the concentration dependence of the activity coefficients (which are necessary for integration of Eq. (60)) and the accuracy of σ_{LV}, this method gives quite satisfactory agreement of the values of *A* found for different systems.

REFERENCES

Arnold, J. R. (1949). *J. Am. Chem. Soc. 71,* 104.

Berezkina, Yu. F., Kazarjan, S. A., Kurbanbekov, E., Larionov, O. G., and Chmutov, K. V. (1972a). *Zh. Fiz. Khim. 46,* 545; VINITI N 3775-71 Dep.

Berezkina, Yu. F., Kazarjan, S. A., Kurbanbekov, E., Larionov, O. G., and Chmutov, K. V. (1972b). *Zh. Fiz. Khim. 46,* 2166; VINITI N 4525-72 Dep.

Berezkina, Yu. F., Kazarjan, S. A., Kurbanbekov, E., Larionov, O. G., and Chmutov, K. V. (1972c). *Zh. Fiz. Khim. 46,* 2966; VINITI N 4498-72 Dep.

Berezkina, Yu. F., Kazarjan, S. A., Larionov, O. G., and Chmutov, K. V. (1973). VINITI N 4569-72 Dep.

Bering, B. P. (1957). Dissertation. Moskva.

Bering, B. P. (1972). In "Fizicheskaya adsorbtsiya iz mnogokomponentnykh faz," p. 106. Nauka, Moskva.

Bering, B. P., and Pokrovskii, N. L. (1969). *Izv. Akad. Nauk SSSR, Ser. Khim.,* p. 263.

Bering, B. P., and Serpinskii, V. V. (1970). *Izv. Akad. Nauk SSSR, Ser. Khim.,* p. 1232.

Bering, B. P., and Serpinskii, V. V. (1971). *Izv. Akad. Nauk SSSR, Ser. Khim.,* p. 2641.

Bering, B. P., and Serpinskii, V. V. (1972a). *Izv. Akad. Nauk SSSR, Ser. Khim.,* 166.

Bering, B. P., and Serpinskii, V. V. (1972b). *Izv. Akad. Nauk SSSR, Ser. Khim.,* p. 169.

Bering, B. P., and Serpinskii, V. V. (1972c). *Izv. Akad. Nauk SSSR, Ser. Khim.,* p. 171.

Bering, B. P., Myers, A. L., and Serpinskii, V. V. (1970). *Dokl. Akad. Nauk SSSR 193,* 119.

Bering, B. P., Serpinskii, V. V., and Surinova, S. I. (1973a). *Izv. Akad. Nauk SSSR, Ser. Khim.,* p. 3.

Bering, B. P., Serpinskii, V. V., and Surinova, S. I. (1973b). *Izv. Akad. Nauk SSSR, Ser. Khim.*, p. 7.

Blackburn, A., Kipling, J. J., and Tester, D. A. (1957). *J. Chem. Soc.*, p. 2373.

Bülow, M., Grossmann, A., and Schirmer, W. (1972). *Z. Chem. 12,* 161.

Butler, J. A. V. (1932). *Proc. Roy. Soc. 135A,* 348.

Coltharp, M. T., and Hackerman, N. (1973a). *J. Colloid Interface Sci. 43,* 176.

Coltharp, M. T., and Hackerman, N. (1973b). *J. Colloid Interface Sci. 43,* 185.

Cornford, P. V., Kipling, J. J., and Wright, E. H. M. (1962). *Trans. Faraday Soc. 58,* 74.

Davis, K. M. C., Deuchar, J. A., and Ibbitson, D. A. (1973). *J. Chem. Soc., Faraday Trans. 1, 69,* 1117.

Elovich, S. Yu., and Larionov, O. G. (1962a). *Izv. Akad. Nauk SSSR, Otd. Khim. Nauk,* p. 209.

Elovich, S. Yu., and Larionov, O. G. (1962b). *Izv. Akad. Nauk SSSR, Otd. Khim. Nauk,* p. 216.

Eltekov, Yu. A. (1970). *In* "Surface Area Determination," p. 291. Butterworth, London.

Eltekov, Yu. A., Khopina, V. V., and Kiselev, A. V. (1972). *J. Chem. Soc., Faraday Trans. 1, 68,* 889.

Everett, D. H. (1964). *Trans. Faraday Soc. 60,* 1803.

Everett, D. H. (1965). *Trans. Faraday Soc. 61,* 2478.

Kagiya, T., Sumida, Yu., and Tachi, T. (1971). *Bull. Chem. Soc. Japan 44,* 1219.

Kazarjan, S. A., Larionov, O. G., Chmutov, K. V., and Yudilevich, M. D. (1973a). *Zh. Fiz. Khim. 47,* 1619; VINITI N 5533-73 Dep.

Kazarjan, S. A., Larionov, O. G., Chmutov, K. V., and Yudilevich, M. D. (1973b). *Zh. Fiz. Khim. 47,* 2171; VINITI N 6049-73.

Khopina, V. V. (1972). In "Fizicheskaya adsorbtsiya iz mnogokomponentnykh faz," p. 106. Nauka, Moskva.

Kipling, J. J. (1965). "Adsorption from Solutions of Nonelectrolytes," p. 41. Academic Press, London and New York.

Kiselev, A. V. (1945). *Acta Physicochim. URSS 20,* 947.

Kiselev, A. V. (1946). *Zh. Fiz. Khim. 20,* 239.

Kiselev, A. V., and Pavlova, L. F. (1962). *Izv. Akad. Nauk SSSR, Otd. Khim. Nauk,* p. 2121.

Klinkenberg, A. (1959). *Rec. Trav. Chim. 78,* 83.

Kurbanbekov, E., Larionov, O. G., Chmutov, K. V., and Yudilevich, M. D. (1973a). *Zh. Fiz. Khim. 47,* 1617; VINITI N 5006-72 Dep.

Kurbanbekov, E., Larionov, O. G., Chmutov, K. V., and Yudilevich, M. D. (1973b). *Zh. Fiz. Khim. 47,* 1617; VINITI N 5025-72 Dep.

Kurbanbekov, E., Larionov, O. G., Chmutov, K. V., and Yudilevich, M. D. (1973c). *Zh. Fiz. Khim. 47,*1618; VINITI N 5532-73 Dep.

Larionov, O. G. (1966). *Zh. Fiz. Khim. 40,* 1796.

Larionov, O. G. (1970). *Zh. Fiz. Khim. 44,* 2963.

Larionov, O. G. (1972). In "Fizicheskaya adsorbtsiya iz mnogokomponentnykh faz," p. 123. Nauka, Moskva.

Larionov, O. G. (1973). In "Rasshirennye tezisy dokladov na IV Vsesoyuznoi konferentsii po teoreticheskim voprosam adsorbtsii," p. 111. Moskva.

Larionov, O. G., and Myers, A. L. (1971). *Chem. Eng. Sci. 26,* 1025.

Larionov, O. G., and Kurbanbekov, E. (1972). In "Fizicheskaya adsorbtsiya iz mnogokomponentnykh faz," p. 85. Nauka, Moskva.

Larionov, O. G., Tonkonog, L. G., and Chmutov, K. V. (1965). *Zh. Fiz. Khim. 39,* 2226.

Larionov, O. G., Chmutov, K. V., and Yudilevich, M. D. (1967a). *Zh. Fiz. Khim. 41,* 1011.

Larionov, O. G., Chmutov, K. V., and Yudilevich, M. D. (1967b). *Zh. Fiz. Khim. 41,* 2616.

Lu, B., and Lama, R. F. (1967). *Trans. Faraday Soc. 63,* 727.

McClellan, A. L., and Harnsberger, H. F. (1967). *J. Colloid Interface Sci. 23,* 577.

Matayo, D. R., and Wightman, J. P. (1973). *J. Colloid Interface Sci. 44,* 162.

Myers, A. L. (1968). *Ind. Eng. Chem. 60,* 45.

Myers, A. L., and Sircar, S. (1972a). *J. Phys. Chem. 76,* 3412.

Myers, A. L., and Sircar, S. (1972b). *J. Phys. Chem. 76,* 3415.

Myers, A. L., and Prausnitz, J. M. (1965). *AIChE J. 11,* 121.

Parfitt, I. D., and Thompson, P. C. (1971). *Trans. Faraday Soc. 67,* 3372.

Pavlutchenko, N. M. (1969). VINITI N 1088-69 Dep.

Payne, K., Sturdevant, G. A., and Leland, T. W. (1968). *Ind. Eng. Chem. Fund. 7,* 363.

Prigogine, J., and Marechal, J. (1952). *J. Colloid Sci. 7,* 122.

Rudzinski, W. (1973). *Chem. Phys. Lett. 19,* 221.

Rusanov, A. I. (1960). "Termodinamika poverkhnostnykh yavlenii," p. 92. Izd.LGU, Leningrad.

Rusanov, A. I. (1967a). "Fazovye ravnovesiya i poverkhnostnye yavleniya," p. 13. Khimiya, Leningrad.

Rusanov, A. I. (1967b). "Fazovye ravnovesiya i poverkhnostnye yavleniya," p. 84. Khimiya, Leningrad.

Rusanov, A. I. (1972). In "Fizicheskaya adsorbtsiya iz mnogokomponentnykh faz," p. 74. Nauka, Moskva.

Schay, G. (1970). In "Surface Area Determination," p. 273. Butterworth, London.

Schay, G. (1971a). *J. Colloid Interface Sci. 35,* 254.

Schay, G. (1971b). *Monatsh. Chem. 102,* 1419.

Schay, G. (1973). *J. Colloid Interface Sci. 42,* 478.

Schay, G., and Nagy, L. Gy. (1972a). *J. Colloid Interface Sci. 38,* 302.

Schay, G., and Nagy, L. Gy. (1972b). *In* "Fizicheskaya adsorbtsiya iz mnogokomponentnykh faz," p. 96. Nauka, Moskva.

Schay, G., Nagy, L. Gy., and Szekrenyesy (1960). *Periodica Polytechnica 4,* 95.

Schay, G., Nagy, L. Gy., and Szekrenyesy (1962). *Periodica Polytechnica 6,* 91.

Scherbakova, K. D., and Kiselev, A. V. (1947). *In* "Sbornik rabot po fizicheskoi khimii," p. 225. *Izd. Akad. Nauk SSSR*, Moskva-Leningrad.

Schuchowitzky, A. (1938). *Acta Physicochim. URSS 8*, 531.

Serpinskii, V. V. (1972). *In* "Fizicheskaya adsorbtsiya iz mnogokomponentnykh faz," p. 124. Nauka, Moskva.

Sircar, S., and Myers, A. L. (1970). *J. Phys. Chem. 74*, 2828.

Sircar, S., and Myers, A. L. (1971). *AIChE J. 17*, 186.

Sircar, S., and Myers, A. L. (1973). *AIChE J. 19*, 159.

Sircar, S., Novosad, J., and Myers, A. L. (1972). *Ind. Eng. Chem. Fund. 11*, 249.

Šprow, F. B., and Prausnitz, J. M. (1966). *Trans. Faraday Soc. 62*, 1105.

Wright, E. H. M. (1966). *Trans. Faraday Soc. 62*, 1275.

Wright, E. H. M. (1967). *Trans. Faraday Soc. 63*, 3026.

Wright, E. H. M., and Powell, A. V. (1972). *J. Chem. Soc., Faraday Trans. 1, 68*, 1908.

PHASE TRANSITIONS IN TWO-DIMENSIONAL AMPHIPHILIC SYSTEMS

J. Francois Baret

Departement de Physique des Liquides
Université de Provence
Marseille, France

PROGRESS IN SURFACE AND MEMBRANE
SCIENCE, VOL. 14

ISBN 0-12-571814-4

I. INTRODUCTION

At the interface between two nonmiscible fluids A and B there is a boundary zone. Mixtures between molecules A and B are energetically unfavored since the two fluids are not miscible but entropy imposes some degree of mutual penetration. Molecule A in the bulk is surrounded symmetrically by other A molecules, whereas in the boundary layer it may have some B molecules as neighbors. The interactions are no longer isotropic, an uncompensated force component, directed for molecules A toward the bulk of A, creates the anisotropy of the interfacial zone. To increase the interfacial area, new molecules have to move from the interior to the surface, this requires some energy to be spent and the interfacial tension has to be overcome. The more abrupt the crossing from one phase to the other, the more anisotropic are the interaction forces and the larger is the interfacial tension. An amphiphile has two parts, one that interacts positively with phase A (when A is water this part is called hydrophilic), and a second that interacts positively with phase B (when phase B is oil, it is called lyophilic). Since A and B are not miscible, these interactions are in general exclusive of one another, the hydrophilic part is lyophobic and the lyophilic part is hydrophobic. If an amphiphile C is dissolved in one of the two phases, it will tend to concentrate in the interface in order to minimize its free energy by immersing its A-phile part in A and its B-phile part in B. Doing so, it reduces the abruptness of the boundary between the two phases and the interfacial tension decreases. The adsorbed layer plays the role of a buffer between the two fluids, it diminishes the possibility of direct repulsive contacts between A and B, to replace them by A-C or B-C interactions that, because of the ambivalent character of C, are both attractive.

Unavoidably one part of an amphiphile molecule in solution should be in repulsive interaction with the solvent; this is the

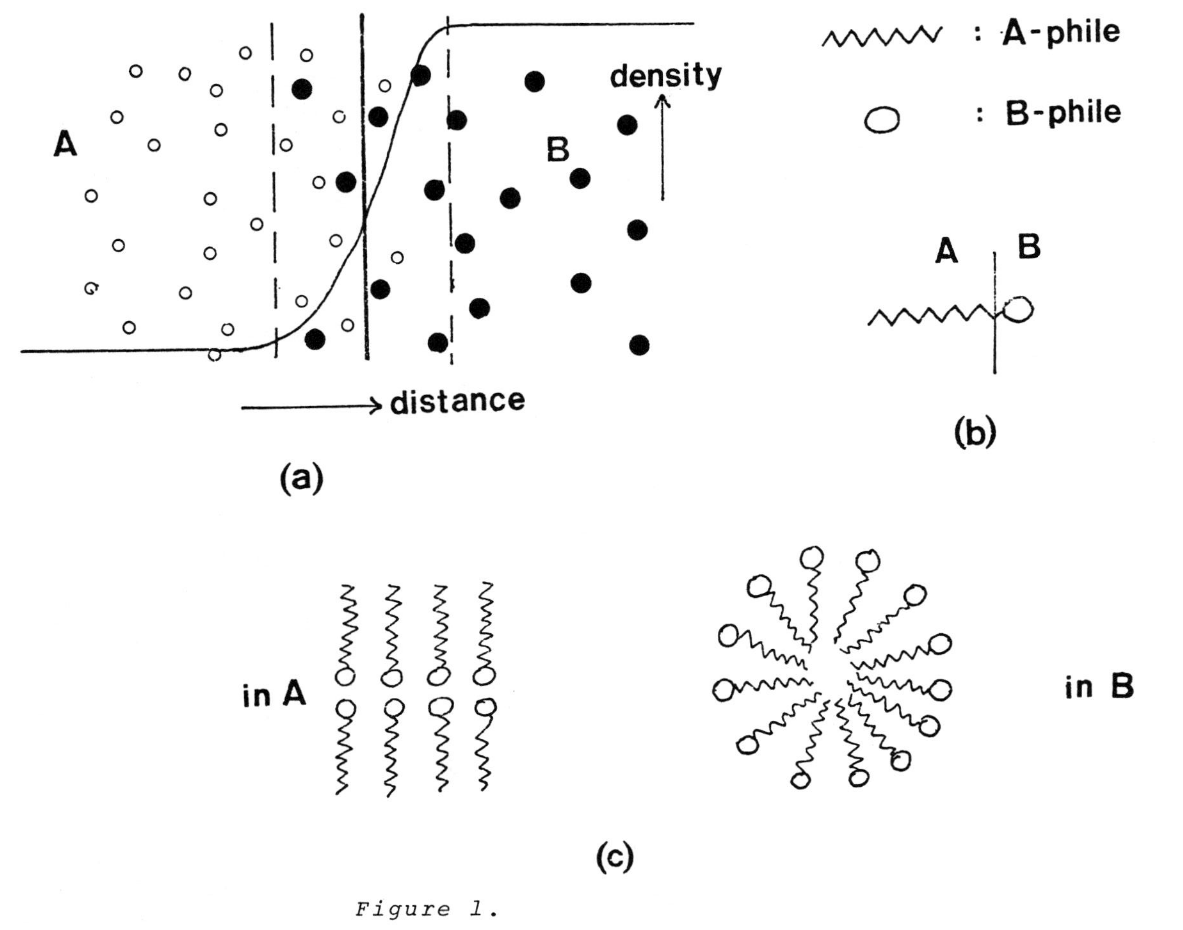

Figure 1.

cause of the appearance of mesomorphic structures. Indeed, beyond a certain concentration the amphiphiles are going to aggregate in order to put their solvent-repulsive parts out of the solvent [see Fig. 1(c)]. The first structure obtained is the spherical micelle. When the concentration increases it becomes cylindrical, then lamellar. If amphiphiles do not find an interface they will create one. The repulsion-attraction ambivalence of this kind of molecule confers to the surface (two-dimensional) a lower free energy state than the solution (three-dimensional). Therefore it is without surprise that we shall note the important role played by these compounds each time a two-dimensional structure bounds a liquid medium. Thus, the basic components of the biological membranes are the phospholipids that have a strong amphiphilic character because of their two hydrocarbon chains and their glycerol-derived polar head, organized in a double layer. The compounds with a single chain like soaps, fatty alcohols, or amines are important in many industrial processes including detergency, solubilization, or lubrication. To these polar compounds with fatty chains one should add now the amphiphilic polymers that have a wide range of uses because of the synthetic possibly of compounds with variable hydrophilic-hydrophobic balances either sequenced or grafted.

The function of the two-dimensional amphiphilic system at the surface where it is adsorbed, or at the one it has formed, will depend on the intermolecular organization, i.e., the interaction state of the molecules. Moreover, depending on the values of the thermodynamical variables this organization will change and phase transitions will occur. If a membrane bilayer can change from a dense state to a more loosely packed state, this should affect the transport properties of the membrane as well as the enzymatic activities of the bound proteins. Similarly, if a monolayer, spread on a water pool to retard evaporation, can exist in two states with different densities, evaporation will be better suppressed by the denser state. Monolayers and bilayers alike may exhibit phase transitions under varying thermodynamical conditions. These phase transitions, in spite of their differences, have similar characters whose comparison may help in elucidating their nature. We shall analyze separately the two two-dimensional amphiphilic systems that received the most attention: the monolayer spread at the air-water interface and the phospholipid bilayer. From there we shall see that the theoretical interpretation of their phase transitions is complex and that, up to this day, no universally accepted model allows us to describe both bilayer and monolayer transitions. Incidentally, we shall examine the much less-studied system of the layer adsorbed at an oil-water interface and we shall find there a clue to resolve the hiatus between bi- and monolayer data.

II. BILAYERS

When mixing a phospholipid with water, depending on the concentration and the temperature, different types of structures are obtained (Fig. 2). The more ordered phase, the gell, has a lamellar structure with the all-trans chains closely packed. In the liquid-crystalline phase (mesomorphic), the molecules are also organized in alternating layers of water and lipid but here the chains resemble those in a liquid hydrocarbon, i.e., with several gauche bonds (Williams and Chapman, 1970). These lamellar structures have received much attention because the lipid is organized in bilayers as we expect it to be in biological membranes. In particular the gel-crystal liquid transition has been extensively studied by different techniques we shall examine now.

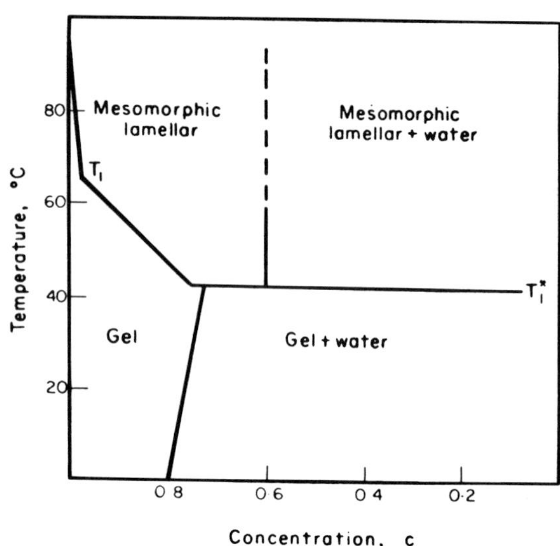

Fig. 2. Phase diagram of the DPPC-water system (Williams and Chapman, 1970).

A. Techniques

1. Differential Scanning Calorimetry (DSC). In this technique two compartments are used, one for the sample and the other for the reference material. In order to maintain the two compartments at the same temperature, different energies have to be furnished. It is this differential energy that is recorded (van Dijck, 1977). DSC has been used extensively to characterize thermotropic process in bilayer systems (Ladbrooke and Chapman, 1969). Most diacyl phospholipids organize in bilayers upon hydration. When these bilayers are subjected to DSC analysis, a sharp transition peak is observed at a particular temperature with an enthalpy change that is characteristic for the phospholipid under study (Fig. 3). For phosphatidylcholine (PC) another small peak is observed at lower temperature; it corresponds to the so-called pretransition. Gel-to-liquid-crystalline phase transitions have been demonstrated in aqueous dispersions of PC (Keough and Davis, 1979;

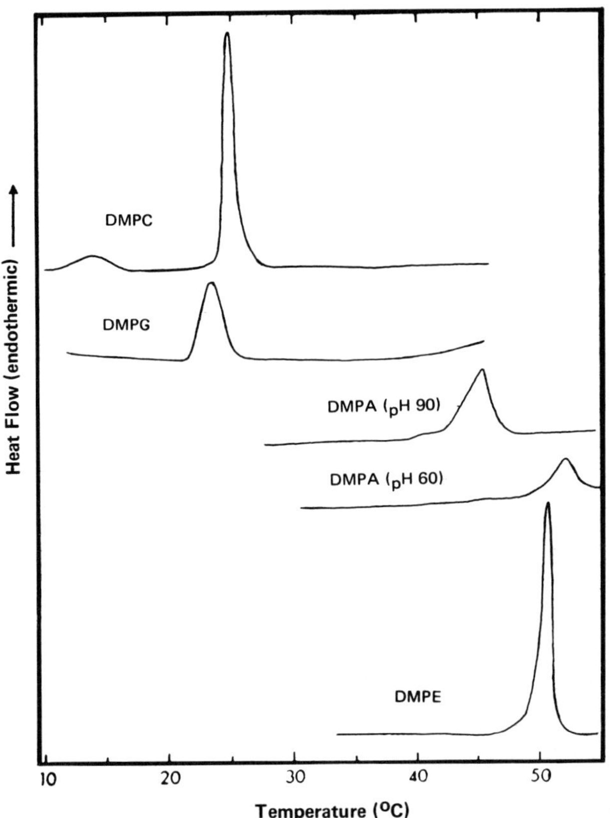

Fig. 3. DSC thermograms of various dimyristoyl phospholipids (Papahadjopoulos, 1977).

Hinz and Sturtevant, 1972; Barton and Gunstone, 1975), phosphatidic acid (PA) (Trauble and Eibl, 1974; Jacobson and Papahadjopoulos, 1975), phosphatidylethanolamine (PE) (Shimshick and McConnell, 1973; Lee, 1975b; Blume and Ackerman, 1974; Chapman *et al.*, 1974), phosphatidylglycerol (PG) (Verkleij *et al.*, 1974; Kimelberg and Papahadjopoulos, 1974; Findlay and Barton, 1978), phosphatidylserine (PS) (van Dijck *et al.*, 1975; MacDonald *et al.*, 1976), phosphatidylinositol (PI) (Schnepel *et al.*, 1974), etc. Data for some of these transitions are collected on Table I (from Lee, 1977a; van Dijck, 1977).

Table I

Phospholipid		Transition temperature (°C)	Latent heat (kcal/mole)
DSPC		54-58	10.7
DPPC		41-42	8.6-9.7
DMPC		23-24	6.2-6.8
DPPE		60-65	8-8.5
DMPE		47.5-51	6.5
dioleoyl-PC	(cis)	(-22) - (-14)	7.6-11.2
dioleoyl-PE	(cis)	-16	4.5
dielaidoyl-PC	(trans)	9.5	7.3
dielaidoyl-PE	(trans)	35	7.0

For the same head group, lipids with longer chains have higher transition temperatures than shorter ones, unsaturated chains have lower transition temperatures than saturated ones and cis-unsaturated chains have lower transition temperatures than trans-unsaturated ones (Oldfield and Chapman, 1972). PC and PG both have much lower transition temperatures than PA and PE. This is surprising since PC and PE are zwitterionic, whereas PG and PA may only carry the charge of the phosphate group. It was proposed that hydrogen bonding may explain this behavior (Jacobson and Papahajopoulos, 1975). There is also an even-odd effect of the chain length (see Fig. 4).

The DSC studies mentioned all concern multibilayer dispersion. By a proper sonication, unilamellar vesicles may be obtained. The thermotropic behavior of these vesicles is different from that of multilayers (Suurkuusk *et al.*, 1976; Papahadjopoulos *et al.*, 1975; Kantor *et al.*, 1977). The peculiar results obtained with the single-walled vesicles could be a consequence of a fusion process that enlarged the diameter of the vesicles in the preparation, since it seems that the radius of curvature of the bilayer has some effect on the phase transition (van Dijck *et al.*, 1978).

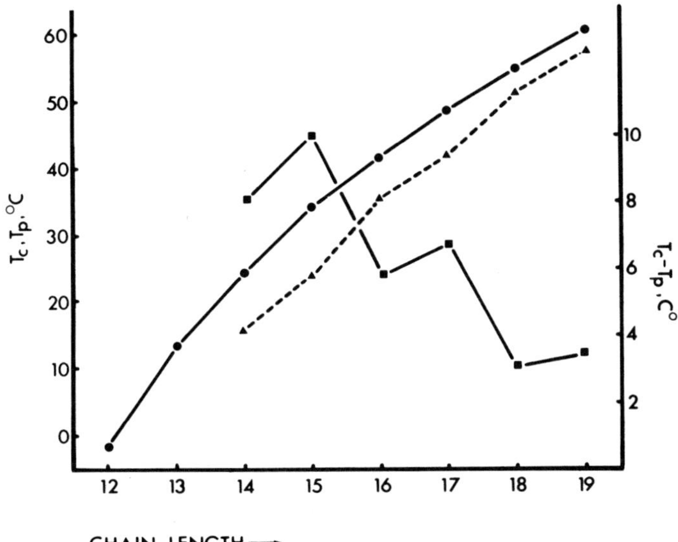

Fig. 4. Main transition and pretransition temperature versus chain length for diacyl phosphatidyl choline. Main transition (●), pretransition (▲), difference (▪) (Silvius et al., 1979).

2. Electron Spin Resonance (ESR). The measurements are done with a nitroxide spin label that has a different spectrum when solvated in water than in lipid. Thus, it is possible to obtain the partition coefficient of the label between the two phases. Tempo has been the most widely used spin label to study the gel-to-liquid crystal transition of phospholipids in aqueous dispersion because of its high solubility in the fluid liquid crystal and its low solubility in the gel (Hubbell and McConnell, 1971). As can be seen on Fig. 5, the experiments give two amplitudes H and P of the high-field nitroxide hyperfine signals. To a first approximation H is proportional to the amount of spin label dissolved in the bilayer and P is proportional to the amount dissolved in water. The change in the relative values of H and P as a function of temperature for DPPC reflects the fact that this phospholipid undergoes a transition from the gel state to the liquid crystalline state. In Fig. 6 is plotted the fraction of

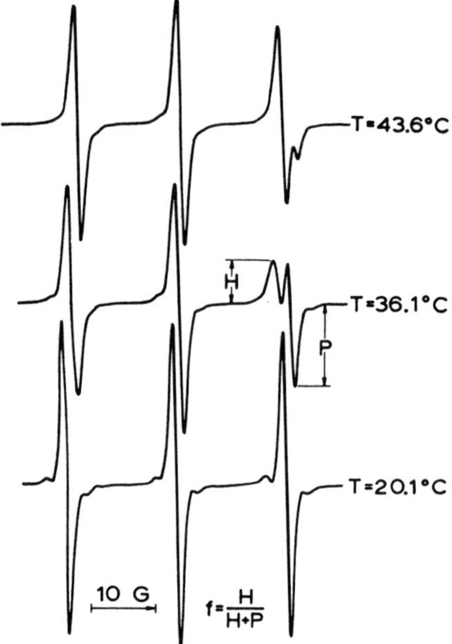

Fig. 5. Paramagnetic resonance spectra of Tempo dissolved in an aqueous dispersion of DPPC (Shimshick and McConnell, 1973).

spin label dissolved in the lipid bilayer

$$f = H/(H+P)$$

as a function of temperature for four phospholipids. These curves exhibit abrupt changes in the magnitude of the spectral parameter

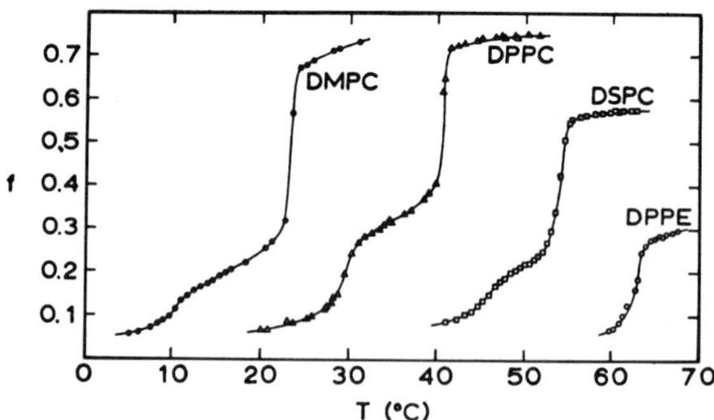

Fig. 6. The Tempo spectral parameter f versus temperature for aqueous dispersion of different phospholipids (Shimshick and McConnell, 1973).

at temperatures corresponding to the calorimetrically measured phase transition temperature.

	DMPC	*DPPC*	*DSPC*	*DPPE*
esr (Tempo)	*23.2*	*40.5*	*54.0*	*63.0*

This is confirmed by Marsh (Marsh *et al.*, 1976) for DMPC from the fractional partition of Tempo but also with a spin-labeled DPPC intercallated between the bilayer molecules and whose outer splitting ($A_{||}$) in the spectrum is a measure of the amplitude of motion of the phospholipid chains (Fig. 7). Other authors have determined the transition temperatures of various lipid bilayer by esr (Trudell *et al.*, 1974a; Sackmann *et al.*, 1973). In a review by Träuble (1972) on the optical and spin label methods applicable to the study of lipid phase transition, there is an analysis of the information that may be obtained from esr data, in particular, with respect to the dynamical parameters of the membrane. Marsh

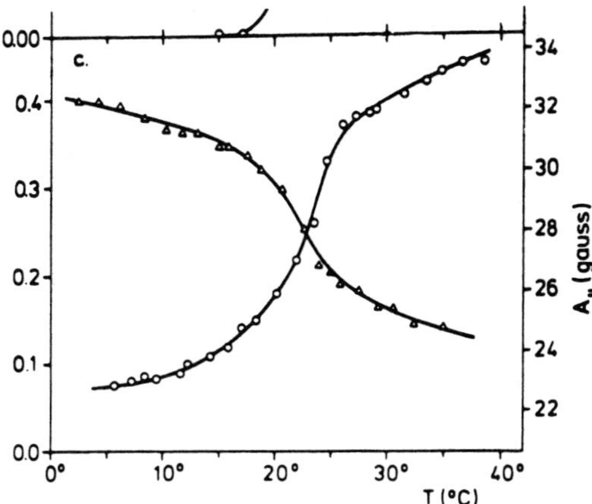

Fig. 7. Temperature dependence of spin-label indicators of the phase transition of DMPC. (○) Fractional partition of Tempo, (△) $A_{||}$ hyperfine splitting of DPPC spin-label (Marsh et al., 1976).

(1974) has used the method of interacting spin label pairs to measure the lateral expansion in phospholipid bilayers at the transition temperature.

3. *Nuclear Magnetic Resonance (NMR).* NMR is a well-adapted technique for studying molecular mobility in bilayer systems. Proton magnetic resonance is difficult to use because of the relatively broad and overlapping signals obtained, but nuclei with smaller gyromagnetic ratio like ^{13}C or ^{31}P can be of interest (Frenzel *et al.*, 1978). Trahms and Boroska (1979) using pulse NMR have studied the transitions of DPPC and they show that the second moment of the proton absorption line displays significant changes at 42°C and at about 35°C.

4. *Fluorescence.* Several fluorescence probes may be used to detect phase transitions of lipid bilayers. Some, like *N*-phenylnapththylamine are supposed to penetrate into the hydrophobic core of the bilayer, whereas some other like 1-anilino-8-napthalenesulfonate (ANS) are expected, due to their amphiphilic nature, to bind at the interface with the polar group in the aqueous phase (Träuble, 1972; Oldfield and Chapman, 1972). Eibl and Blume (1979) have shown the influence of charge on phosphatidic acid bilayer phase transition by measuring the fluorescence intensity of *N*-phenylnaphthylamine (Fig. 8). Other probes like chlorophyll or dehydroergosterol or *n*-(9-antroyloxy) fatty acids, have also been used (Lee, 1978; Rogers *et al.*, 1979; Haigh *et al.*, 1979). Suurkuusk *et al.* (1976) have measured the fluorescence depolarization of 1,6-diphenyl-1,3,5-hexatriene (DPH), which is a fluorescent probe specific for the hydrophobic region (Shinitzky and Barenholz, 1974). The incident beam is polarized and the rotational motion of the molecules reduces this polarization. Hence, the value of the fluorescence intensity perpendicular to the plane of polarization of the excitation beam characterizes the fluidity of the lipid environment of the DPH. As can be seen on Fig. 9, the correspondence between calorimetric and fluorescence data is

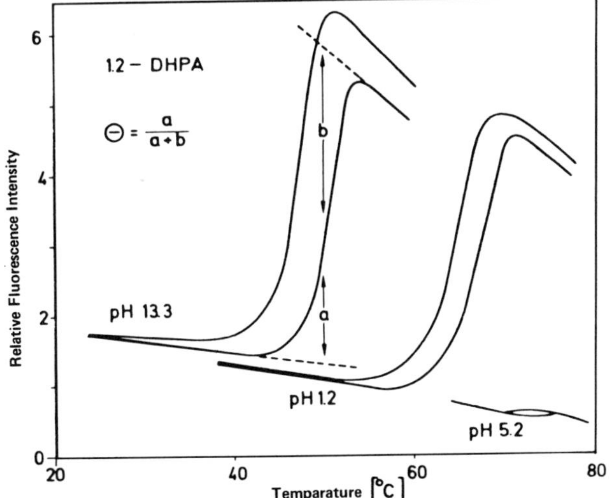

Fig. 8. Fluorescence indication of the phase transition of 1,2-dihexadecyl-sn-glycerol-3-phosphoric acid at different degree of ionization (Eibl and Blume, 1979).

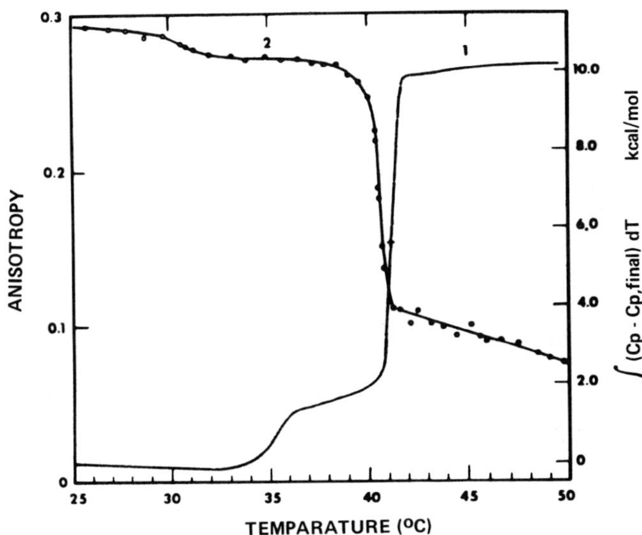

Fig. 9. Comparison of calorimetric and fluorescence results for DPPC multilamellar vesicles: (curve 1) excess enthalpy; (curve 2) fluorescence anisotropy of DPH (Suurkuusk et al., 1976).

excellent in the region of the main phase transition but not so good for pretransition. The main transition temperatures are, respectively, 41.2 and 41.1, the pretransition temperature are, respectively, 25.2-33.9 and 33.2-36.7. Lakowicz *et al.* (1979) have extended the measurement of depolarization fluorescence of DPH to DPPC and DSPC bilayers.

5. *Light scattering and adsorption.* Light is scattered by the fluctuations of the medium and therefore at a phase transition a change in the scattered intensity is expected. The intensity of the light scattered by lipid bilayers shows an increase when going from the fluid phase to the gel phase (Overath and Träuble, 1973; Eibl and Blume, 1979; van Dijck *et al.*, 1978). The change in turbidity accompanying the phase transition that allow light scattering measurements can be followed by light absorption determination (Kamaya *et al.*, 1979). It is also possible to follow the change in an absorption band of a label (BTB) that binds to the bilayer. The number of binding sites increases sharply at the phase transition and the optical density increases accordingly (see Fig. 10). The figure shows that the different optical methods give concordant results even though the transition temperature with BTB is somewhat lower ($\approx 40.5^{\circ}C$), indicating that this label causes some perturbations of the system. This kind of perturbation is the usual drawback of any techniques using chemical labels.

6. *X-Ray Diffraction.* At the main transition temperature of lipid multilayer, the sharp line seen on their x-ray diffraction pattern at about 4.2 Å moves to a larger, more diffuse line (≈4.5 Å) for DPPE (Harlos, 1978), for DPPC (Gottlieb and Eanes, 1974; Janiak *et al.*, 1976), for PA (Jähnig *et al.*, 1979), and the small angle reflections change toward smaller lamellar spacing indicating that the lipid is in the liquid-crystal state. Below the main transition temperature the wide angle reflections show a temperature dependence. For instance, for DSPE at low temperature a strong line at 4.18 Å is followed by a weaker reflection at 3.38 Å which moves,

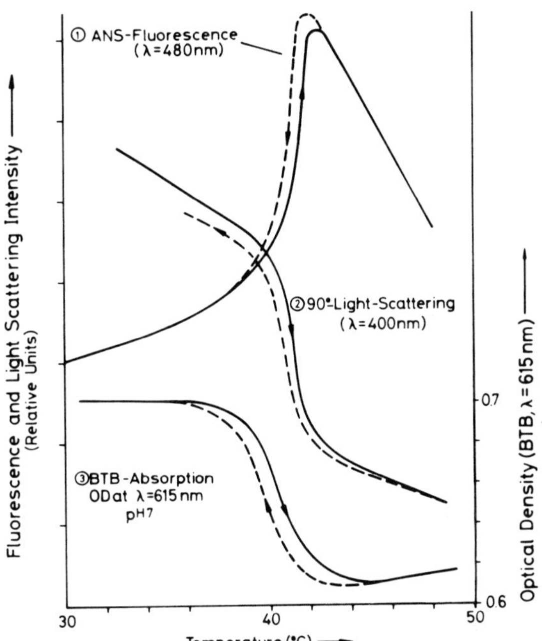

Fig. 10. Phase transition of sonicated DPPC dispersion detected by different optical techniques (Träuble, 1972).

with increasing temperature, toward the main band (Fig. 11). This change in chain packing would be characteristic of the pretransition. For phosphatidylcholine at temperature above the pretransition, the hydrocarbon chains assume also an hexagonal configuration but the structure is slightly different (Tardieu *et al.*, 1973) mainly because of an important tilt angle (28°) that has not been reported for phosphatidylethanolamine.

From a membrane of Mycoplasma laidlawii, Engelman (1970) has obtained x-ray diffraction patterns showing that the lipid chain experiences a phase transition. Below 40°C there is a strong sharp ring near 4.2 Å and above, a broad strong ring near 4.6 Å.

7. *Dilatometry*. The transition involves an increase in the fluidity of the hydrocarbon interior of lipid bilayers and therefore a simultaneous increase in specific volume. For instance, for DPPC there is a change of 0.0350 ml/g at 41.8°C (MacDonald, 1978). In addition to the abrupt thermal expansion accompanying

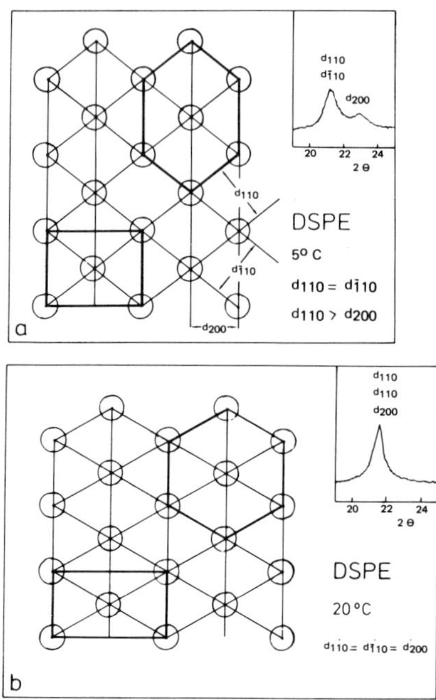

Fig. 11. Model for the hydrocarbon chain packing of DSPE below (a) and above (b) the pretransition. Inserts show the corresponding wide angle x-ray diffraction line (Harlos, 1978).

the main transition, there is another marked change in slope at a temperature a few degree below that corresponds to the pretransition [Fig. 12 (Laggner and Stabinger, 1976)].

8. Raman and infrared spectroscopies. Vibrational spectroscopy techniques allow us to identify which molecular bonds are involved in a conformational change if the spectral bands have been previously assigned. Therefore it may be expected that a wealth of information is to be obtained from Raman and infrared spectra of lipid bilayers experiencing a phase transition (Lippert and Peticolas, 1971, 1972; Spiker and Levine, 1975, 1976; Mendelsohn and Maisano, 1978; Cameron and Mantsch, 1978; Casal *et al.*, 1979). Verma and Wallach (1976) show that intense bands at 2850

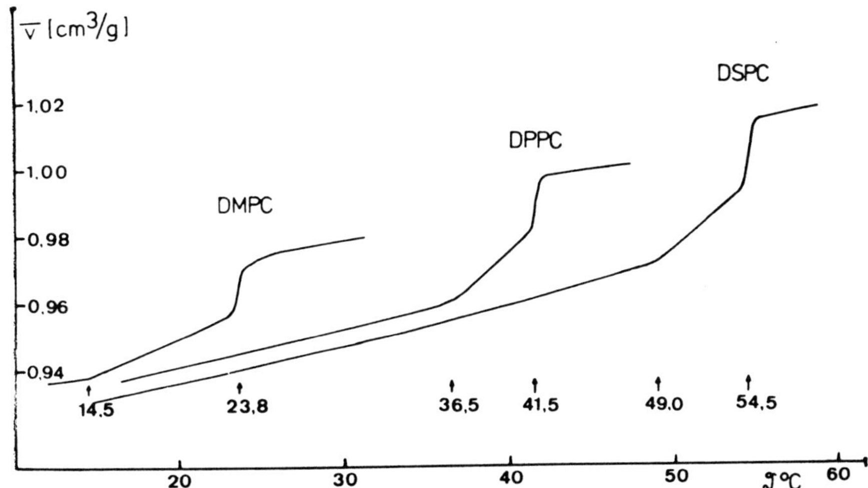

Fig. 12. Partial specific volume versus temperature curves for different lecithins (Laggner and Stabinger, 1976).

and 2885 cm^{-1} in the DMPC liposomes spectra, arising from symmetric and antisymmetric CH stretching of methylene group, have a ratio that drops sharply at 21°C. Mendelsohn *et al.* (Mendelsohn and Maisano, 1978; Mendelsohn *et al.*, 1976) determine the trans-gauche population ratio by comparing the mode at 1130 cm^{-1}, assigned to a C-C vibration of the carbon chains in the all-trans conformation, to the 1080 cm^{-1}, which is assigned to the C-C stretching vibration in chain-containing gauche rotamers. They also obtain a better determination of the ratio of the C-H stretching vibration (2880 and 2850 cm^{-1}) by using deuterated derivatives.

B. Bilayers in Interactions

With the techniques described above it has been possible to study not only phase transitions of single-component bilayers but also of more complex systems. Indeed, the aim of most of the studies on bilayers is to gain insight about biomembranes, and therefore it is of much interest to study model systems that contain several components of the biological membrane. So, the second step, in this attempt of modelization, after the study of single-component bilayers, will be obviously to add a second component. This second component may be another phospholipid or cholesterol, but it may be also a drug or a protein.

1. Phase transitions of phospholipid mixtures. In a bulk phase the melting of a single component proceeds by a sharp transition from the solid to the liquid state, whereas for a binary mixture the melting of solid occurs over a broad temperature range. For single-component bilayers the transition from gel to liquid-crystal takes place over a range of 1-2°C. This is apparently in contradiction with the concept of first-order phase transition that is underlying all the works of the field. Indeed, a classical first-order phase transition (like three-dimensional melting) occurs at a well-defined temperature that does not change from the appearance of the first drop of liquid until the whole sample is melted. In such a transition the two phases involved are separate and distinct. But in lipid bilayer one phase is necessarily generated in the matrix of the other and the two phases cannot be considered to be independent (Lee, 1977b). The coexistence of two lipid phases, gel and liquid-crystal, within the same bilayer, creates for the boundary molecules unusual states of packing. If the size of the distinct patches is small, this interfacial effect is likely to be important and, as Marsh *et al.* (1976) show, may account for the width of the transition. We shall return to this point when examining the theoretical models; for the time being, the width of the transition of a single-component bilayer will be considered as a negligible second-order effect. On the other hand, with bilayers made of a phospholipid mixture the range over which the transitions occurs is an important feature of the system that allows us to draw the phase diagram, that is, the temperature-concentration plot of the lower temperature at which the transition begins and the upper temperature at which it ends. Thus, two curves are defined, liquidus and solidus. Above the liquidus curve the lipid mixture is in the liquid-crystalline state, below the solidus curve it is in the gel state,

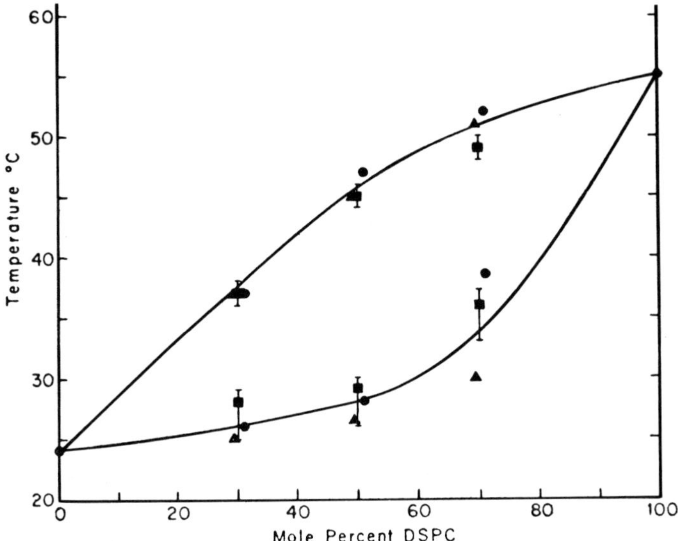

Fig. 13. Phase diagram of the DMPC-DSPC mixture obtained by NMR, compared to calorimetry and fluorescent data (Gent and Ho, 1978).

between the two curves will be a mixture of gel and liquid-crystal of different compositions.

There has been several experimental determinations of the phase diagram of aqueous dispersion of phospholipid mixtures. For instance, in Fig. 13 the phase diagram of the DMPC-DSPC mixture obtained by NMR of fluorine 19 fatty acid incorporated in the bilayer is compared with those obtained by other techniques (Gent and Ho, 1978). This diagram is typical of ideal behavior. Studies with other binary phospholipid dispersions in water, such as studied by DSC (Findlay and Barton, 1978; Blume and Ackerman, 1974; Stewart *et al.*, 1979) or by ESR (Trudell *et al.*, 1974b; Shimshick and McConnell, 1973) or by fluorescence (Lee, 1975a; Lentz *et al.*, 1978; Sklar *et al.*, 1979), may reveal nonideal be-behavior. In fact, ideal mixing in both phases occurs only with similar polar head groups and close chain lengths (Oldfield and Chapman, 1972; Phillips *et al.*, 1970).

2. The effect of cholesterol. Cholesterol is an important component of many biological membranes; it is present in most eucaryotic membranes but its role is still not clearly understood. It is therefore of interest to study the effect of cholesterol on phospholipid bilayers. When mixed with phospholipid dispersion in water, the most striking effect of cholesterol is that it removes completely the phase transition on the thermogram for a concentration above 33 mole % cholesterol (Fig. 14) (Ladbrooke *et al.*, 1968; Hinz and Sturtevant, 1972b; Estep *et al.*, 1978, 1979; Calhoun and Shipley, 1979). Although it is not anymore detectable by DSC, the transition still takes place as it can be seen on Raman spectra (Lippert and Peticolas, 1971). What is lost is the cooperative character of the transition, there is no latent heat associated with the transition, the transition of a phospholipid-cholesterol dispersion in water is not of the first-order type. The hypothesis that a 1:2 stoichiometric complex between cholesterol and phospholipid

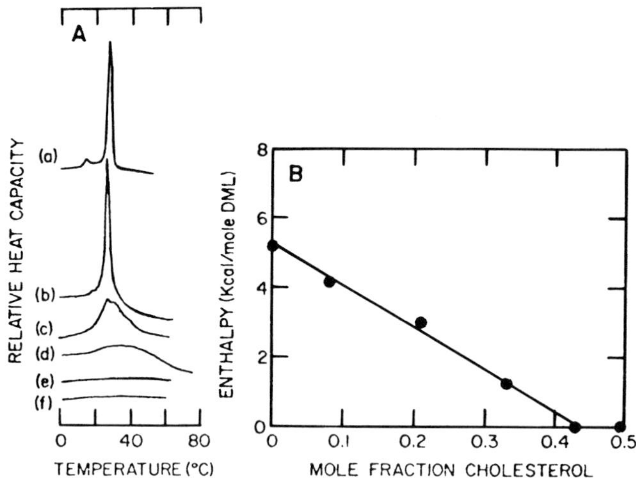

Fig. 14. Effect of cholesterol on DMPC thermograms. (a) 0.00, (b) 0.083, (c) 0.21, (d) 0.34, (e) 0.43, (f) 0.49 mole fraction of cholesterol (Calhoun and Shipley, 1979).

can be formed seems well supported (de Kruijff *et al.*, 1972; Ladbrooke *et al.*, 1968; Hinz and Sturtevant, 1972b; Engelman and Rothman, 1972). Below 33 mole % cholesterol two phases would be present: a pure phospholipid phase and a mixed phase at a molar ratio of 1:2 (cholesterol:phospholipid). But several data are inconsistent with this conclusion, thus NMR experiments (Phillips and Finer, 1974) suggest that in mixtures containing less than equimolar amount of cholesterol, discrete regions of 1:1 lipid-cholesterol complex separate out. Some fluorescence experiments seem consistent with a 1:1 complex (Lee, 1976) whereas some other ones (Rogers *et al.*, 1979) seem to support the 1:2 complex (Fig. 15). However all authors agree on the dual role of cholesterol in bilayer: It reduces the fluidity of the lipid chain [at least the mobility of the first eight carbon atoms (Hubbell and McConnell, 1971)] in the liquid-crystalline state and increases its fluidity in the gel state (Lee, 1976). When cholesterol interacts with

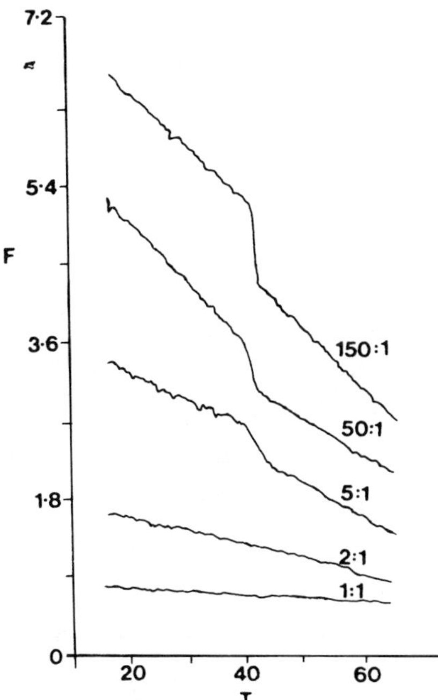

Fig. 15. Plot of fluorescence intensity versus temperature for dehydroergosterol-DPPC mixture (Rogers, 1979).

binary phosphatidylcholine mixtures showing two separated peaks on DSC thermograms, low concentration of cholesterol abolishes only the transition peak of the component with the lowest transition temperature, cholesterol is preferentially excluded from the gel phase regions of the bilayer. In the case of PC-PE mixtures, cholesterol show a specific preference for PC even though the PC has a higher transition temperature (van Dijck, 1977).

3. Lipid-protein interactions. Biomembranes may be considered to be a complicated lipoprotein system. Many studies have attempted to shed some light on the complex interactions between proteins and lipids, among them several are concerned with the phase transition of the lipid bilayer. This is done in two ways, either the effect of lipid phase transitions on the activity of a membrane-bound enzyme is studied or the influence of bound proteins on bilayer phase transitions is observed.

Enzymatic activity and phase transition. The membrane is both a border and link for a cell or an organelle with respect to its environment. Therefore most of the enzymes bound to the membranes play a role in some kind of exchange between inside and outside. The best documented type of exchange is transport membrane and it is well established now that correlations exist between the rate of transport and the physical state of the lipid bilayer (Overath and Träuble, 1973; Linden *et al.*, 1973; Esfahani *et al.*, 1971; Machtiger and Fox, 1973). It is possible to change the composition of a natural membrane by feeding the animal with a special diet, thus the transition temperature of the bilayer is changed as it can be seen on the plots of the succinate oxidase activity (a membrane-bound enzyme) shown on Fig. 16. Similarly, ATPases that are also membrane-bound enzymes, show an activity that changes drastically at the transition point. This has been shown either directly from calorimetric studies of membrane suspensions (de Kruijff *et al.*, 1973, 1974) or with reconstituted

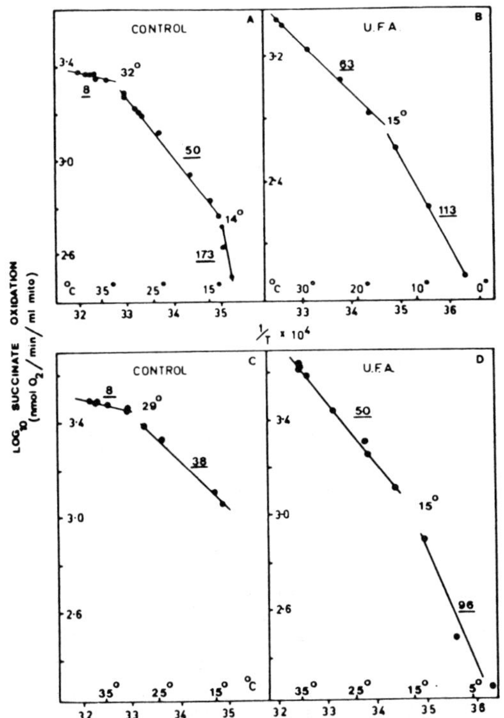

Fig. 16. Succinate oxidase activity versus temperature for different lipid composition of the membranes, from sheep liver (A and B) and kidney (C and D). The numbers underlined, adjacent to the straight lines, are Arrhenius plot activation energy in kilojoules per mole. Plots A and C are from control sheep and plots B and D are from sheep fed with the unsaturated fatty acid (UFA)-rich diet (McMurchie and Raison, 1979).

systems from synthetic phospholipids. For instance, Kimelberg and Papahadjopoulos (1974) have found breaks in the activity plots of $(Na^{+}+K^{+})$-ATPase at temperatures close to the phase transition temperature of the phospholipids used to form the bilayer. This has been confirmed by Bruni *et al.* (1975) for F_1-ATPase.

Protein effect on the thermotropic behavior of bilayers. Generally, when a protein interacts with a bilayer the transition temperature and the transition enthalpy are changed. The amount and the sign of this change depend on the type of interaction involved between lipid and protein. If the interaction is electrostatic, there is a small increase in temperature and enthalpy. If the protein has a large hydrophobic region and is therefore embedded in the hydrophobic part of the membrane, it has little effect on the transition temperature but leads to a decrease in the enthalpy value. There is a third class of proteins representing an intermediate situation, proteins of this category adsorb at the bilayer interface, i.e., neutralize the surface charges and penetrate partially into the hydrophobic region of the bilayer, in this case both transition temperature and enthalpy are decreased (Verkleij *et al.*, 1974; Papahadjopoulos *et al.*, 1975; London *et al.*, 1973; Janoff *et al.*, 1979). The effect of a synthetic polypeptide on a bilayer depends strongly on the polar character of the aminoacids used (Bach *et al.*, 1978; Susi *et al.*, 1979).

4. The effect of drugs, anesthetics and pressure. It is now well documented that anesthetics and pressure counteract each other's effect (Truddell *et al.*, 1975; Halsey and Wardley-Smith, 1975). This is true in particular for bilayer phase transitions. The anesthetics decrease the phase transition temperature, whereas the pressure increases it. This has been verified with DPPC liposomes interacting with diethyl ether, trichloroethylene, and methylflurane (MacDonald, 1978) and with halothane and lidocaine by Kamaya *et al.* (1979) (see Fig. 17). Anesthetics are not the only compounds expected to have a fluidizing effect on the membrane. There is evidence that some antidepressant-active drugs may induce the gel-liquid-crystalline transition (Frenzel *et al.*, 1978). Lee (1978) shows also that tetracaine and procaine decrease the transition temperature of DPPC bilayers.

To summarize this first section, I would like to note some of the features of the bilayer phase transition that appear as fundamental. The transition is sharp, i.e., well defined in temperature, when the phospholipid is pure. Adding another phospholipid species, or cholesterol, or a drug may shift the transition temperature but principally increases the temperature range

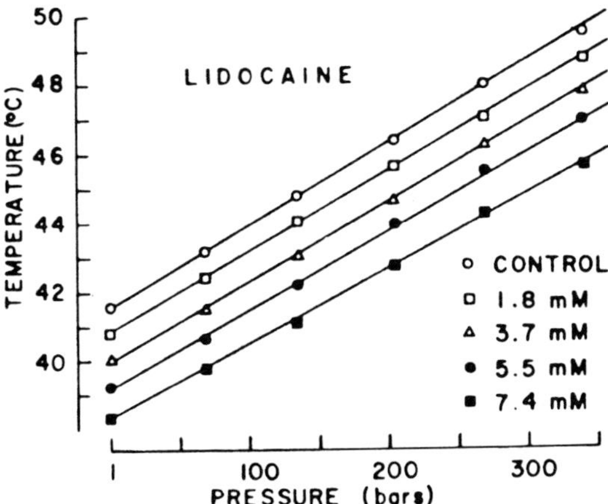

Fig. 17. The effect of pressure on the phase transition temperature of DPPC vesicles with lidocaine at different concentrations (Kamaya et al., 1979).

over which the transition takes place. The introduction of a second component reduces the cooperativity of the transition. There is another transition at lower temperature that implies much less energy and seems associated with a change in the molecular packing. This pretransition appears only with single-component bilayers.

III. MONOLAYERS

That a monolayer is half of a bilayer may seem a rather trivial statement. However, as we shall see, taking this assertion too literally may be misleading. Indeed, bilayers are made of two interacting monolayers and because of this interaction the two systems behave differently in many ways. With regard to the phase transitions there is one basic difference between the two systems: the bilayer phase transition is of the first-order type (with a latent heat of transition), whereas, in most of the cases, monolayer phase transitions are second-order (with no change in area at the transition point). We shall discuss this point in the theoretical section, for the time being we shall examine the data.

A. Air-Water Interface

When deposited at the air-water interface an amphiphilic substance spreads sponteneously to form a monomolecular layer. If the substance is neither soluble in water nor volatile, this monolayer is stable. By a simple but delicate device the area of the monolayer may be varied and the interfacial tension recorded. Such a two-dimensional compressing apparatus is generally known as a Langmuir trough. As a matter of fact experiments on monolayers began even before Langmuir and may be traced back to Agnes Pockels (Raleigh and Pockels, 1891). So, it is not in my intention to give an exhaustive review of the data on monolayers since that time; this would be a tremendous job! Fortunately, we have at our disposal Gaines' book (Gaines, 1961) which is a comprehensive analysis of the works in this field up to 1965. Therefore my intent is to limit myself to the data that recently have been published. Most of the early experiments were dealing with relatively simple compounds like fatty acids or fatty alcohols. Since the number of substances of this kind that can be spread as monolayers is limited, almost everything that could be done with pure single-chain fatty substances has been done before World War II and has been since then reviewed several times. Even though, now and then, a paper is published with isotherms of a single-chain, single-component monolayer (Goddard and Kung, 1973), by and large, in the recent years the activity in the monolayer field has been concentrated on more complex substances or on mixture of single-chain substances.

1. Two-chain amphiphiles. As we saw in Section I the basic components of membranes are phospholipids that are two-chain amphiphiles. Therefore, since they are available in a pure enough form to give reproducible data, phospholipids have been spread on water to form monolayers (Fig. 18). The compressional isotherms of these monolayers have the same basic features as the fatty acid isotherms; viz. when the area decreases, the surface pressure (i.e., the difference between the pure water interfacial tension and the mono-

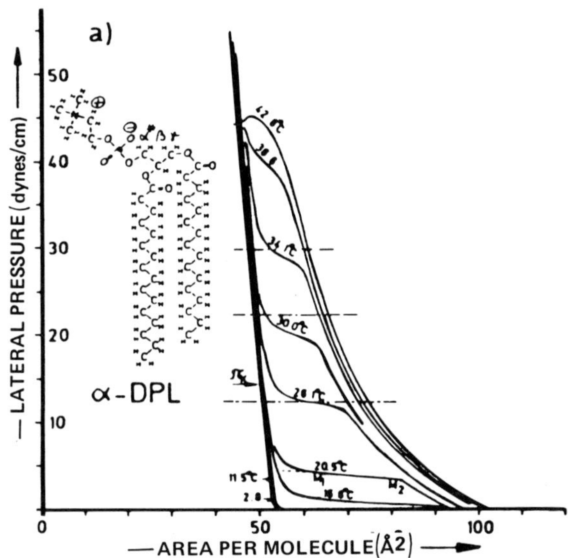

Fig. 18. π-A *isotherms of DPPC on pure water. The arrow at* π_k *indicates another break that would correspond to a change of tilt angle of the chains.* M_1M_2 *could be the coexistence plateau (Albrecht et al., 1978).*

layer interfacial tension) increases and the compressibility decreases until a transition point is reached. At this point the compressibility increases sharply and then monotonously decreases again. The transition indicated by the slope discontinuity occurs at higher surface pressure and smaller area when the temperature increases. Similarly with longer chain lengths the transition occurs for less dense monolayers than it does for those with shorter chains (Phillips and Chapman, 1968; Albrecht *et al.*, 1978). The difference introduced by the polar head is substantial as evidenced by the data on phosphatidyethanolamine (Standish and Pethica, 1968; Demel and Joos, 1968). By the surface tension measurements only one transition is observed and it may in some way be identified with the main transition observed in bilayer thermograms. However when looking more closely at monolayers by other techniques like viscosimetry or light scattering, a second transition can be observed that is associated with the pretransition of the bilayers (Albrecht *et al.*, 1978; Birecki and Amer, 1979).

2. Cyclic and bipolar amphiphiles. As we shall see below several theories assume that orientational effects are involved in monolayer phase transitions; and molecules that can remain flat or at least lying on the interface are by their geometrical characters good candidates to exhibit a high degree of orientation in the interface plane. Unfortunately, the experiments done with this kind of molecules have been disappointing: There is no evidence for any orientation. For instance, with sterols (Cadenhead and Phillips, 1967) bearing two polar heads like β-estradiol diacetate there is a break in the isotherm for an area of 96 $Å^2$/molecule and a limiting value of 38 $Å^2$/molecule (Fig. 19). The molecules in the dilute state are lying flat, until, in this position, the close-packed state is reached, then they begin to stand up. Since this molecular rearrangement shows little temperature dependence, it is likely that only small interaction energies between the molecules in both the lying and standing states are implied. Transitions, apparently also due to a lifting up of the

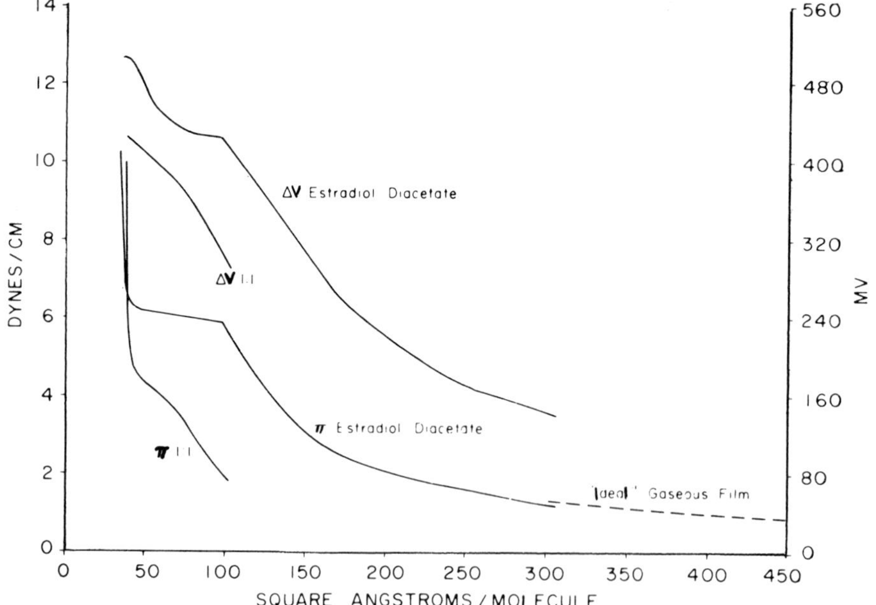

Fig. 19. Surface pressure and surface potential versus area plots for β-estradiol diacetate and 1:1 molar ratio mixture with cholesterol (Cadenhead and Phillips, 1967).

adsorbed molecules, are observed with other polycyclic compounds (Gaines *et al.*, 1978). On the other hand, it seems that if the immersion requirement of the polar groups strongly maintains the molecule flat on the interface there is no break in the isotherm (Miyasaka *et al.*, 1978; Chifu *et al.*, 1975, 1978; Kellner and Cadenhead, 1978a).

Other good candidates to orient horizontally in the monolayer plane are the bipolar molecules. With diols, if the chain is shorter than C_{20}, the molecules in the condensed state are mostly standing up with one hydroxyl group in water and one away from it, whereas if the chain is longer than C_{20} it may bend so that, even in the condensed state, both polar heads dip in the water (Ueno *et al.*, 1975). In contrast, diacids (Jeffers and Dean, 1965; Pal *et al.*, 1975) do not orient vertically, the two carboxyl groups remain in the water and the hydrocarbon chain bows in the condensed phase. With diesters the opposite situation occurs: even with C_{12} chains the molecules orient vertically (Dubault *et al.*, 1978). Obviously, all these results may be understood if the different polar heads are arranged in descending order of affinity to water:carboxyl > hydroxyl > ester group. In the case where the two polar heads are not identical, for instance, with hydroxyoctadecanoic acid (Tachibana and Hori, 1977; Tachibana *et al.*, 1979) or hydroxyhexadecanoic acid (Kellner and Cadenhead, 1978b, 1979), the less hydrophilic group will leave the water.

3. *Two-components monolayers*. Since it has been clearly established that cholesterol suppresses the gel-liquid-crystal transition of phospholipid bilayer, it is interesting to show the effect of cholesterol on phospholipid monolayers. Usually, investigators have reported that cholesterol and phospholipid occupy a smaller molecular area in the mixed system than in monolayers of the pure material (Demel *et al.*, 1967; Shah and Schulman, 1967; Chapman *et al.*, 1969; Tinoco and McIntosh, 1970). This condensing effect of cholesterol is interpreted either as the result of

specific interactions between the two lipids or as due to the formation of cavities between phospholipid molecules (Shah and Schulman, 1967). For Gershfeld and Pagano (1972) there is no surface mixing between cholesterol and DPPC whatsoever, whereas for Zatz and Cleary (1975) these two lipids behave like a two-dimensional ideal solution. In fact it appears that the surface miscibility is dependent on the surface pressure value (Cadenhead *et al.*, 1976). By studying the collapse of monolayers of DPPC and cholesterol mixtures, Joos and co-workers (Snik *et al.*, 1978) noticed no squeeze-out of cholesterol molecules.

It has been demonstrated that addition of anesthetics, like methoxyflurane or halothane, to a phospholipid monolayer shifts the transition toward more condensed area and higher surface pressure, a result that is consistent with the melting effect observed on bilayers (Ueda *et al.*, 1974).

The interaction between β-casein and lecithin promotes condensation due to the formation of a shell of relatively restricted lecithin molecules around the periphery of the protein molecule (Phillips *et al.*, 1975). With synthetic polypeptides in most of the cases studied, there is no deviation from the additivity rule of the molecular areas (Yamashita *et al.*, 1978; Shibata *et al.*, 1978; Gabrielli, 1975; Gabrielli and Maddii, 1978).

There have been many studies on mixture of two single-chained components like a fatty acid and a fatty alcohol, as shown on Fig. 20 (Matuo *et al.*, 1978; Roberts *et al.*, 1976), or a fatty acid and a fatty ester (Matuo *et al.*, 1979a; Motomura *et al.*, 1979; Matuo *et al.*, 1979b), or two different fatty esters (Matuo *et al.*, 1976c), or a fatty alcohol and a fatty sulfate (Costin and Barnes, 1975). The change in the position of the transition point in the compressional isotherms with composition allows the two-dimensional phase diagrams of these different mixtures. Depending on the chain length difference and the polar heads there may be positive or negative azeotropy types, eutectic type, or more complicated types of phase diagrams.

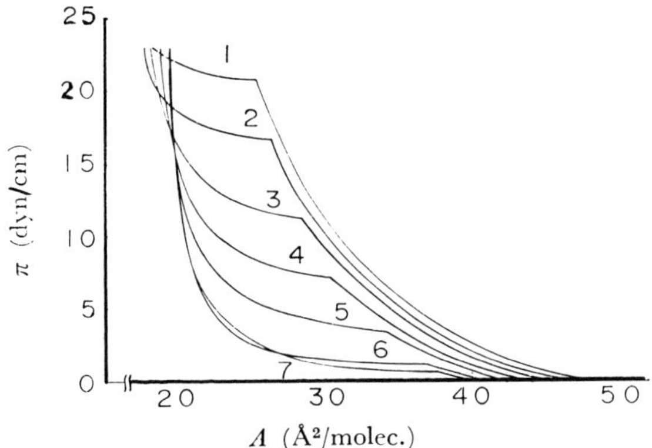

Fig. 20. Surface pressure versus mean area curves of tetradecanoic acid-1-tetradecanol monolayers: 1, 1(tetradecanoic acid); 2, 0.9; 3, 0.7; 4, 0.5; 5, 0.3; 6, 0.1; 7, 0(1-tetradecanol) mole fraction of tetradecanoic acid (Matuo et al., 1978).

B. Oil-Water Interface

Since an amphiphilic molecule is by definition partly hydrophilic and partly hydrophobic, and that hydrophobicity is oftentimes synonymous with lyophilicity, it may find at the oil-water interface the same kind of potential well it finds at the air-water interface. However the experimental handling of a monolayer is much easier with the latter system than with the former. Indeed, in order to make a stable monolayer at the air-water interface the substance has to be insoluble in water and nonvolatile; this is easily achieved with many fatty substances having long enough hydrocarbon chains. At the oil-water interface the amphiphilic substance should be insoluble in both phases in order to get a stable monolayer. The Unilever group at Port Sunlight has, with much success, studied with a Langmuir's type surface balance the behavior of several phospholipids at an oil-water interface. The isotherms obtained (Fig. 21) are quite different from those obtained at the air-water interface, mainly because there is a really flat region following the transition break. Even though there is no second singular point at the end (low areas) of these plateaus, it remains that this kind of isotherms are much more similar to what is expected for classical first-order transition. These authors have shown that the decrease in chain length has a similar effect to

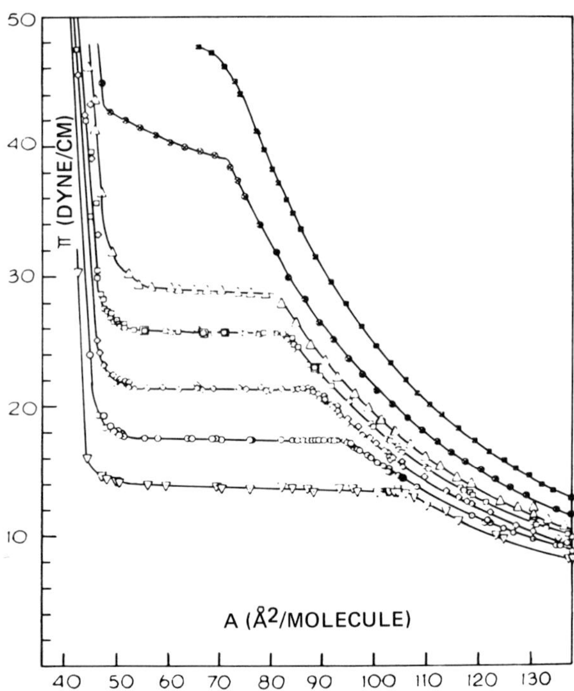

Fig. 21. Surface pressure versus area isotherms for DSPC at the n-heptane/water interface: ▽ *= 5°; 0 = 10°;* ◇ *= 15°;* □ *= 20°;* △ *= 23.4°;* ⊗ *= 30°;* ⊕ *= 32.7°;* □ *= 40.4° (Pethica et al., 1976).*

an increase in temperature (Pethica *et al.*, 1976; Taylor *et al.*, 1973) and that lipids with an ethanolamine headgroup condense more easily than those with a choline headgroup (Llerenas and Mingins, 1976).

There is another technique for studying monolayers at the oil-water interface: It is the amphiphile adsorption from the solution onto the interface. This technique is not as straightforward as the surface balance, since Gibbs' equation must be used in order to calculate the number of adsorbed molecules. The adsorption at the oil-water interface has been studied for many substances but there have been little data to suggest that a phase transition occurs (Hutchinson and Randall, 1952; Ambwani *et al.*, 1973). Lutton *et al.*, 1969) have evidenced transitions of the adsorbed layers of several monoglycerides and fatty alcohols at a triglyceride-water interface. The interfacial tension versus temperature curves show a break separating two straight portions of different slopes. At lower temperatures the slopes are quite steep and positive, at higher temperatures the slopes are smaller. The transition temperature increases with concentration and with chain length. Other studies by Lin *et al.* (Lin, 1975, 1979; Lin *et al.*, 1979) and by Matubayasi *et al.* (1978) on better-defined systems have confirmed the reality of these two interfacial states that can be obtained at the oil-water interface. The condensed state is obtained at high concentration, high hydrostatic pressure and low temperature (see Figs. 22, 23). The mechanism underlying this transition should be some sort of solidification of the amphiphile layer during which the alkyl chains go from a gauche state to an all-trans state. This stiffening of the chains is accompanied by a reduction of the molecular area that excludes the solvent from the

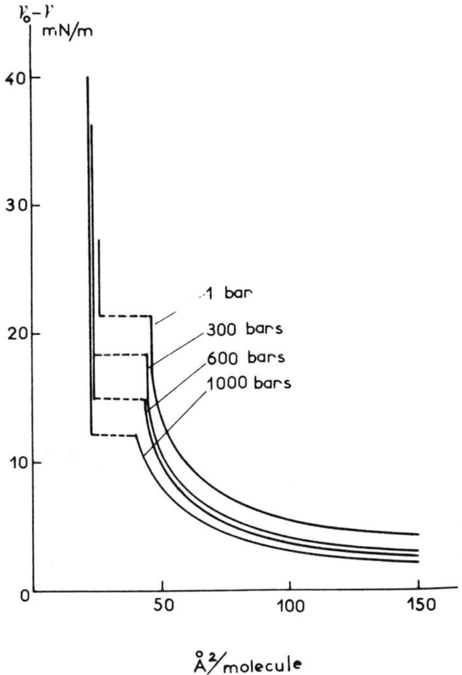

Fig. 22. Interfacial pressure versus molecular area for tridecanol at a paraffinic oil-water interface under various pressures (Lin et al., 1979).

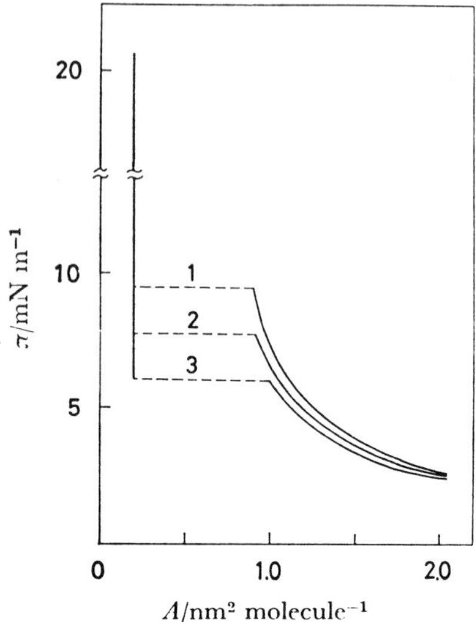

Fig. 23. Interfacial pressure versus molecular area for octadecanol at the hexane-water interface, under 800 bars, at various temperatures: 1-30°C; 2-25°C; 3-20°C (Matubayaschi et al., 1978).

adsorbed layer. The spatial structure created by the adsorption facilitates the solute-solute interaction as compared to the solute-solvent interaction, which is not favored by any ordered structure.

C. The Gas-Liquid Expanded Transition

In the traditional classification (Gaines, 1966), monolayer spread at the air-water interface may display three different types of phase transitions: at very low surface pressure, the first known as the gas-to-liquid-expanded (G-LE) transition, characterized by an area discontinuity; then at medium range of surface pressure, a second transition called liquid expanded-liquid condensed (LE-LC), where the surface compressibility sharply increases; and then at high pressure, the liquid condensed-solid (LC-S) transition, where the surface compressibility sharply decreases. All the data reviewed above are concerned with transitions of the second type, the liquid expanded-liquid condensed type. As a matter of fact, very few papers are concerned with the two other transitions. Now and then it is possible to find in recent literature an isotherm exhibiting the LC-S transition, but it is incidental (Vochten and Petre, 1975). With respect to the G-LE transition, the situation is quite different, even though the data are also very scarce. Indeed, it is not because of lack of interest that few works are done on this transition but because it implies a real experimental tour de force. The chain length and temperature range for which the G-LE transition may be observed is very narrow, and therefore it is not surprising that the only studies concerned with this transition are all dealing with the same substance: pentadecanoic acid (C_{15}) (Hawkins and Benedek, 1974; Kim and Cannell, 1976). These authors show that the two-dimensional "gas" formed by the acid molecules moving along the water surface undergoes a "gas-liquid" phase transition at very low surface pressure. They find, as with any van der Waals condensation, a coexistence curve and a critical point. Therefore they may deduce critical exponents, i.e., the power-law for the variation of parameters like compressibility and density versus temperature. These critical exponents play a very important role in modern theories of critical phenomena (Wilson, 1979) and their values are used to test the validity of various models. Curiously, those obtained from Hawkins' and Kim's experiments are in agreement with the predictions of mean-field theories, whereas there is generally disagreement between these theories and three-dimensional experiments. This should be understood as resulting from a difference in the intermolecular force ranges, in the two-dimensional system where each molecule would move in the mean field due to all the other molecules (Stanley, 1971).

IV. THEORETICAL MODELS

Before examining the different models that have been proposed to describe the transitions observed in two-dimensional amphiphilic systems, there is one question that should be raised: Is the break observed in the isotherms of spread monolayers a real thermodynamical event or an artefact? This is a much more serious question that it may seem at first sight. The main argument against the reality of the LE-LC transition has been put forward very recently first by Gershfeld (1976) and then by Zografi (Jalal and Zografi, 1979). The equilibrium-spreading pressure (i.e., the surface pressure obtained by placing a crystal or a drop of an amphiphilic substance on a water surface) of a monolayer coincides for many substances with the transition pressure. It is then suggested that the liquid-condensed state is a metastable state that will slowly collapse to form a bulk phase. This argument is supported by the fact that transitions appear only to occur when temperature is below the melting point of the fatty acid used. Therefore it would seem that what is called the LE-LC transition is in fact merely the slow transformation from a two-dimensional phase to a three-dimensional crystal. There are many things that may be advocated against this critique. First, the coincidence noticed by Zografi between transition pressures and equilibrium-spreading pressures does not occur for all temperatures (Müller-Landau *et al.*, 1980). Second, what is true for fatty acids is not true for phospholipids; in contradiction to Zografi's and Gershfeld's papers, for most of the phospholipids the equilibrium-spreading pressure is much above the transition pressures (Compare data from Albrecht *et al.* (1978) and Chapman *et al.* (1969) with those of Phillips and Hauser 1974).) Third, the values of the molecular areas attained by the liquid-condensed phase when it is compressed is roughly 20 Å^2 for a fatty acid and 50 Å^2 for a phospholipid. These figures are more likely those of a close-packed lattice of vertical molecules than the transition areas that are much larger.

However there is no reason for a monolayer, whose molecules are strongly bound to the water surface, to collapse before the close-packing of the molecules is achieved.

Thus, once convinced of the reality of the LE-LC transition, we may proceed by reviewing the different models proposed to describe this transition as well as those occurring in bilayers and in adsorbed layers. But we shall first examine the basic concepts underlying the notion of phase transition. A phase transition is a qualitative change in the state of the system. By qualitative change is meant the disappearance of a character of the system that allows the clear separation of the state of the system that bears a given character (state I), from that which does not (state II). Both states are equilibrium states, i.e., corresponding to a free energy minimum. The system has two ways to get a free energy minimum: either the enthalpy decreases or the entropy increases. Every time the external conditions are modified, the system should find a new stable position, a position with the lowest possible free energy. In most of the cases this can be achieved continuously; with the continuous variation of an external parameter there is a continuous response of the conjugate variable. When the pressure of an isothermal gas is continuously increased, the volume continuously decreases. This is up to a point where there is condensation; to a quantitative change in pressure corresponds a quantitative change in density from the gas to the liquid density value. At the onset of condensation the system has found a new response to the increase of pressure; instead of merely reducing the PV term by a decrease of volume, it drastically decreases the energy by condensing the molecules close together in small droplets. Indeed, a typical intermolecular potential has a repulsive part with a vertical asymptote at very short distances as well as an attractive tail. In between there is a minimum. In this position a molecule will minimize its energy, and the system will tend to form a bound state. Of course the entropy will tend to oppose this bound state and in order to minimize its free energy, the system will have to compromise between the energetic need to get the molecule in a potential well and the entropic need to scatter them in a volume as wide as possible. When the volume is too small the number of accessible states is not large enough to compete with the tendency, driven by the energy, to have a bound state, and the system then condenses. If the volume is decreased again, the population of the bound state will increase up to a point where all the molecules have condensed; then the pressure will rise again.

This picture of the gas-liquid condensation is typical of the so-called first-order phase transition. Apart from the gas-liquid transitions, examples of this kind of transition are solid-liquid transitions, phase separations of binary mixture, or ferromagnetic transition. A first-order transition is a two-state process, an all-or-nothing change of the system: the spin in a ferromagnet is either up or down, a molecule is either liquid or gaseous. These transitions imply a discontinuity, at the transition point, of quantities like enthalpy or volume that are first-order derivatives of the free energy. Other types of transitions may be encountered in physical systems for which first derivatives of free energy are continuous and only second-order derivatives like compressibility or heat capacity are discontinuous; these transitions are called "second-order." The molecular process involved in this kind of transition is not an all-or-nothing change but a gradual change. For instance, elongated molecules may minimize their free energy by lining up in one preferred direction. A qualitative change occurs at the transition point: The isotropic system (molecules at random) becomes oriented (lined up molecules) but orientation is a gradual quality, molecules are more or less oriented. At the onset of the anisotropic phase, the departure from randomness is very slight but important enough to destroy the symmetry that the isotropic system had. The departure from randomness monotonously increases. It is not the number of molecules belonging to the "oriented state" but the average degree of orientation of the system that increases. The oriented molecules do not separate out to form a new phase (at least, if the interaction energy favoring the clustering of parallel molecules remains low enough, otherwise the system exhibits a first-order phase transition) but they are interspersed among random molecules. In a second-order phase transition the cooperativity is too weak to allow the formation of cluster of molecules all in the same state.

In a debate about the nature of the LE-LC transitions occurring in two-dimensional lipid systems there are two sides: the first-order and the second-order. Supporters of the former view are more numerous, many authors are convinced that in both bilayers and monolayers the transitions are first-order transitions. In the case of bilayers, at least for the main transition, this assumption seems well grounded since there is an important latent heat associated with the phase change. However, in the case of the LE-LC transition there is a difficulty with this assumption, since all the data evidenced a single break in the compressional isotherms of spread monolayers. On the other hand an area discontinuity has been clearly shown with monolayers adsorbed at the oil-water interface. Hence, any attempt to understand these transitions as a single first-order phase transition would have to explain why the LE-LC transition does not fit into the classical scheme. The reason second-order supporters are few is because the calorimetric data on phospholipid dispersions clearly indicates the first-order character of the gel-liquid crystal phase transion of bilayers.

A. Order-Disorder Models

Several models have been proposed, which analyzed the phase transitions observed in two-dimensional amphiphilic systems as resulting from the disordering of the hydrocarbon chains. The carbon-carbon bond may exist in three rotational states, one ground state called trans and two excited states called gauche. An ordered hydrocarbon chain is all-trans, whereas the number of gauche bonds gives the degree of disorder of the chain. Therefore in these models an essential ingredient is the temperature-dependent balance between the energetically preferred ordered trans state and the high entropy of the disordered state in which the chains enjoy a great conformational diversity with many gauche bonds.

In order to compute the number of states of the system, Nagle (1975, 1976) uses a two-dimensional lattice perpendicular to the interface plane. The chains that are supposed infinite have their methylene groups on consecutive vertices of the lattice. In the case of a triangular lattice, trans bonds are represented as vertical segments whereas gauche bonds are represented as nonvertical segments. Therefore, an all-trans chain would be a vertical straight line and an all-gauche chain a zigzag line. The excluded volume interaction is taken into account by the requirement that two chains cannot occupy the same lattice site. An attractive potential is added in the form of a power-law of density. By arguments derived from those used by van der Waals for three-dimensional gas, Nagle proposed a 3/2 exponent, even though such a value overestimates long range part of interactions; an alternative value could be 5/2 if only nearest-neighbor interactions are taken into account. Then, to calculate the partition function, this author built an isomorphism between his lattice and the two-dimensional dimer lattice, which in the close-packed state can be exactly solved. Horizontal dimers correspond to gauche bonds whereas vertical dimers correspond to trans bonds. Thus the dimer density in each direction gives the gauche-to-trans ratio. The results are in qualitative agreement with bilayer experimental data. When a lateral pressure term is added to the energy, π-A isotherms are obtained; they do not display the basic features of monolayer isotherms (Fig. 24):

(i) the high area portion of the curves is concave instead of convex;

(ii) the increase of compressibility at the onset of the transition

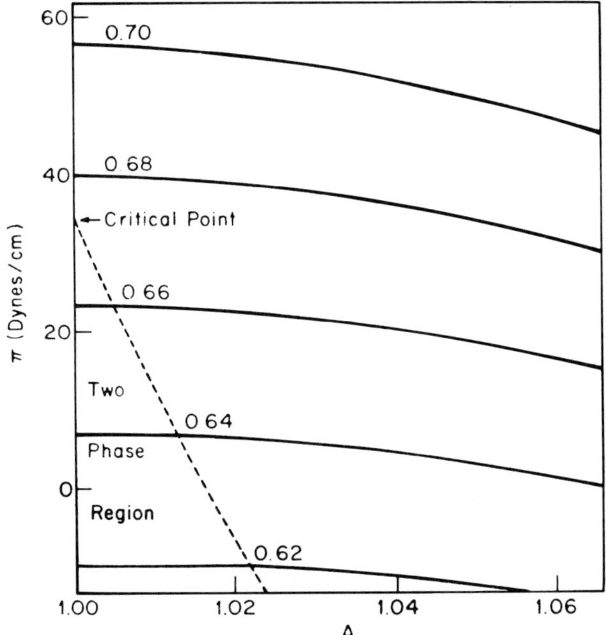

Fig. 24. π-*A isotherms calculated for different values of kT. A = 1 corresponds to the close-packed all-trans state (Nagle, 1975).*

is very slight;

(iii) there is a flat portion;

(iv) the condensed phase appears abruptly at the end of the coexistence plateau instead of showing a slow increase in compressibility toward the limiting area observed experimentally.

Nagle attempted to improve the model by introducing a lattice spacing which increases with temperature; this separates the isotherms in the ordered phase portion. Moreover he includes a headgroup dipolar interaction and a soft core between the chains and between the headgroups, but nevertheless the data are still poorly reproduced.

As Scott (1974) points out, Nagle has been obliged to use an oversimplified model in order to make exact calculations. He notes that the crucial assumption of an infinite chain length is unrealistic and he devises a model in which the chains still lie in a vertical plane but are of finite length with their end attached on a line (Scott, 1975a,b). Here again a two-dimensional lattice is used, each methylene group of the chain occupies a site and no site can be occupied by more than one group. A limited number of rotational isomers are considered [in the more elaborated case 21 states are considered for two chains of 15 units each (Scott, 1975b)]. The fractions of molecules that are in one of these configurations are computed by maximizing the entropy. No transition occurs if a van der Waals' interaction is not included. Scott chooses the same 3/2 exponent power-law of density. Then he obtains isotherms with a discontinuity in density. This is to be expected since his model is basically a van der Waals' gas. In subsequent papers (Scott, 1975b; Scott and Cheng, 1977), Scott attempts to eliminate the flat portions of his isotherms by including an extra surface pressure due to the headgroups. This tacitly supposes the hydrocarbon region and the water surface to be thermodynamically isolated from each other, which assumption is very questionable. Moreover the results are deceiving, if there is no other flat portion the isotherms still have two slope discontinuities (Fig. 25).

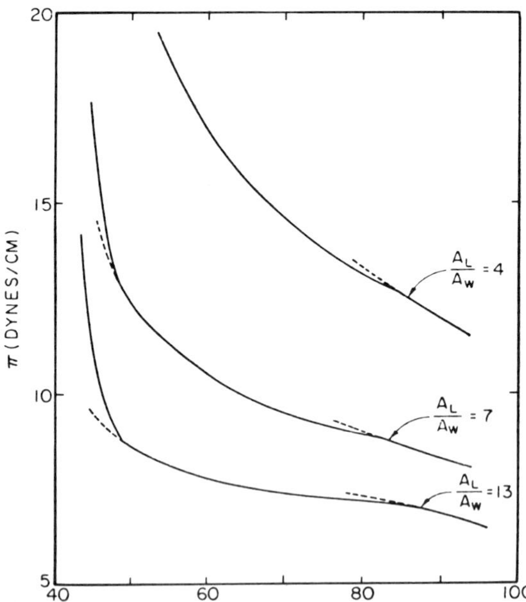

Fig. 25. Effect of different headgroup size on the π-A isotherms. A_L/A_W is the ratio of the molecular surface area of lipid to that of the unbound water (Scott, 1975b).

Marčelja (1974) has proposed a model that treats all the possible conformations of a single chain in the field created by all the other chains. Here there is no need for a lattice, the steric repulsion are taken into account through a lateral pressure contribution. The attractive interaction has the form of an anisotropic van der Waals' potential, that is, a potential that depends on the bond angle with the normal to the interface plane. The all-trans conformation is therefore energetically favored both because of the lowest energy of trans bonds and because of the anisotropic intermolecular potential. The isotherms resulting from this model are satisfactory except for the existence of flat portions (Fig. 26 should be compared with Fig. 18).

In a related way Jacobs *et al.* (1975) give a model that includes, beside gauche-trans rotation energy, an interchain attraction and an excluded volume interaction for the chains and for the headgroups. The headgroups are supposed to be packed in a two-dimensional layer as hard disks. The excluded volume interaction is expressed as in Marčelja's model as a lateral pressure term, but here the area increase is calculated from the degree of bend-

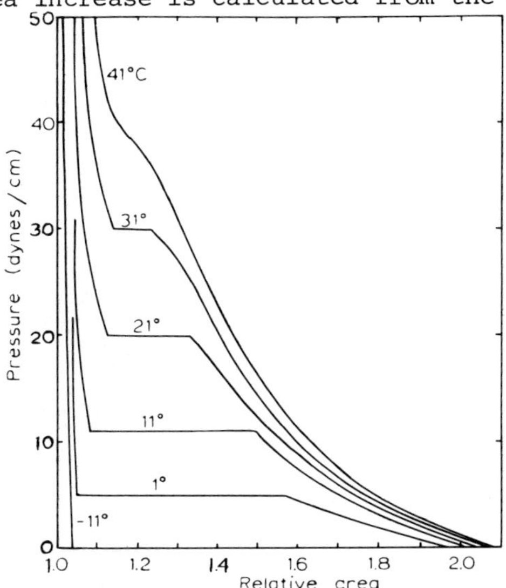

Fig. 26. Isotherms calculated at different temperatures. The area is measured relative to its value in the perfectly ordered state (Marcelja, 1974).

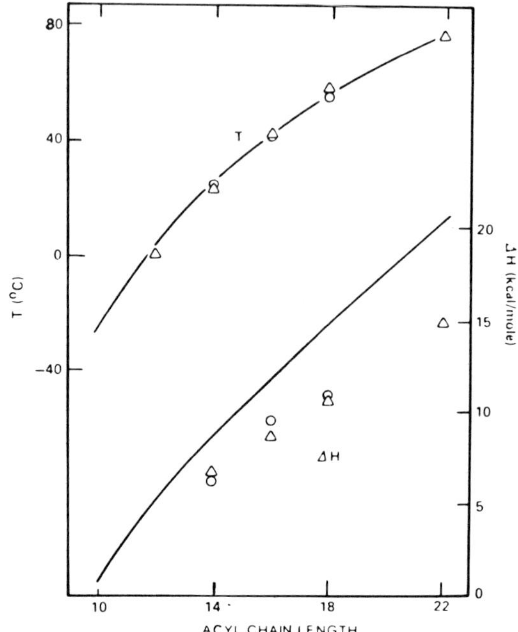

Fig. 27. Transition temperature and enthalpy changes for phosphatidyl cholines. The solid lines are calculated; triangles and rectangles are experimental data from Ladbrooke and Chapman (1969) and Hinz and Sturtevant (1972), respectively (Jacobs et al., 1975).

ing of the chain. Their results are compared with experimental data on Fig. 27.

Bothorel and his collaborators have calculated explicitly the interaction energy and the melting entropy for two hexane chains enclosed in an hexagonal box (Belle and Bothorel, 1972; Belle *et al.*, 1974; Bothorel *et al.*, 1972, 1974). To do this they introduced an intramolecular energy, which is the energy difference between the rotational isomers, and an intermolecular energy computed from a Lennard-Jones potential. Satisfactorily, the interaction energy between the two chains has a minimum for a distance of 4.6 Å and the entropy increases sharply at distances above 4.9 Å. To refine their model these authors have replaced the hexagonal box by eight other molecules and they increase the chain length up to 14 carbon atoms. Beside the minimum corresponding to the close-packed state they find two other minima for molecular areas between 30 and 50 $Å^2$ (Bothorel *et al.*, 1974).

In this last model the chains are supposed to have simultaneously the same conformation and this may seem at first sight quite a drastic assumption. However, as McCammon and Deutch (1975) put it, the all-trans molecules are virtually immiscible with kinky chains (a kink is two consecutive and opposite gauche bonds). There are two coexisting phases constituted by coherently organized clusters during the phase transition. Thus they set two partition functions, one for the solid all-trans phase and the other for fluid kinky phase. They reproduce quite well the transition temperature of pure lecithin bilayers and the phase diagram of the DPPC-DSPC mixture bilayers.

The assumption that kinks are the essential contribution to disorder is also made by Jackson (1976). This author considers two kinds of kinks; a chain which is kinked into a vacancy is said to have a falling kink, its activation energy is twice the energy for the formation of a gauche bond; a chain which kinks by pushing into neighbors is said to have a forcing kink, the energy associated with a forcing kink depends on the position of the kink along the chain and on the packing of the chains.

All these models have to look for the root of the phase transition in the rotational isometry of the chains. Of course each author places this gauche-trans change in different frameworks but the results are very coherent: These order-disorder models give good results for bilayers but are rather impotent to describe monolayer isotherms especially because they all give isotherms with two slope discontinuities. Certainly many reasons may be advocated to explain this discrepancy between the shapes of theoretical and experimental isotherms. It is possible that impurities in the monolayer disrupt the cooperativity of the transition by perturbing the interaction between hydrocarbon chains and thereby produce the rounded isotherms experimentally observed. But it should be remarked however that transitions with a single slope discontinuity have been observed in a number of monolayers, over many years, with no known exception. The absence of flat portion in experimental isotherms cannot be an artefact. Nagle (1976) gives an alternative explanation: The walls of the trough containing the monolayer are rough and the molecules near the walls could be expected to disorder out of the all-trans ordered conformation at a higher surface pressure than the molecules in the interior. This possibility could be tested by increasing the size of the trough. For Albrecht *et al.* (1978) the finite slope seems to be an intrinsic property of the monolayer that has to be attributed to the limited size of the cooperative units. A similar explanation has been advanced by Marsh (Marsh *et al.*, 1976) to justify the finite temperature range through which occurs a bilayer transition. This argument is very fundamental since it questions the differentiation between first- and second-order phase transition. In the condensation process of a gas to surface tension acts as a driving force for coalescence of the small drops in bigger ones. If in two-dimensional systems such a process does not exist or is too slow, the small clusters of melted chains may remain scattered among clusters of all-trans chains; there is no phase separation. The properties of the system may depend not only on the number of melted molecules but also on the number of clusters, since each cluster is surrounded by boundary molecules that are in a peculiar interaction state. There is a very simple explanation: experimental isotherms display a single slope discontinuity because the LE-LC phase transition exhibited by spread monolayers is second-order. Curiously very few authors have adopted this position.

B. Orientational Models

To introduce the different models reviewed in this section we shall first consider Ohki's model (1967). It does not yield a second-order transition and does not consider molecular orientation, but it is typical of the method used in the more sophisticated models described below. As a matter of fact it is merely a two-dimensional lattice gas traced from Ising lattice of spins. Instead of having spins up or down, each lattice site is occupied either by a molecule or by a hole. The surface pressure is related to the magnetic field and the area to the magnetization. Then, using the well-known results of Lee and Yang, the π-A isotherms are obtained; they show an area discontinuity. In such a model there is of course an interaction energy between nearest neighbors, but no internal molecular energy.

Kaye and Burley (1974) insert an infinite repulsion between nearest neighbors so that no two molecules can be nearest neighbors. The model shows separation into two sublattices and this ordering transition is of the liquid-solid type. By adding an attractive tail of potential, up to the ninth neighbors, they obtain simultaneously a gas-liquid and a solid-liquid transition which is second-order (Fig. 28). The salient feature of these results is the fact that second-order phase transition may be obtained without any conformational change and, with isotropic molecules merely

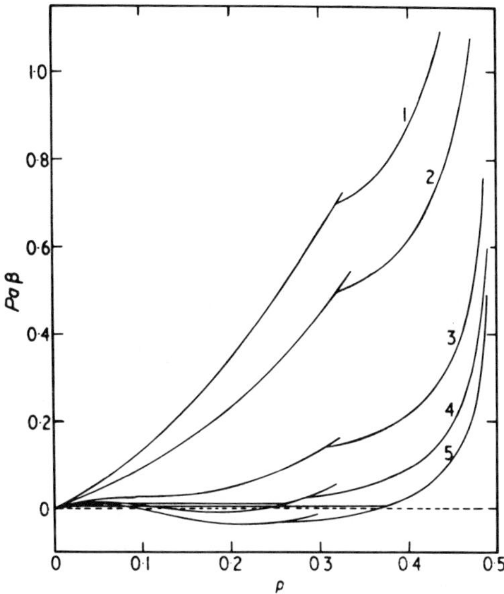

Fig. 28. Two-dimensional pressure-density diagram showing the van der Waals (first-order) and the ordering (second-order) transitions for various temperatures. The interaction are extended up to ninth neighbors (Kaye and Burley, 1974).

represented as points, the only requisite being the first neighbor repulsion.

Bell *et al.* (1978) use also a lattice gas but they considered two molecular orientations, labeled 1 and 2. The nearest-neighbor pairs of types 11 or 22 have a lower interaction energy than the nearest-neighbor pairs of type 12. There is therefore a preferential interaction between similarly oriented molecules. In fact three species are involved because holes have to be added, and for each species the free energy, calculated from the energy and the number of configurations is minimized. The isotherms display a discontinuity of slope at the transition point and the compressibility is lower on the high area side of the transition than on the low area side (Fig. 29). Moreover, by giving to the interaction energy large-enough negative values, phase separation occurs in the disordered region.

For Caillé and Agren (1975) molecules on the lattice may have three different orientations, two in the interface plane and one perpendicular to it. The molecules perpendicular to the interface

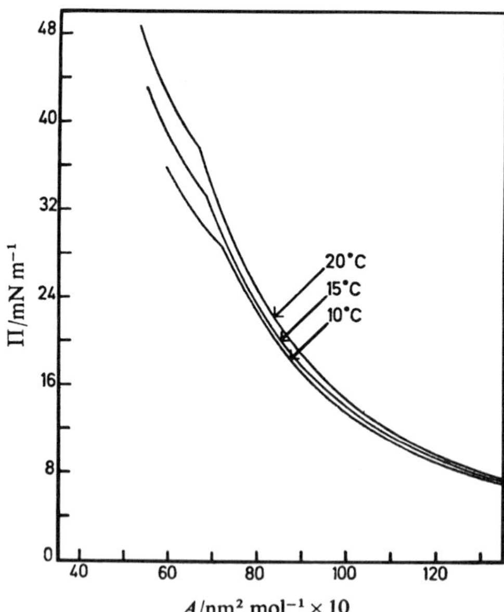

Fig. 29. Theoretical surface pressure against area for DPPC at the heptane-water interface (Bell et al., 1978).

occupy one cell whereas the molecules lying on the interface occupy several cells. When the density is increased two-dimensional nematic order appears, i.e., one direction of the interface plane is privileged. This transition from an isotropic liquid to an anisotropic one is a second-order phase transition. In a subsequent paper, Caillé and collaborators (Caillé *et al.*, 1978) consider a two-state model; in the first state the molecule occupies a single lattice site, in the second state it occupies several sites. The first state corresponds to an upright orientation of the molecule in an all-trans rotational conformation, whereas the second state may correspond either to a molecule lying on the interface or to the projection of a molecule having gauche bonds. Since there is no other orientational state in the interfacial plane, no nematic ordering will show up. An intramolecular energy is introduced to take into account the trans-gauche chain melting as well as an intermolecular energy between the upright molecules. Then these authors consider a multistate model in which a variety of intramolecular conformations are possible, this expressed by letting the number of sites occupied by a molecule projection vary. They obtain a coexistence curve corresponding to the area discontinuities characteristic of first-order phase transitions.

In the same vein, Firpo *et al.* (1978) have described a model where a rectangular molecule occupies several contiguous sites in one of the two perpendicular directions of the interface plane; beside the two orientational species, a third one is constituted by the holes that were not included in Caillé's models. The isotherms obtained exhibit, when the degree of anisotropy of the projection is large enough, two transitions, one at low density which is first-order and another one at high density which is first-order at low temperature and becomes second-order at high temperature (Fig. 30). Therefore these authors have evidenced a so-called tricritical point where the transition shifts from the first-order to the second-order. This very interesting feature may express the ambivalence of the LE-LC transition, which could be first-order at

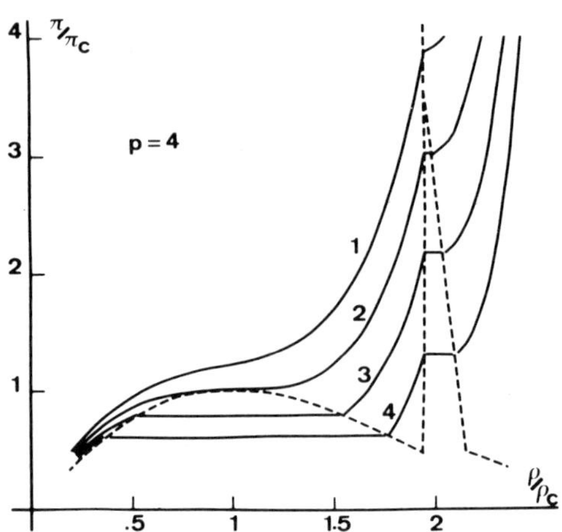

Fig. 30. Calculated isotherms and coexistence curves showing a transition of condensation (first-order) at low density and an orientational transition (with tricritical point) at high density. p is the degree of anisotropy of the molecular projection (Firpo et al., 1978).

low temperature (see for instance Fig. 18). They modify their model by differentiating not only the directions on the interface plane but also the sense. This induces in addition to the nematic ordering a two-dimensional smectic ordering that appears on the phase diagram as another transition line. Thus, the schematic phase diagram proposed by Albrecht *et al.* (1978) is partially reproduced. These authors seem to have evidenced four different phases (isotropic, anisotropic, crystal tilted, crystal not tilted), whereas the model of Firpo *et al.* yields only three phases (isotropic, nematic, smectic) (Dupin *et al.*, 1979).

V. CONCLUSION

In physical systems as well as in biological systems interfaces are both boundaries and links between the two phases they separate. Some molecular species cannot cross the interface, others can. Amphiphilic molecules are specialized compounds for the interfaces. Their schizophrenic chemical structure allow them to enjoy a free energy minimum when adsorbed at an interface. Therefore, the exchange properties of the interface may be monitored through the amphiphiles adsorbed. The amphiphilic layer lying at an interface may be in different states of organization, and with changes of the thermodynamic parameters of the system it may display phase transitions. These phase transitions are of different kinds that do not seem reducible to a single model. The bilayer main transition exhibits an important latent heat associated with the idea of first-order phase transition, whereas the LE-LC transition of spread monolayers shows a discontinuity of compressibility with no area discontinuity, as expected for second-order phase transition. The molecular mechanism underlying the bilayer main transition seems well understood, it is the hydrocarbon chain melting, i.e., the transformation of an all-trans state to a rotamer containing gauche bonds. On the contrary, the molecular mechanism occurring during the LE-LC transition appears to be highly controversial, even though an orientation in the plane of the interface seems able to yield the second-order transition experimentally observed. When the hydrocarbon chains of amphiphile molecules are solvated, the transition observed is again first-order, and this leads to the conclusion that monolayers adsorbed at the oil-water interface are much more similar to bilayers than are monolayers spread at the air-water interface. But the study of the adsorbed layer transitions has been, up to now, too superficial to get a complete understanding of its relations with both the bilayer transitions and the spread monolayer transitions. However, it seems clear that it is also due to a chain melting process. The bilayer pretransition is likely the result of an orientational process, but rather different from that involved in the LE-LC transition. In the former it is the collective orientation of the molecules that would change (tilt angle or ripples), whereas in the latter it is the degree of alignment of individual molecules in the plane of the interface that would change at the transition point. Anyway, chain melting and orientation are complementary processes since molecular asymmetry, on one hand, and tilt angle, on the other hand, depend on the stiffness of the molecule. Hence it is important now to elucidate, by more theoretical and experimental works, how these two phenomena are connected.

References

Albrecht, O., Gruler, H., and Sackmann, E. (1978). *J. Physique 39*, 301.

Ambwani, D. S., Jao, R. A., and Fort, T., Jr. (1973). *J. Colloid Interface Sci. 42*, 8.

Bach, D., Bursuker, I., Eibl, H., and Miller, I. R. (1978). *Biochim. Biophys. Acta 514*, 310.

Barton, P. G., and Gunstone, F. D. (1975). *J. Biol. Chem. 250*, 4470.

Bell, G. M., Mingins, J., and Taylor, J. A. G. (1978). *J. C. S. Faraday I. 74*, 223.

Belle, J. and Bothorel, P. (1972). *C. R. Acad. Sci. Paris 275*, 355.

Belle, J., Bothorel, P., and Lemaire, B. (1974). *FEBS Lett. 39*, 115.

Birecki, H., and Amer, N. M. (1979). *J. Physique 40*, C3-433.

Blume, A., and Ackerman, T. (1974). *FEBS 43*, 71.

Bothorel, P., Lussan, C., Lemaire, B., and Belle, J. (1972). *C. R. Acad. Sci. Paris 274*, 1541.

Bothorel, P., Belle, J., and Lemaire, B. (1974). *Chem. Phys. Lipids 12*, 96.

Bruni, A., van Dijck, P. W. M., and de Gier, J. (1975). *Biochim. Biophys. Acta 406*, 315.

Cadenhead, D. A., and Phillips, M. C. (1967). *J. Colloid Interface Sci. 24*, 491.

Cadenhead, D. A., Kellner, B. M. J., and Phillips, M. C. (1976). *J. Colloid Interface Sci. 57*, 224.

Caillé, A., and Agren, G. (1975). *Can. J. Phys. 53*, 2369.

Caillé, A., Rapini, A., Zuckermann, M. J., Cros, A., and Doniach, S. (1978). *Can. J. Phys. 56*, 348.

Calhoun, W. I., and Shipley, G. G. (1979). *Biochemistry 18*, 1717.

Cameron, D. G., and Mantsch, H. H. (1978). *Biochem. Biophys. Res. Commun. 83*, 886.

Casal, H. L., Smith, I. C. P., Cameron, D. G., and Mantsch, H. H. (1979). *Biochim. Biophys. Acta 550*, 145.

Chapman, D., Owens, N. F., Phillips, M. C., and Walker, D. A. (1969). *Biochim. Biophys. Acta 183*, 458.

Chapman, D., Urbina, J., and Keough, K. M. (1974). *J. Biol. Chem. 249*, 2512.

Chifu, E., Tomoaia, M., and Ioanette, A. (1975). *Gazz. Chim. Ital. 105*, 1225.

Chifu, E., Tomoaia, M., Nicoara, E., and Olteanu, A. (1978). *Rev. Roumaine Chimie 23*, 1163.

Costin, I. S., and Barnes, G. T. (1975). *J. Colloid Interface Sci. 51*, 94.

Demel, R. A., and Joos, P. (1968). *Chem. Phys. Lipids 2*, 35.

Demel, R. A., van Deenen, L. L., and Pethica, B. A. (1967). *Biochim. Biophys. Acta 135*, 11.

van Dijck, P. (1977). Thesis, Utrecht.

van Dijck, P. W., Ververgaert, P. H., Verkleij, A. J., van Deenen, L. L., and de Gier, J. (1975). *Biochim. Biophys. Acta 406*, 465.

van Dijck, P. W. M., de Kruijff, B., Aarts, P. A. M. M., Verkleij, A. J., and de Gier, J. (1978). *Biochim. Biophys. Acta 506*, 183.

Dubault, A., Casagrande, C., Veyssié, M., Caillé, A., and Zuckermann, M. J. (1978). *J. Colloid Interface Sci. 64*, 290.

Dupin, J. J., Firpo, J. L., Albinet, G., Bois, A. G., Casalta, L., and Baret, J. F. (1979). *J. Chem. Phys. 70*, 2357.

Eibl, H., and Blume, A. (1979). *Biochim. Biophys. Acta 553,* 476.

Engelman, D. M., and Rothman, J. E. (1972). *J. Biol. Chem. 247,* 3694.

Epand, R. M. (1978). *Biochim. Biophys. Acta 514,* 185.

Esfahani, M., Limbrick, A. R., Knutton, S., Oka, T., and Wakil, S. J. (1971). *Proc. Nat. Acad. Sci. 68,* 3180.

Estep, T. N., Mountcastle, R. L., Biltonen, R. L., and Thompson, T. E. (1978). *Biochemistry 17,* 1984.

Estep, T. N., Mountcastle, R. L., Berenholz, Y., Biltonen, R. L., and Thompson, T. E. (1979). *Biochemistry 18,* 2112.

Findlay, E. J., and Barton, P. G. (1978). *Biochemistry 17,* 2400.

Firpo, J. L., Dupin, J. J., Albinet, G., Bois, A., Casalta, L., and Baret, J. F. (1978). *J. Chem. Phys. 68,* 1369.

Frenzel, J., Arnold, K., and Nuhn, P. (1978). *Biochim. Biophys. Acta 507,* 185.

Gabrielli, G. (1975). *J. Colloid Interface Sci. 53,* 148.

Gabrielli, G., and Maddii, A. (1978). *J. Colloid Interface Sci. 64,* 19.

Gaines, G. L., Jr. (1966). *In* "Insoluble Monolayers at Liquid-Gas Interfaces," p. 177. Interscience, New York.

Gaines, G. L., Jr., Behnken, P. E., and Valenty, S. J. (1978). *J. Am. Chem. Soc. 100,* 6549.

Gent, M. P. N., and Ho, C. (1978). *Biochemistry 17,* 3023.

Gershfeld, N. L., and Pagano, R. E. (1972). *J. Phys. Chem. 76,* 1244.

Gershfeld, N. L. (1976). *Ann. Rev. Phys. Chem. 27,* 349.

Goddard, E. D., and Kung, H. C. (1973). *J. Colloid Interface Sci. 43,* 511.

Gottlieb, M. H., and Eanes, E. D. (1974). *Biophys. J. 14,* 335.

Haigh, E. A., Thulborn, K. L., and Sawyer, W. H. (1979). *Biochemistry 18,* 3525.

Halsey, M. J., and Wardley-Smith, B. (1975). *Nature London 257,* 811.

Harlos, K. (1978). *Biochim. Biophys. Acta 511*, 348.

Hawkins, G. A., and Benedek, G. B. (1974). *Phys. Rev. Lett. 32*, 524.

Hinz, H. J., and Sturtevant, J. M. (1972a). *J. Biol. Chem. 247*, 3697.

Hinz, H. J., and Sturtevant, J. M. (1972b). *J. Biol. Chem. 247*, 6071.

Hubbell, W. L., and McConnell, H. M. (1971). *J. Am. Chem. Soc. 93*, 314.

Hutchinson, E., and Randall, D. (1952). *J. Colloid Interface Sci. 7*, 151.

Jackson, M. B. (1976). *Biochemistry 15*, 2555.

Jacobs, R. E., Hudson, B., and Anderson, H. C. (1975). *Proc. Nat. Acad. Sci. U.S. 72*, 3993.

Jacobson, K., and Papahadjopoulos, D. (1975). *Biochemistry 14*, 152.

Jalal, I., and Zografi, G. (1979). *J. Colloid Interface Sci. 68*, 196.

Jähnig, F., Harlos, K., Vogel, H., and Eibl, H. (1979). *Biochemistry 18*, 1459.

Janiak, M. J., Small, D. M., and Shipley, G. G. (1976). *Biochemistry 15*, 4575.

Janoff, A. S., Haug, A., and MacGroarty, E. J. (1979). *Biochim. Biophys. Acta 555*, 56.

Jeffers, P. M., and Dean, J. (1965). *J. Phys. Chem. 69*, 2368.

Kamaya, H., Ueda, I., Moore, P. S., and Eyring, H. (1979). *Biochim. Biophys. Acta 550*, 131.

Kaneshina, S., Shibata, O., and Nakamura, M. (1979). *Bull. Chem. Soc. Jpn. 52*, 42.

Kantor, H. L., Mabrey, S., Prestegard, J. H., and Sturtevant, J. M. (1977). *Biochim. Biophys. Acta 466*, 402.

Kaye, R. D., and Burley, D. M. (1974). *J. Phys. A: Math. Nucl. Gen. 7*, 1303.

Kellner, B. M. J., and Cadenhead, D. A. (1978a). *Biochim. Biophys. Acta 513,* 301.

Kellner, B. M. J., and Cadenhead, D. A. (1978a). *J. Colloid Interface Sci. 63,* 452.

Kellner, B. M. J., and Cadenhead, D. A. (1978b). *Chem. Phys. Lipids 23,* 41.

Keough, K. M. W., and Davis, P. J. (1979). *Biochemistry 18,* 1453.

Kim, M. W., and Cannell, D. S. (1976). *Phys. Rev. 13,* 411.

Kimelberg, H. K., and Papahadjopoulos, D. (1974). *J. Biol. Chem. 249,* 1071.

Kruijff de, B., Demel, R. A., and Van Deenen, L. L. M. (1972). *Biochim. Biophys. Acta 255,* 331.

Kruijff de, B., van Dijck, P. W. M., Goldbach, R. W., Demel, R. A., and van Deenan, L. L. H. (1973). *Biochim. Biophys. Acta 330,* 269.

Kruijff de, B., Gerritsen, W. J., Oerlmans, A., van Dijck, P. W. M., Demel, R. A., and van Deenen, L. L. M. (1974). *Biochim. Bio-Phys. Acta 339,* 44.

Ladbrooke, B. D., and Chapman, D. (1969). *Chem. Phys. Lipids 3,* 304.

Ladbrooke, B. D., Williams, R. M., and Chapman, D. (1968). *Biochim. Biophys. Acta 150,* 333.

Lagaly, G. (1976). *Angew. Chem. Int. Ed. Engl. 15,* 575.

Laggner, P., and Stabinger, H. (1976). *In* "Colloid and Interface Science" (M. Kerker, ed.), Vol. 5, p. 91. Academic Press, New York.

Lakowicz, J. R., Prendergast, F. G., and Hogen, D. (1979). *Biochemistry 18,* 508.

Lee, A. G. (1975a). *Biochemistry 14,* 4397.

Lee, A. G. (1975b). *Biochim. Biophys. Acta 413,* 11.

Lee, A. G. (1976). *FEBS Lett. 62,* 359.

Lee, A. G. (1977a). *Biochim. Biophys. Acta 472,* 237.

Lee, A. G. (1977b). *Biochim. Biophys. Acta 472,* 285.

Lee, A. G. (1978). *Biochim. Biophys. Acta 514,* 95.

Lentz, B. R., Freire, E., and Biltonen, R. L. (1978). *Biochemistry 17,* 4475.

Lin, M. (1975). *C. R. Acad. Sci. Ser. B. 281,* 619.

Lin, M. (1979). *J. Chim. Phys. Phys. Chim. Biol. 76,* 61.

Lin, M., Firpo, J. L., Mansoura, P., and Baret, J. F. (1979). *J. Chem. Phys. 71,* 2202.

Linden, C. D., Wright, K. L., McConnell, H. M., and Fox, C. F. (1973). *Proc. Nat. Acad. Sci. U.S. 70,* 2271.

Lippert, J. L., and Peticolas, W. L. (1971). *Proc. Nat. Acad. Sci. U.S. 68,* 1572.

Lippert, J. L., and Peticolas, W. L. (1972). *Biochim. Biophys. Acta 282,* 8.

Llerenas, E., and Mingins, J. (1976). *Biochim. Biophys. Acta 419,* 381.

London, Y., Demel, R. A., Geurts van Kessel, W. S. M., Vossenberg, F. G. A., and van Deenen, L. L. M. (1973). *Biochim. Biophys. Acta 311,* 520.

Lutton, E. S., Stauffer, C. E., Martin, J. B., and Fehl, A. J. (1969). *J. Colloid Interface Sci. 30,* 283.

McCammon, J. A., and Deutsch, J. M. (1975). *J. Amer. Chem. Soc. 97,* 6675.

Mcdonald, A. G. (1978). *Biochim. Biophys. Acta 507,* 26.

McDonald, R. C., Simon, S. A., and Baer, E. (1976). *Biochemistry 15,* 885.

Machtiger, N. A., and Fox, C. F. (1973). *Annu. Rev. Biochem. 42,* 575.

McMurchie, E. J., and Raison, J. K. (1979). *Biochim. Biophys. Acta 554,* 364.

Marčelja, S. (1974). *Biochim. Biophys. Acta 367,* 165.

Marsh, D. (1974). *Biochim. Biophys. Acta 363,* 373.

Marsh, D., Watts, A., and Knowles, P. F. (1976). *Biochemistry 15,* 3570.

Matubayasi, N., Motomura, K., Aratono, M., and Matuura, R. (1978). *Bull. Chem. Soc. Jpn. 51,* 2800.

Matuo, H., Hiromoto, K., Motomura, K., and Matuura, R. (1978). *Bull. Chem. Soc. Jpn. 51,* 690.

Matuo, H., Motomura, K., and Matuura, R. (1979a). *J. Colloid Interface Sci. 69,* 192.

Matuo, H., Yosida, N., Motomura, K., and Matuura, R. (1979b). *Bull. Chem. Soc. Jpn. 52,* 667.

Matuo, H., Motomura, K., and Matuura, R. (1979c). *Bull. Chem. Soc. Jpn. 52,* 673.

Mendelsohn, R., and Maisano, J. (1978). *Biochim. Biophys. Acta 506,* 192.

Miyasaka, T., Watanabe, T., Fujishima, A., and Honda, K. (1978). *J. Amer. Chem. Soc. 100,* 6657.

Motomura, K., Yano, T., Ikematsu, M., Matuo, H., and Matuura, R. (1979). *J. Colloid Interface Sci. 69,* 209.

Müller-Landau, F., Cadenhead, D. A., and Kellner, B. M. J. (1980). *J. Colloid Interface Sci. 73,* 264.

Nagle, J. F. (1975). *J. Chem. Phys. 63,* 1255.

Nagle, J. F. (1976). *J. Membrane Biol. 27,* 233.

Ohki, S. (1967). *J. Theoret. Biol. 15,* 346.

Oldfield, E., and Chapman, D. (1972). *FEBS Lett. 23,* 285.

Overath, P., and Träuble, H. (1973). *Biochemistry 12,* 2625.

Pal, R. P., Chatterjee, A. K., and Chattoraj, D. K. (1975). *J. Colloid Interface Sci. 52,* 46.

Papahadjopoulos, D. (1977). *J. Colloid Interface Sci. 58,* 459.

Papahadjopoulos, D., Moscarello, M., Eylar, E. H., and Isac, T. (1975). *Biochim. Biophys. Acta 401,* 317.

Pethica, B. A., Mingins, J., and Taylor, J. A. G. (1976). *J. Colloid Interface Sci. 55,* 2.

Phillips, M. C., and Chapman, D. (1968). *Biochim. Biophys. Acta 163,* 301.

Phillips, M. C., Ladbrooke, B. D., and Chapman, D. (1970). *Biochim. Biophys. Acta 196,* 35.

Phillips, M. C., and Finer, E. G. (1974). *Biochim. Biophys. Acta 356,* 199.

Phillips, M. C., and Hauser, H. (1974). *J. Colloid Interface Sci.* *49*, 31.

Phillips, M. C., Evans, M. T. A., and Hauser, H. (1975). *Adv. Chem. Ser. 144*, 217.

Rayleigh, Lord, Pockels (1891). *Nature (London) 43*, 437.

Roberts, K., Österlund, R., and Axberg, C. (1976). *J. Colloid Interface Sci. 55*, 563.

Rogers, J., Lee, A. G., and Wilton, D. C. (1979). *Biochim. Biophys. Acta 552*, 23.

Sackmann, E., Träuble, H., Galla, H. J., and Overath, P. (1973). *Biochemistry 12*, 5360.

Schnepel, G. H., Hegner, D., and Schummer, U. (1974). *Biochim. Biophys. Acta 367*, 67.

Scott, H. L. (1974). *J. Theor. Biol. 46*, 241.

Scott, H. L. (1975a). *J. Chem. Phys. 62*, 1347.

Scott, H. L. (1975b). *Biochim. Biophys. Acta 406*, 329.

Scott, H. L., and Wood-Hi Cheng (1977). *J. Colloid Interface Sci. 62*, 125.

Shah, D. O., and Schulman, J. H. (1967). *J. Lipid Res. 8*, 215.

Shibata, A., Yamashita, S., and Yamashita, T. (1978). *Bull. Chem. Soc. Jpn. 51*, 2757.

Shimshick, E. J., and McConnell, H. M. (1973). *Biochemistry 12*, 2351.

Shinitzky, M., and Barenholz, Y. (1974). *J. Biol. Chem. 249*, 2652.

Silvius, J. R., Read, B. D., and McElhaney, R. N. (1979). *Biochim. Biophys. Acta 555*, 175.

Sklar, L. A., Miljanich, G. P., and Dratz, E. A. (1979). *Biochemistry 18*, 1707.

Snik, A. F. M., Kruger, A. J., and Joos, P. (1978). *J. Colloid Interface Sci. 66*, 435.

Spiker, R. C. Jr., and Levin, I. W. (1975). *Biochim. Biophys. Acta 388*, 361.

Spiker, R. C. Jr., and Levin, I. W. (1976). *Biochim. Biophys. Acta 433*, 457.

Standish, M. M., and Pethica, B. A. (1968). *Trans. Faraday Soc. 64,* 1113.

Stanley, E. (1971). *In* "Introduction to Phase Transition and Critical Phenomena,: p. 76. Clarendon Press, Oxford.

Stewart, T. P., Hui, S. W., Portis, A. R., and Papahadjopoulos, D. (1979). *Biochim. Biophys. Acta 556,* 1.

Susi, H., Sampugna, J., Hampson, J. W., and Ard, J. S. (1979). *Biochemistry 18,* 279.

Suurkuusk, J., Lentz-B.R. Barenholz, Y., Biltonen, R. L., and Thompson, T. E. (1976). *Biochemistry 15,* 1393.

Tachibana, T., and Hori, K. (1977). *J. Colloid Interface Sci. 61,* 398.

Tachibana, T., Yoshizumi, T., and Hori, K. (1979). *Bull. Chem. Soc. Jpn. 52,* 34.

Tardieu, A., Luzzati, V., and Reman, F. C. (1973). *J. Mol. Biol. 75,* 711.

Taylor, J. A. G., Mingins, J., Pethica, B. A., Tan, B. Y. J., and Jackson, C. M. (1973). *Biochim. Biophys. Acta 323,* 157.

Tinoco, J., and McIntosh, D. J. (1970). *Chem. Phys. Lipids 4,* 72.

Trahms, L., and Boroska, E. (1979). *Biochim. Biophys. Acta 552,* 189.

Träuble, H. (1972). *Biomembranes 3,* 197.

Träuble, H., and Eibl, H. (1974). *Proc. Nat. Acad. Sci. U.S. 71, 214.*

Trudell, J. R., Payan, D. G., Chin, J. H., and Cohen, E. N. (1974a). *Biochim. Biophys. Acta 373,* 141.

Trudell, J. R., Payan, D. G., Chin, J. H., and Cohen, E. N. (1974b). *Biochim. Biophys. Acta 373,* 436.

Trudell, J. R., Payan, D. G., Chin, J. H., and Cohen, E. N. (1975). *Proc. Nat. Acad. Sci. U.S. 72,* 210.

Ueda, I. M. D., Shieh, D. D., and Eyring, H. (1974). *Anesthesiology 41,* 217.

Ueno, M., Kawanabe, M., and Meguro, K. (1975). *J. Colloid Interface Sci. 51,* 32.

Verkleij, A. J., de Kruijff, B., Ververgaert, P. H., Tocanne, J. F., and van Deenen, L. L. (1974). *Biochim. Biophys. Acta 339*, 432.

Verma, S. P., and Wallach, D. H. H. (1976). *Biochim. Biophys. Acta 426*, 616.

Vochten, R., and Petre, G. (1975). *J. Colloid Interface Sci. 51*, 360.

von Dreele, P. H. (1978). *Biochemistry 17*, 3939.

Williams, R. M., and Chapman, D. (1970). "Progress in the Chemistry of Fats and Other Lipids.: 11, 1.

Wilson, K. (1979). *Sci. Am. 241*, 158.

Yamashita, T., Shibata, A., and Yamashita, S. (1978). *Bull. Chem. Soc. 51*, 2751.

Zatz, J. L., and Cleary, G. W. (1975). *J. Pharmaceutical Sci. 64*, 1534.

THE WETTABILITY OF SOLIDS BY LIQUID METALS

Ju. V. Naidich

Institute of Material Problems
Academy of Sciences, Ukranian S.S.R.
Kiev, USSR

ISBN 0-12-571814-4

I. INTRODUCTION

Contact between a metal melt and the surface of a solid high-melting substance or material occurs in many physical phenomena and some important technological processes. They include the processes of crystal nucleation and growth from the metal melt, various metallurgical processes-melting, casting, moulding of steel and alloys, soldering and welding, processes of powder metallurgy mineral production by liquid-phase sintering or impregnation of a porous high-melting framework by a liquid-metal binder. Contact systems involving metal melts take place in a number of operating plants where a metallic liquid acts as a working fluid-heat exchanger with liquid-metal heat carriers, nuclear fuel systems where the active substance is distributed as a stable, finely dispersed suspension in the liquid metal, etc.

The contact systems under consideration are usually characterized by developed interphase surfaces, a highly dispersed state of the solid phase, cavities, channels and spaces of small cross-section (capillaries) present in the solid phase. In such systems, as the process advances, the velocity and the direction, as well as the properties of the product obtained, depend on the state and properties of the interface, the degree of wettability of the solid phase by the liquid metal and the contact adhesion strength.

Let us consider several well-known formulae for capillary pressure:

$$p = \frac{k\sigma_{lv} \cdot \cos\theta}{r},$$

for the height of the liquid in the vertical capillary

$$H = \frac{2\sigma_{lv} \cdot \cos\theta}{rg\rho}$$

for the velocity of the movement of the liquid along the horizontal slot

$$L^2 = \frac{D\sigma_{lv} \cdot \cos\theta}{6\eta}\tau$$

for the adhesion strength of the liquid and the solid body

$$W_a = \sigma_{lv} \cdot (1 + \cos\theta)$$

where σ_{lv} is the surface tension at the interface of the liquid and gaseous phases, θ the wetting angle of the solid substance wettability by the liquid, r the capillary radius, D the capillary slot width, η the viscosity of the liquid, g the acceleration of the gravity, and τ the time.

Liquid penetration into pores and thin capillary channels of the solid phase and liquid spreading over solid surfaces are elementary processes that determine the state and properties of disperse capillary systems: solid body, liquid, gas (pores).

From the formulas cited above it follows that the progress of these processes is determined by the most important parameters of the system: the wetting angle of the solid body wettability by the liquid (θ) and the surface tension of the latter (σ_{lv}).

Bottom boiling (nucleation of carbon monoxide bubbles and steel decarbonization in the open-hearth bath, growth and removal of non-metallic products of deoxidation) depend respectively on the degree of wettability of the furnace bottom and the non-metallic inclusion of the liquid metal.

With a high degree of wettability (strong adhesion and the liquid to the solid surface), nucleation of bubbles at the interface steel-bottom is hampered or stops at all well wetted non-metallic inclusions with strong adhesion to the metal. The bubbles hardly coalesce and are not removed from the steel bath.

At a low degree of wettability of the solid body by the liquid metal ($\theta > 90^\circ$) it is impossible to realize the process of liquid phase sintering or impregnation of mineral bodies. Here liquid metal displacement from the solid framework pores is taking place. This is known in sintering technology as "sweating."

The degree of wettability and adhesion of the metallic binder to the high-melting component of cermet determines many properties of the material obtained: the degree of phases dispersion, character and uniformity of their distribution, the material strength, its physical and other characteristics.

While using the liquid metal as a heat carrier, film boiling occurs in the case of poor wetting of heat exchanger walls which sharply lowers heat removal. The solder badly wetting the surfaces being soldered does not penetrate into the gap between the parts to be connected and cannot form a strong bond between them. The process of soldering proves to be impossible.

Measuring the degree of wettability of solid bodies surfaces by liquid metals makes it possible to determine the energetic parameter of interaction between the pair of substances, i.e., the work of adhesion. For systems with a rather low degree of interaction (solid bodies insoluble in the liquid metal and nonwettable by it) it is practically impossible to obtain such characteristics by any other method (calorimetric, for instance). The method of measuring wettability for such purposes becomes the only one.

One can measure the degree of wettability, which depends on the liquid and the solid phase composition, and construct a composition-property diagram using the method of wettability by the liquid as the means of physico-chemical analysis.

Thus the study of contact and capillary properties of metal melts, the processes of wetting solid bodies surfaces by them, and metal adhesion to the solid surface constitutes one of the most important aspects of modern metallurgy, physics and physical chemistry of solid substances and metallic liquids.

The phenomena of wettability and adhesion have been studied during the 14th through the beginning of the 20th century in low-temperature contact systems formed by various solid bodies and organic liquids, water, mercury, among others. Solid body wettability by molten metals at high temperatures under the conditions of chemical interphasic reactions possesses a number of purely specific features and presents an independent problem.

The present paper deals with the problems of physico chemical wetting, contact and capillary phenomena in metal melts at high temperatures.

II. GENERAL LAWS OF THE WETTING OF SOLIDS BY LIQUIDS. WETTABILITY BY LIQUID METALS

The main thermodynamic relations of the wetting theory are the equations for wetting angle (θ) and the adhesion work (w_a):

$$\cos\theta = \frac{\sigma_{sv} - \sigma_{sl}}{\sigma_{lv}} \tag{1}$$

$$w_a = \sigma_{lv}\cdot(1 + \cos\theta) \tag{2}$$

where σ_{sv}, σ_{sl}, σ_{lv}, are interphasic specific surface energies at the interfaces solid body-gas, solid body-liquid and liquid-gas (Fig. 1).

Both of the above equations were proposed in 1804 by T. Young [1] without strict proof. These equations were then obtained theoretically by different methods [2-9] ranging from the mechanical equilibration of forces at the three-phase boundary to the use of variational methods [2,8,13] or from thermodynamic concepts [3,9]. Only lately has experimental confirmation of the essential validity of the above relations been obtained [10]. In this work an analysis and a discussion of the validity of both equations is given. It is, however, difficult to experimentally verify these relations due to the fact that until recently, there were no reliable methods of determining the interfacial tension at the boundary of a solid body.

The above equations have been verified by comparing the values of the work of adhesion. These values were determined experi-

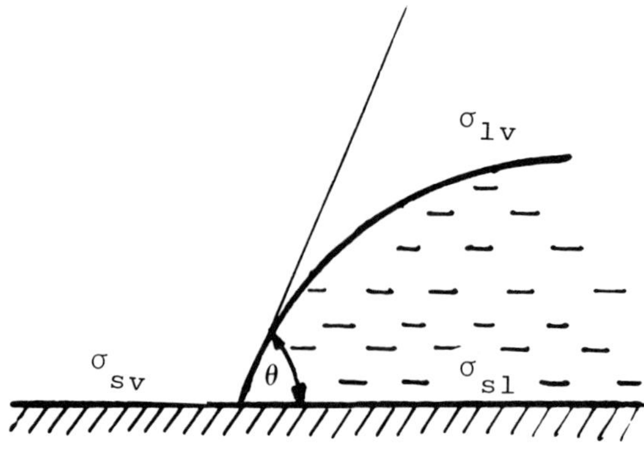

Fig. 1. The contact angle (θ) of wetting the solid surface by the liquid.

mentally by measuring critical slope angles of the solid body surface at which the liquid drop begins to roll off [8,11, and 12] which were calculated from equation (2) and by the experimental data for the contact angle and the liquid surface tension. Connection between the work of adhesion and the slope angle of the solid body surface (α) at which the liquid drop with the mass m and the base radius r begins to roll off is presented by the formula [8,11]:

$$w_a = \frac{mg \sin \alpha}{2r}$$

which has been verified experimentally.

Satisfactory correspondence of the values of the adhesion work on wetting various solid bodies by different liquids (water, mercury, glycerine) confirms the essential correctness of equation (2). Equation (1) can be readily obtained from 2 from the evident relation:

$$w_a = \sigma_{sv} + \sigma_{lv} - \sigma_{sl} \tag{3}$$

(Dupre equation). That is why the correctness of equation (2) confirms the validity of equation (1).

At the same time a number of phenomena complicate the process of wetting and spreading and make an impression of apparent unsuitability and incorrectness of the equations being considered. These complicating factors include hysteresis of wetting, the solid body surface slope in special cases of wetting (a liquid drop at the inclined surface), deformability of the solid surface, its non-uniformity and roughness and liquid vapor adsorption at the solid surface which decreases the surface tension of the latter. The effect of these factors is considered in [10].

A. *Interfacial Surface Energy and Contact Angles of Wetting in Equilibrium and Non-Equilibrium Systems*

When a liquid is in contact with the surface of a solid body, the system can be either in thermodynamic equilibrium or not in equilibrium during frequently large time periods. In particular, some high-temperature systems containing metal melts as a liquid phase which can react intensively with solid phases belong to the latter class. Thus, even in non-equilibrium systems, there are interfaces and interfacial surface energies connected with them. Differentiation of various classes of systems has been given little attention in literature up to now. Meanwhile, contact surface processes take place in various systems with inherent peculiarities. Equation (1) and (2) are derived for "equilibrium conditions" (the term which is to be specified in this case). There arises a question about the possibility of applying these equations to non-equilibrium systems.

Classification of the various systems into those in equilibrium and nonequilibrium and analyses of wetting phenomena in each type of system are given in [15].

In equilibrium systems, solid and liquid contacting phases are under conditions of thermodynamic equilibrium; the chemical potential of each component, and the temperature and pressure in each of the phases are the same. The contact of a liquid of a composition with a solid body of *b* composition at temperature T for a system having a phase diagram of the "cigar" type (Fig. 2a) can serve as an example for two-component systems as well as for the contact of the pure liquid A with the pure solid body B for the case of mutually immiscible components (Fig. 2b).

If the solid and the liquid phases, which correspond to an equilibrium composition at the given temperature and pressure are

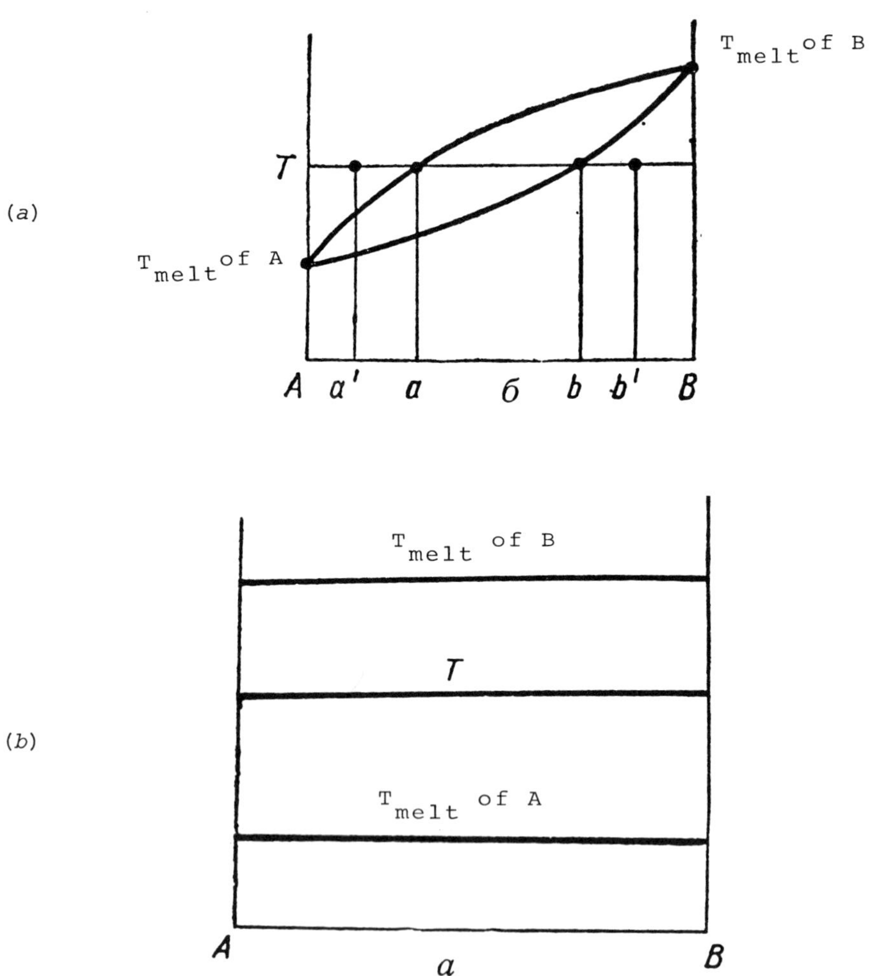

Fig. 2a,b. Types of contact systems (explained in the text)

brought into contact, then the interface between these faces is characterized by a certain interfacial energy and the system remains stable and in equilibrium. When contact between the liquid and the solid body is made, the process of adsorption can be observed at the interface, i.e. transition of some part of the substance into the surface or back but not through the interface. These processes are characterized by a finite velocity which can be rather small in the case of the solid phase. In the present treatment, instantaneous adsorption is assumed and any non-equilibrium of the system is due to the kinetics of a simple adsorption process and is ignored. At high enough values of the interfacial energy and the liquid phase surface tension, a certain non-zero contact angle of wetting of the solid phase by the liquid is established. The possibility of applying equation (1) to such systems is obvious. The equilibrium angle of contact in this case is accepted to be the wetting angle corresponding to the value determined by this equation.

The interfacial energy σ_{sl} in equilibrium systems depends on nature and structure of the contacting phases and decreases as their properties become increasingly similar.

For non-equilibrium systems one usually considers the case where the system is in thermal and mechanical equilibrium but not in chemical equilibrium. This means that temperature and pressure in each of the phases are the same, but that the components chemical potentials in the solid and the liquid phases may not be equal. Such a situation for a two-component system is presented by the contact of the liquid of composition a and the solid body of composition b (Fig. 2a). After the contact surface appears in such a system, processes will take place which will result in the equalization of the component's chemical potentials in the liquid and solid phases. Such a process could be the transition of one of the components from one phase into another (dissolution, diffusion) or one of several chemical transformations. Whatever the process, the system will shift to an equilibrium state. The interfacial energy σ_{sl}* of such a system will change with time and will depend on irreversible processes of chemical interaction at the interface. The action of these processes can result in a substantial interfacial tension decrease. Such a statement has been formulated in a variety of different ways by many authors. Thus A. M. Levin [14], who had studied the stability of refractories in liquid steel and formulated conditions for vigorous wetting of the solid phase by liquid metal, considered that it is necessary for the liquid and solid phases to be as far as possible from equilibrium contact conditions. In the work of Kingery *et al.* [17,18] and in our study [16,19] it has been found that a metallic liquid will moisten a solid metallic body if a sufficiently intensive chemical reaction is taking place between them.

Let us assume that liquid is chemically adsorbed at the solid body surface. The energy of the bonds formed between the liquid and the solid body is the adhesion work in accordance with the definition of the latter. Binding energy (the work of adhesion) released during the absorption reaction should be equal to the energy of the system, i.e., from the Dupre equation:

$$\sigma_{sl} = \sigma_{sv} + \sigma_{lv} - w_a ,$$

it follows directly that growth of the binding energy of the liquid and the solid body (w_a) results in an interfacial tension (σ_{sl}) decrease. Proceeding from this, energy released in the course of the reaction between the solid body and the liquid can be equated to the work of adhesion.

Jordane and Lane [20] pointed to an interfacial energy decrease on wetting solid metals by a liquid phase under nonequilibrium conditions, as a result of a chemical reaction and dissolution. Ono and Kondo [21], have applied the thermodynamics of irreversible processes and obtained the generalized Gibbs formula for interfacial tension under non-equilibrium systems. Apart from the usual factors

In the general case it is also σ_{lv}.

determining the variation of tension under equilibrium conditions (adsorption and temperature change) the equation includes factors determining interfacial tension change connected with non-equilibrium in the system, for example with the irreversible process of component transition through the interface.

A number of works have been devoted to research into the wettability of solid bodies by metal melts under given deviations of the system (difference of the component's chemical potentials in the liquid and the solid phases) from the equilibrium state [23,24].

Zhukhovitsky *et al.* [22] have calculated variations of interfacial tension for stationary processes (where a constant difference of the component's chemical potentials in the solid and the liquid phases is maintained):

$$\Delta\sigma = M\Delta\mu$$

where M is a constant. According to this simple relation, the dynamic variation of the interfacial tension is proportional to the difference of the component chemical potentials in the contacting phases. As the system approaches equilibrium, the difference of chemical potential decreases while the interfacial tension increases and stabilizes on reaching equilibrium. In [15] it is shown that for the systems where the solid phase is partially soluble in the liquid phase but the reverse is not true, there should be a minimum in the curve of interfacial tension vs. contact time if pure substances are initially in contact (Fig. 3). For systems with a layer of a new chemical compound forming at the interface, the melting point of which is higher than the temperature at which the process is taking place, a part of the first downward branch is realized. Evidently, the process is slowed down sharply and is already determined by the component rate of diffusion through the new boundary layer.

Aksay *et al.* [25] have considered a number of concrete cases of wettability accompanied by chemical interface reactions. The

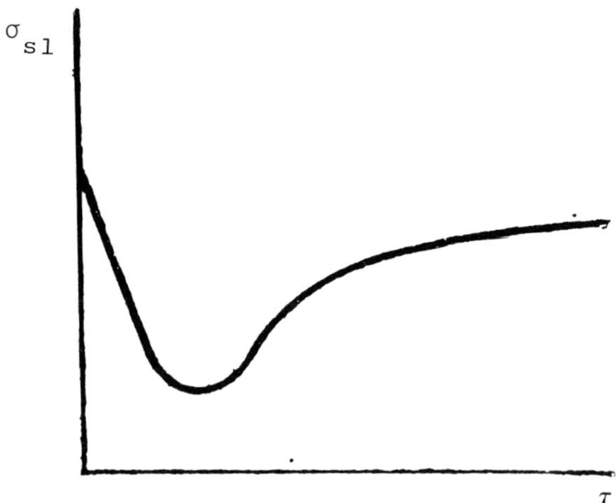

Fig. 3. The curve of interfacial tension as a function of contact time for non-equilibrium systems when the solid phase is partially soluble in the liquid.

curves of interfacial energy change in non-equilibrium systems where the solid phase is dissolving are similar to those in reference [15], and are presented in Fig. 3.

The adhesion work between the two phases 1 and 2 in the general case may be presented in the form

$$W_a = W_{a(equil)} + W_{a(non\text{-}equil)} \tag{4}$$

Interfacial tension is expressed as

$$\sigma_{1,2} = \underbrace{\sigma_1 + \sigma_2 - W_{a(equil)}}_{\sigma_{equil}} - \underbrace{W_{a(non\text{-}equil)}}_{\Delta\sigma_{non\text{-}equil}}$$

$$\sigma_{1,2} = \sigma_{equil} - \Delta\sigma_{non\text{-}equil} \tag{5}$$

where σ_{equil} is a function of the difference in the nature and properties of contacting phases P_1 and P_2: $\sigma_{equil} = f(P_1 - P_2)$ and $\Delta\sigma_{non\text{-}equil}$ is the function of the systems' deviation from the equilibrium state (the differences of the component chemical potentials) $\Delta\sigma_{non-equil} = \phi(\mu_i^1 - \mu_i^2)$.

Differences of the phase properties are evidently the original reason for a positive interfacial tension. The interfacial tension ($\sigma_{1,2}$) is equal to the value of σ_{equil} for the case when the phases 1 and 2 are under conditions of thermodynamic equilibrium at their contact, i.e. the component chemical potentials in these phases are equal and $\Delta\sigma_{non-equil}$ ($W_{a_{non-equil}}$) is zero.

When the phases properties approach each other, $P_1 \to P_2$, the tension σ_{equil} decreases and, when the phases become identical in the limit, the interfacial tension disappears and $W_{A(equil)_2}$ equates to the work of cohesion between the phases. At $\mu_i \neq \mu_i$ the interfacial tension is less than $\sigma_{sl_{equil}}$ by the value of energy of the chemical reaction taking place between the phases, $\Delta\sigma_{non-equil} = W_a$.

It should be noted that the tension decrease due to intensive chemical interaction at the interface can be high, i.e., that $\Delta\sigma_{non-equil}$ interfacial tension can become negative, i.e. an energy decrease can be observed at the interface. Such a system is considered to be thermodynamically unstable. However if one of the contacting phases is solid, the slugishness of mass transfer processes in the solid phase can prolong such a situation for a long time (metastable equilibrium). Dependence of σ_{sl}, θ and W_a on time is the characteristic feature of non-equilibrium systems.

The possibility of applying equations (1) and (2) in non-equilibrium systems where σ_{sl}, θ and W_a are the functions of time is discussed in [10] and [15].

One may recognize non-equilibrium spreading by the fact that the initial angle of contact will differ from that at equilibrium as determined by equation (1) (usually $\theta_{init.} > \theta_{equil.}$), but gradually assumes the equilibrium value. The dynamic force of spreading is

$$F = \sigma_{lv}(\cos\theta_{equil} - \cos\theta\tau) \tag{6}$$

where $\theta\tau$ is the value of wetting angle at the given moment of time.

Liquid inertia and viscosity are the forces which inhibit spreading. For liquid metals with low viscosity the process of spreading takes place at high speed since the variation of the angle of contact from about 180-160° to 30-50° occurs during approximately 10^{-2} seconds [26]. During the whole period of spreading, the application of (1) and (2) is evidently meaningless. For similar systems a process of slow spreading is also possible ("equilibrium" or quasi-static spreading). This can be frequently observed in metal melts* due to the change with time of the interfacial tension (σ_{sl}). The change in σ_{sl} is in turn due to the chemical contact reaction or diffusion at which the inertial forces and the liquid phase viscosity are small and equilibrium of surface forces is gradually realized during the process. For this latter case equations (1) and (2) are valid and by measuring the instantaneous contact angle value one can calculate the corresponding value of the interfacial energy and the work of adhesion.

B. The Nature of Adhesive Forces Determining the Wettability of a Solid by a Metal Melt

The wettability of a solid body by a liquid and the bond between the liquid and the solid body is determined, just as in adsorption, by two types of forces acting between the phases: (1) physical interactions which combines polarization and dispersion forces; (2) chemical forces, both ionic and homopolar. The basic difference, of significance here, between these two types of forces is their magnitude. The bond energies of physical (van der Waals) forces are functions or units of a kcal/mole. The energies of chemical interaction equal tens and hundreds of kcal/mole.

*****The authors [55] have observed the reduction of the contact angle of a copper-aluminum liquid alloy wetting uranium dioxide from 140 to 80° over a period of several hours. The contact angle of graphite wetting by stannus-titanium melts decreases to a stable value over 2-4 hours [10].*

Taking into account that the surface tension of most metals (excluding alkali metals) is of the order of 10^3 erg/cm^2, for spreading, or a significant degree of wettability, by a liquid metal on any solid body, the energy of interaction (the work of adhesion of the metal with the solid body) is of the same order of magnitude, i.e., 10^3 erg/cm^2. Taking the adhesion work to have this value

$$W_{a(mol)} \cong W_a \cdot S_M, \; S_M = \left[\frac{M}{\rho} \right]^{2/3} N^{1/3} \qquad (7)$$

(where M is the solid material molecular weight, ρ is density, N is Avogardo number, S_M is the area occupied by a mole of substance if it is extended into a monoatomic film) we shall obtain for W_a (mol) a value of tens of kcal/mole. Physical forces cannot provide such energy; this energy can be obtained by chemical interaction. Physical (dispersion) forces determine the wettability of liquids with low surface tension (water, organic solvents and so on) for which σ_{lv} and W_a constitute tens of dyne/cm. The distinguishing feature of "chemical" wetting, apart from the large value of interphase forces, is a comparatively strong dependence of the degree of wettability on temperature. Such dependence is often characterized by a "wetting threshold", i.e. by the presence of a temperature beyond which the contact angle begins to decrease sharply and the work of adhesion begins to increase. The character of the change of $W_a = f(T)$ can evidently in some cases be a characteristic by which this or that system can be related to a certain class. For systems with physical wetting it is possible to establish the connection between the degree of wetting and other properties of contacting materials, such as polarizability, ionization potential, dielectric constant, etc.

In the case of wetting determined by chemical interaction, it is necessary to allow for variables characterizing the chemical affinity of atoms from different phases, the difference of the component chemical potentials in phases, the reaction equilibrium contact, the free energy of formation of the compounds, etc. In connection with the characteristic feature of wetting in non-equilibrium systems (namely, its irreversibility) only the wetting angle of infiltration makes any sense.

The process of irreversible wetting can be described in turn within the framework of the thermodynamics of irreversible processes. The latter is developed today only for partially suitable cases (small deviations of the system from the equilibrium state, stationary processes). That is why in the present work, the usual thermodynamic interpretation of wetting as a chemical process reaching equilibrium (pseudo-equilibrium) in the near-contact layer is accepted and applied as far as possible.

1. The Determination of W_a (non-equil) values

The interaction energy W_a (non-equil) can be approximately calculated using thermodynamic concepts. Let us consider this approach for the contact between the solid body and the liquid.

Let the atomic plane of the solid phase A with a surface density of N_A moles/cm^2 be in contact with the atomic plane of the liquid B containing N_B moles/cm^2. Let the reaction

$$mA + nB \rightleftarrows A_mB_n$$

take place in the system A - B.

Keeping in mind that the surface energy is localized within a very short distance of the immediate surface of the substance, let us restrict ourselves to considering the interaction of atoms of only the first contact planes of the substance A and B which form. In accordance with Guggenheim [28,35] treatment, this is the surface phase. Let us calculate the energy released by the reaction taking place in this phase between n_A moles of A and n_B moles of B as the initial state (a mixture of n_A moles of A and n_B moles of B) to form the final equilibrium state in this phase characterized by the presence of all three substances A, B, and A_mB_n.

The concentrations of these substances, at any moment of time, can be expressed as functions of n_A, n_B and α where α, the degree of reaction conversion, is the number of A_mB_n moles formed.

As it is easy to see for the given reaction equation

$$C_A = \frac{n_A - m\cdot\alpha}{n_A + n_B + \alpha(1-m-n)} ; \qquad C_B = \frac{n_B - n\cdot\alpha}{n_A + n_B + \alpha(1-m-n)}$$

$$C_{A_mB_n} = \frac{\alpha}{n_A + n_B + \alpha(1-m-n)} .$$

In the initial state $\alpha = 0$, at equilibrium $\alpha = \alpha_0$ the energy variation accompanying the transition from any intermediate state at a given α to the equilibrium one, so that an infinitesimal number of reactants will take part in the process and an infinitesimal number of reaction products will be formed, is evidently expressed in the following way (reaction isotherm):

$$dZ = \left(RT \ln \frac{C_{A_mB_n}}{C_A^m C_B^n} - RT \ln \frac{C^0_{A_mB_n}}{C_A^{0m} C_B^{0n}} \right) d\alpha \ , \tag{8}$$

where the superscript 0 refers to the equilibrium state.

The total work of reaction equal to the adhesion work is expressed as:

$$W_a = \Delta Z = \int_0^{d_0} dZ \tag{9}$$

After integrating we shall obtain an expression for ΔZ as a function of α_0. The former is found from the expression for the constant of the reaction equilibrium

$$\Delta Z^0 = -RT \ln K = RT \ln \frac{C^0_{A_mB_n}}{C_A^{0m} C_B^{0n}} \tag{10}$$

where ΔZ^0 is the standard change of isobaric potential.

Thus for calculating $W_{a\,(\text{non-equil})}$ it is necessary to know the following characteristics of the system: n_A, n_B, the approximate mechanism of reaction and how the standard isobaric potential ΔZ^0

changes. Calculation of these formulae for different systems has been done in [10] and is also presented below.

2. The Value of $W_{a(equil)}$

The value of $W_{a(equil)}$ for systems with a high energy of interatomic bonding in each of the phases and a high surface energy (metals, high melting chemical compounds) can be written in two separate parts:

$$W_{a(equil)} = W_{a(chem.equil)} + W_{a(VDV)}$$

Here $W_{a(chem.equil)}$ is the cohesive energy of the solid and the liquid phases due to establishment of chemical equilibrium bonds, i.e. due to the mutual saturation of the free valences of the contacting surfaces. The establishment of such bonds is not accompanied by rupture or partial dissociation of interatomic bonds in each of the phases (the process characteristic for chemical irreversible interaction) taking place in non-equilibrium systems; $W_{a(VDV)}$ is the energy of van der Waals (physical) interaction (dispersion bond).

For many systems where the phases present differ in their nature (liquid metals, non-metal crystals: ionic compounds, covalent substances and compounds) the value of $W_{a(chem.equil)}$ is rather small (there is little or no saturation of surface valences). Against a background of weak chemical interaction, dispersion binding forces become pronounced; the low total value of the experimentally observed work of adhesion for these systems can be explained by the energy of dispersion interaction forces. This energy can be numerically estimated for some systems. For this purpose one can use the expression for the potential or dispersion interaction between a pair of atoms:

$$E \cong \frac{3}{2} \frac{\alpha_1 \alpha_2}{R^6} \cdot \frac{I_1 . I_2}{I_1 + I_2} , \qquad (11)$$

where α_1 and α_2 is polarizability; I_1 and I_2 are the first ionization potentials of atoms 1 and 2 and R is distance between them.

Neglecting entropy contributions and taking into account only interactions between each pair of atoms, we shall obtain $W_{a(equil)} = nE$ where n is the number of substance atoms/cm^2 of the surface area. Calculations of the energies of binding dispersion forces for the contact of solid bodies and liquid metals have been published in works by Benjamin and Weaver [29], MacDonald and Eberhart [30] and the author [10]. Knowing $W_a = W_{a(equil)} + W_{a(non\text{-}equil)}$ one can, with the help of equation (2) find the angle of contact.

III. METHODS OF MEASURING THE WETTABILITY OF A SOLID BODY BY A LIQUID METAL

The main characteristics of the contact capillary system (solid body, liquid and gas) are the angle of contact of the solid phase wetting by the liquid and the surface tension of the latter.

This chapter deals with a number of methods for determining these values applied to high-temperature metal melts. The sessile drop method and the plate weight method are most suitable for this purpose.

A. Sessile Drop Method

The first law of capillarity, a general differential equation determining the form of the liquid surface in the gravity force field for the case of the liquid drop resting on the horizontal solid surface and being the surface of revolution (Fig. 4), can be expressed in the following way [32]:

$$\frac{1}{\rho/b} + \frac{\sin \phi}{x/b} = 2 + Z/b \, \frac{(D-d)gB^2}{\sigma_{lv}} \tag{12}$$

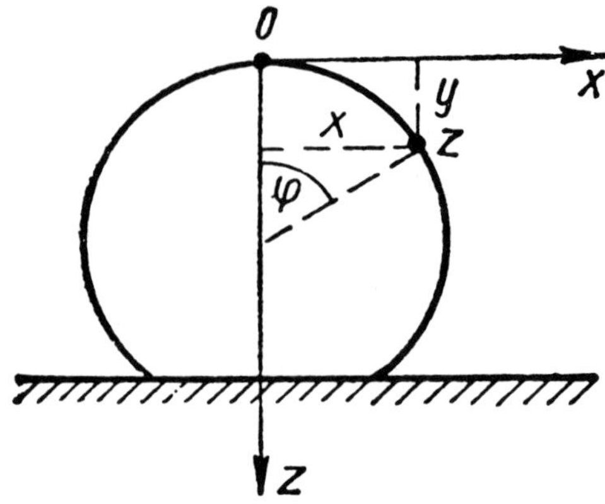

Fig. 4. Coordinates of the point at the liquid drop surface.

where ρ is the radius of the surface curvature in the plane of its meridional section (the plane of the drawing); b is the radius of the drop surface curvature at its top ($R_1=R_2=b$); expression $\frac{(D-d)gb^2}{\sigma_{lv}}$ is denoted by β; $D-d$ is the difference between the densities of the liquid and the medium (the latter can be neglected for a liquid-gas phase system); g is gravitational acceleration; x and z are coordinates of a point on the liquid surface.

The solution of this equation relates the point coordinates on the surface x, z, ϕ with the liquid surface tension at the interface with the medium σ_{lv}, the liquid density (difference of the liquid and medium densities) and the gravitational acceleration. By measuring the point coordinates at the liquid drop surface one can express the liquid tension through its density. From the equation solution one can also find the liquid surface inclination at any point including the point of contact with the solid surface, i.e. the contact angle of wettability.

Thus the sessile drop method allows us to determine the values of σ_{lv} and θ. Having the equation of the liquid surface shape, it is possible to determine the volume enclosed in it, i.e. the drop volume, and hence to determine the liquid density. Though the solution of equation (12) cannot be presented in the definite form, nevertheless its integral can be obtained approximately with a known degree of accuracy.

There are different methods of approximate integration of equation (12) (the list of approximate formulae is given in [4, 31,32], those for graphical integration including [36]). The most accurate solutions of this equation (for figures of revolution) are the tables obtained in the work of Bashfoth and Adams [34].

The formula for calculating σ is of the form

$$\sigma = \frac{g\rho b^2}{\beta}, \tag{13}$$

where b and β are determined from Bashforth's and Adams's tables [34] by the measured coordinates x, z, ϕ. Measurements of x and z are usually carried out at the selected ϕ, most often (if $\theta > 90^\circ$) at $\phi = 90^\circ$. Such measurements are the most accurate ones. Measurements of $\phi < 90^\circ$, which are obtained with good substrate wetting by the liquid, are less accurate, but, on observing special measures, they can also yield good results.

To determine the values of the x, z and ϕ coordinates of the drop contor, the image of the latter is photographed (the print should be of very high quality). Judging by the print, the maximum drop diameter (2x), the distance to the apex z at $\phi = 90^\circ$ are determined with the measuring microscope.

There are other methods of measuring photographs [37]. From the photograph, one can either calculate the angle of contact using Bashforth's and Adams's tables, or directly measure it by drawing a tangent to the drop contor near the wetting perimeter.

Apart from the ordinary method of drawing a tangent to the drop contor, the angle of contact can be calculated from the drop dimension (height h and diameter of the wetting perimeter 2a) by the formulae for spherical segment (assuming that the drop surface is spherical, an assumption which is valid for sufficiently small drops):

$$tg\theta = \frac{2ha}{a^2 - h^2} \qquad (14)$$

or from the drop volume and height

$$\cos\theta = 1 - \frac{3\pi h^3}{3\nu + \pi h^3} \qquad (15)$$

In several variants those methods were applied in the works by Ferguson [33], Volkova [39], Bikerman [40], Wolfram and Weber [41] and Levin [14].

If the drop is rather large so that its curvature at the apex can be neglected, the angle of contact is determined by the expression

$$\cos\theta = 1 - \frac{Dgh}{2\sigma_{lv}} \qquad (16)$$

The above formula is valid only for rather large liquid drops (diameter of 5-10 cm and more) [33].

The sessile drop method has some advantages which are of great importance while working with melted metals. They are principally the simplicity of the method, the need for only small quantities of material for the liquid and solid phases (the liquid phase quantity may sometimes constitute even less than 10^{-2}-10^{-4} g.) and the simple shape of the solid body. This method does not require direct access to the sample during measurements which can be conducted at substantial distances from the object so that the latter can be placed in a special closed apparatus, to be heated and so on. The main disadvantage of the method consists in non-reproducibility of the liquid surface which accumulates contaminants and impurities.

Nevertheless, this disadvantage can be eliminated by observing special precautions and carrying out the test carefully with highly purified materials.

B. The Plate Weight Method

The essence of this method is as follows. The plate, partially submerged in the liquid, is subject to an additional capillary force tangential to the plate surface and normal to the wetting perimeter [38,43]. This force can be expressed by projecting the vector of the liquid surface tension near the wetting perimeter onto the plate plane

$$f = \sigma_{lv} \cdot \cos\ \theta \cdot \ell$$

where ℓ is the wetting perimeter length.

The origin of this force can be understood from the following simple considerations. In Fig. 5 a plate is shown submerged in a liquid. The plane of the diagram intersects the plane normal to the plate plane and that of the liquid surface. The free energy of such a system can be expressed as

$$F(x) = 2b(H-x)\sigma_{sv} + 2bx\sigma_{sl} + S_{lv} \cdot \sigma_{lv} + \phi \qquad (16)$$

where x is the depth of the submerged plate, H is the plate height, b is its width (the thickness is assumed to be negligible), S_{lv} is

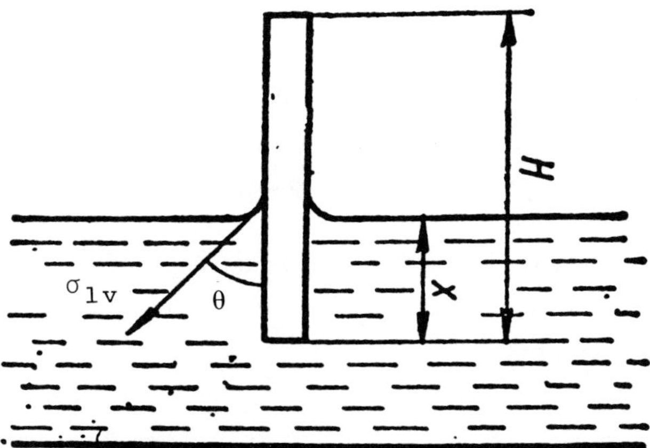

Fig. 5. The derivation of the force equation for the plate immersed into a liquid.

the liquid surface area which borders on the gas phase, ϕ is the volumetric contribution to free energy. The plate weight is not taken into account for simplicity. Then the projection of the capillary force in the *x* direction is determined by

$$f_x = \frac{\partial F(x)}{\partial x} = 2b(\sigma_{sv} - \sigma_{sl}) = -2b\sigma_{lv} \cdot \cos\theta$$

In practice it is necessary to allow for archimedean force and make corrections for edge effects.

Using the plate weight method one can immediately obtain the contact angle value averaged over the whole wetting perimeter, which is rather large in this method (about 2-3 cm).

Really, according to this method, the total force which pulls in or pushes out the plate is

$$f = \sigma_{lv} \sum_i \Delta \ell i \cos\theta_i$$

where $\Delta \ell i$ is the wetting perimeter section along which the angle of contact maintains the constant value θ_i.

The angle of contact being measured is obviously determined by the expression:

$$\cos\theta = \frac{\Sigma\, \Delta \ell i \cos\theta_i}{\ell}$$

where $\ell = \Sigma\, \Delta \ell i$ is the wetting perimeter length.

C. Instrumentation for Measuring Surface Tension and Wetting Angle

The main requirements for measuring surface tension and wetting angles by the sessile drop method in the placement of a symmetrical drop of the melt on the solid surface under study (the substrate) are: a horizontal position of the surface, an enlarged image, as clear as possible, of the melted drop profile and solid substrate parallel to the surface of material being studied. The sample should be in a controlled gaseous atmosphere or in a vacuum at the temperature specified.

These requirements are met by means of devices including the bumping system (vacuum system), the temperature control and the photooptical system. The latter should have a considerable number of degrees of freedom in arranging the components in order to carry out different adjustments of the system.

A standard apparatus for studying the wettability of solid bodies by liquid metals using the sessile drop method is shown in Fig. 6. The vacuum chamber of the apparatus is a steel vessel with a volume of about 15 liters which encloses a horizontal tubular furnace with a sample. The chamber is evacuated down to the pressure of 10^{-5} to 10^{-6} Torr. The pump exhaust is connected to the line of preliminary rarefaction made of stainless steel tubes with 100 mm diameter and evacuated by a rather powerful forevacuum pump removed from the laboratory room to a special compartment. The experience of such forevacuum lines operation demonstrated their reliability and convenience.

Since getting the apparatus into working order (obtaining the needed vacuum at the required high temperature) takes a considerable time, the apparatus includes a device which permits the changing of samples in the furnace at the working temperature without breaking vacuum.

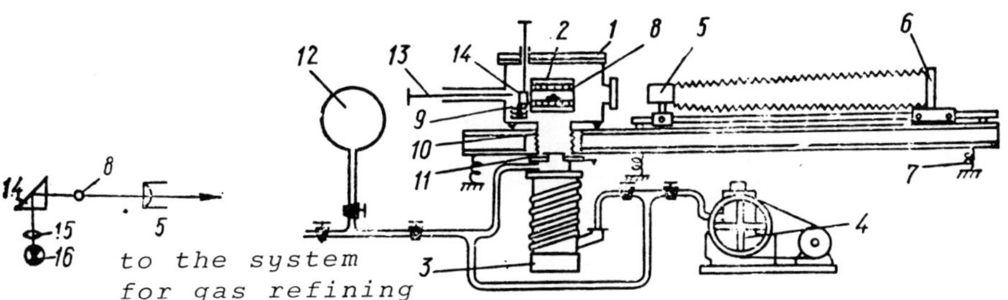

Fig. 6. Scheme of the apparatus for determining surface tension and wetting angles of metallic liquids. 1-vacuum chamber; 2-furnace; 3-steam-oil pump; 4-backing pump; 5-objective; 6-cassette portion; 7-springs; 8-metal sample on the substrate studied; 9-next samples; 10-bellows; 11-vacuum valve; 12-tanks for purified inert gases; 13-horizontal rod; 14-quartz prism; 15-condenser; 16-light source; 17-vertical rod.

Apart from high efficiency for sample handling that such a device provides, it is especially important for studying metals with high chemical activity. Using this device it is possible to heat and degas the apparatus and the furnace components, to obtain the necessary vacuum and temperature (the sample is contained in a special compartment out of the furnace in a cold state and does not interact with residual gases) and only then to introduce the sample under study into the furnace space.

Other modifications provide for batch delivery of the various melts directly to the drop (introducing additives into the drop during the test without breaking vacuum [44]), for the liquid metal drop penetrating the solid surface under study [45], and for touching the drop top (melted on the neutral additional substrate) with the surface of a solid phase under study and establishing the contact angles near this surface [46]. The latter procedure was used to study the kinetics of spreading [26].

Attachment lens of different types permit shooting the spreading process at frame frequencies from 16 to 3000 per second.

The whole assembly (optical bench and camera) is mounted on a solid frame placed on damping springs ("floating" structure) protecting the assembly from vibration which deteriorates the drop contour definition on the print and distorts it. For operation in an inert atmosphere a system for gas refining and drying is incorporated.

The ordinary sessile drop method for determining metal surface tansions is not accurate enough. The measuring error reaches 5 per cent. That is why, while studying such processes as the variation of surface tension with temperature, the magnitude of which is rather small, this method is not applicable.

An improved version of the sessile drop method has been developed and is called the "large drop" method [42]. Here a metal drop of a rather large size is formed at the edge of the flat crucible, "a cup" with sharp edges forming a circle of the required diameter (usually $\sim$10 to 20 mm) (Fig. 7). The base of the drop

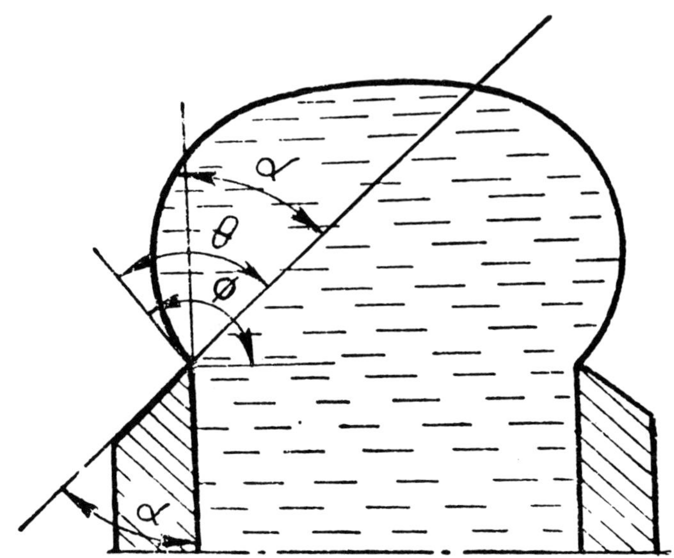

Fig. 7. Liquid drop in the cup ("large drop" method).

takes the form of the cup edge and is symmetric (circular). This produces a strictly symmetrical form of the drop and results in high accuracy and reliability with this method (errors are about 0.3 to 0.5%).

An analysis of the methods of exactly measuring the contact angles and the effect of various factors (posity and roughness of the surface, its slope, etc.) is given in reference [10].

An increase in the surface roughness results in a corresponding increase in the contact angle (infiltration angle). For quartz glass wetting by copper and tin the transition from a polished to a roughly-polished surface for the solid phase causes the contact angle to grow by 17 to 18° [10]. In other work [250] still greater growth of the contact angle was observed for liquid aluminum on the smooth (θ = 53) and rough (θ = 114) surface of titanium nitride and on the smooth (θ = 67) and rough (θ = 129) surface of titanium diboride. In the latter cases evidently an additional effect is superimposed on that of the rough surface. On observing necessary hygienic conditions for the experiment, on having a solid substrate with a high degree of smoothness and maintaining a true horizontal position, errors in determining the contact angle do not exceed 1 to 2°.

IV. THE WETTABILITY OF DIFFERENT CLASSES OF SOLIDS BY LIQUID METALS

While studying and analyzing wettability phenomena in well-defined systems, it is expedient to divide the latter into a number of classes in accordance with the physico chemical nature of the solid phase: (1) Metal melt - ionic compound (high-melting metal oxides, salts), (2) Metal melt - substance or compound with preferably covalent character interatomic bonding (diamond, graphite, carbide, nitride phases), (3) Metal melt - high-melting metals or metal-like compounds. In the first two types of systems, solid bodies are characterized by a closed stable electronic configuration or atoms, and strong saturated interatomic bonding. Interaction of most solid bodies of these types with metals is possible only with partial or complete dissociation of the interatomic bonding in the solid bodies. Such solid bodies are somewhat less reactive than metal melts. With solid bodies, a key feature in systems of the first type is the large size of anion and its high polarizability as compared to cation. In ionic bodies, the surface is formed by anions, and their interaction with metal melts is determined mainly by the interaction with the solid body anions.

In the third type of system, the establishment of equilibrium interphase metal bonds with surface atoms is possible without dissociating bonds within the solid phase. Wettability in these systems is greater than with the first two.

The characteristics of wettability in each type of contact systems are set out below.

A. The Wettability of Ionic Compounds by Metal Melts

Those systems formed by high-melting metal oxides are of the highest practical value. Interatomic bonds in these compounds are essentially of ionic nature.

High-melting metal oxides are as a rule poorly wetted by liquid metals. So contact angles of mercury, tin, lead silver, copper, nickel, cobalt, iron and platinum on high-melting oxides of aluminum, magnesium, silicium and berillium are of the order of 120 to 150°. However even the initial qualitative observations of metal-oxide contact have shown that some metals (titanium, zirconium, barium, aluminum and so on) intensively wet the oxide surfaces [17,47-51]. The factors which govern the phenomenon of oxide wettability by liquid metals have been obscured for a long time.

Weyl [52] considered that metals which are able to form highly charged ions (Sn^{+4}, Al^{+3}) should wet ceramic surfaces better due to an enhanced electrostatic interaction. Kingrey also thought that formation of high charge metal ions would be favorable for metal-ceramic binding. The idea of purely electrostatic metal-oxide binding was also supported by Cabrera and Mott [54] and Livey and Murray [55] who felt that this interaction was essential in metal systems.

While we do not completely neglect such a binding mechanism, it should be noted that it is not the basic one. One can give many examples when it is not possible to explain the wetting phenomena being observed in these terms. Thus, barium (Ba^{+2}) wets and impregnates aluminum ceramics as well as aluminum [47]; sodium (Na^{+1}) spreads over glass [56] and uranium dioxide [55]; silver (Ag^{+1}) spreads over copper oxide [48]. Tin, which according to Wegl has great affinity to oxides, does not actually wet high-melting oxides (Al_2O_3, MgO, ZrO_2, etc.) as accurate quantitative study has shown. According to [54] and proceeding from electrostatic theory, metals in general should effectively wet oxides since the latter are polar. In reality quite an opposite picture is observed.

Brace [49] and Bondi [56] tried to explain the wettability of oxides by some liquid metals. They supposed that the liquid metal reduced the solid oxide surface layer to the metal, and that the latter was well wetted by the metal melt. This explanation is important in that it points out potential wetting process chemistry in oxide-metal systems.

As a result of studying wettability and contact interaction between oxides and various liquid metals the following important features have been established:

(1) On the basis of the works of Kingrey *et al.* [17,18,53,57, 61-65], this author and Eremenko [16,19,95], Popel *et al.* [68,72-76], Armstrong *et al.* [66,67] and other studies [29,59,60,77,79, 80-83,102,103] it can be stated that wettability of oxide by a metal and adhesion in such a system increases with growing affinity of the liquid phase metal for oxygen. This is indicated by variations in the thermodynamic potential of metal oxide formation (ΔZ).

Chemically oxygen-active metals such as titanium, zirconium, aluminium, silicium, manganese and lithium (either pure or in the form of alloys) form small contact angles and spread completely over certain oxide (Al_2O_3, BeO, SiO_2, MgO) surfaces.

In some works [110,111], a natural decrease of the wettability of some oxides (Y_2O_3, Al_2O_3, SiO_2, TiO_2) by iron, nickel, cobalt, aluminium, silicium and tin is accompanied by an increase in the free energy of metal oxide formation. However such dependence allowing only for the properties of one component of the contact pair (the metal) is one sided. It is obligatory to take into account the properties of the other component (the oxide) as well.

(2) On the basis of general considerations, it can be expected that interaction of any substance with a chemical compound will be the less intensive the stronger the intramolecular bonds are. Thus one may consider that the weaker the bonds between the metal and the oxygen in the oxide, the greater the chemical interaction and the wetting of oxides by metals should be.

Research into the wettability of binary oxide phases-solid solutions by metal melts [10,19,71,83] depending on their composition has established a general correlation between the value of the oxide's electrical conductivity (its intrinsic conductivity), its thermodynamic stability or free energy of formation, i.e., the strength of the metal-oxygen bond, and its wettability by a metal. Compounds with a smaller free energy of formation, formed from the metal and oxygen (which have, as a rule, higher electrical conductivities), are better wetted by liquid metals [81,10] (Fig. 9). Similar dependences of wettability on the composition of the oxide solid solutions have been obtained by Tikkanen [71]. Data on the growth of the wetting contact angle of the oxide by the liquid metal with increasing bond ionicity [112] and on the oxide space lattice energy [110,111] are in agreement with the above correlation.

It is, however, necessary to allow for the following circumstances. The surfaces of the highest-melting oxides are formed by oxygen anions of large size. Metal cations of smaller size are displaced from the surface and move to the interior of the oxide. In the process of formation of a fresh ionic crystal surface, ions

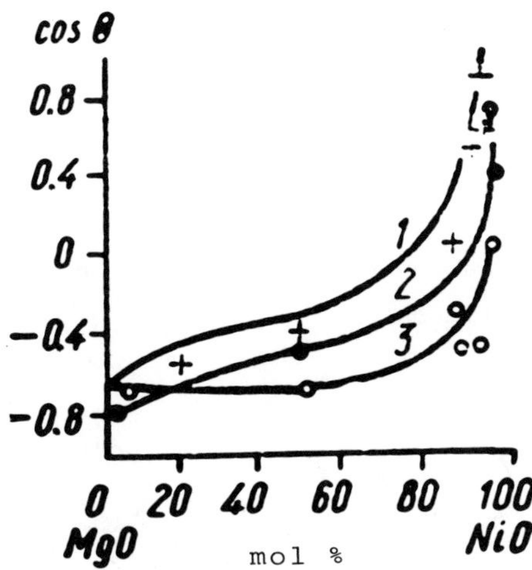

Fig. 8. Wettability by metal melts of binary oxide phases (solid solutions magnesium oxide-nickel oxide) depending on the composition: 1-tin; 1000°C; 2-copper; 1200°C; 3-silver; 1100°C.

of the boundary plane affected by the vector field of the bulk ions move deep into the crystal in a direction normal to its surface. Due to the different size and polarizability of the various kinds of ions the extent of their penetration varies. Madelung [84] was evidently the first to pay attention to this phenomenon which was also discussed in [85]. Metal ions polarizability is much lower than that of oxygen. The polarizabilities of the ions Mg^{+2} and Al^{+3} for instance are $0.094 \cdot 10^{-24}$ and $0.052 \cdot 10^{-24}$ cm^3 (according to Pauling); oxygen ion polarizability (O^{-2}) is $3.88 \cdot 10^{-24}$ cm^3. This circumstance and the relatively large size of the anion lead to the fact that displacement of the surface oxygen ion under the influence of a large crystal field deformation is smaller than that of the metal ion. According to Weyl [86] the oxide surface structure is of the form presented in Fig. 9.

On the basis of these facts, one can state that chemical interaction between a liquid metal and an oxide surface is essentially a metal interaction with the oxygen of the oxide, which can be written as follows:

$$Me'' + M'O \rightleftarrows M' + Me''O$$

The change in the thermodynamic potential due to this reaction is:

$$\Delta Z = \Delta Z'' - \Delta Z'$$

where $\Delta Z''$ and $\Delta Z'$ are the thermodynamic potential changes associated with the oxidizing reaction of the liquid metal and the metal forming the solid oxide.

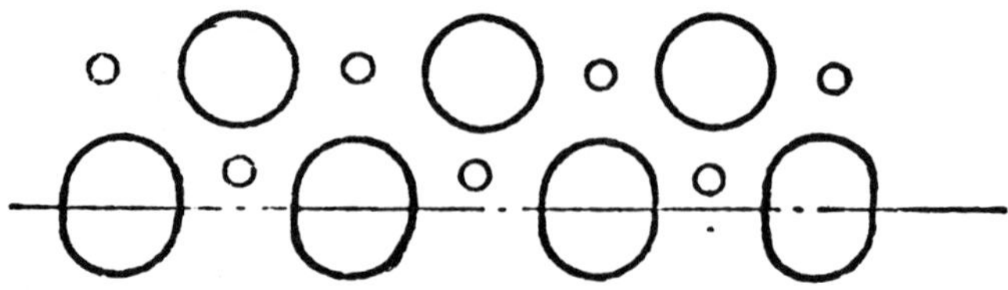

Fig. 9. The surface structure of high-melting oxides.

Just as the degree of solid oxide wettability by the liquid metal is to be associated with ΔZ, the contact angle and the work of adhesion (to be more exact, the non-equilibrium part of the work of adhesion $W_{a(n.-eq.)}$) should also be associated with ΔZ. The results of comparing wettability and thermodynamic properties of contacting components ΔZ are presented in Fig. 10 and Table 1.

In spite of data scattering there is a distinct tendency for θ to decrease with decreasing ΔZ absolute values.

At high positive values of ΔZ, the intensity of metal-oxide chemical interaction is low while contact angles are large. In this case, however, for the more stable inert systems, (Pt-BeO, Pt-Al_2O_3, Au-BeO for instance (Table 1)), angles of contact are other than 180°, i.e. the adhesion work has a finite value and is 10^1 to 10^2 erg/cm^2 (fractions and whole units of kcal/mol).

One can assume that this energy arises from van der Waals dispersion interactions which, being universal, take place in all

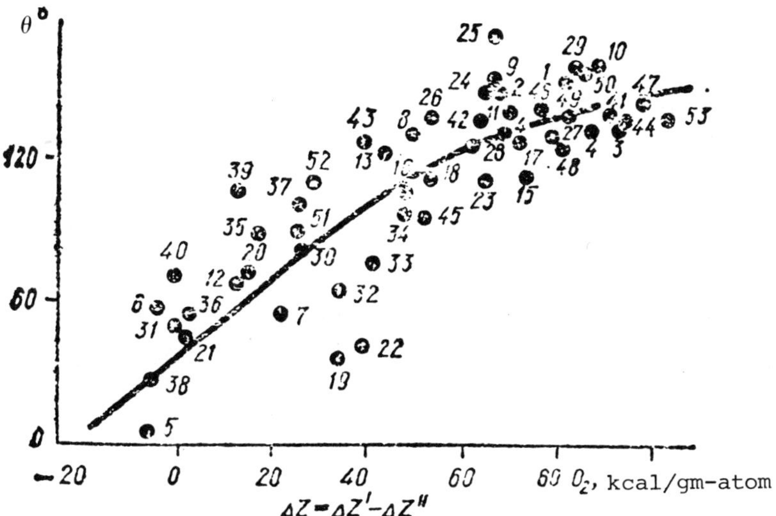

Fig. 10. Correlation between the degree of wettability (wetting angle) and the difference of affinity to oxygen of the liquid phase metal and the solid phase oxide metal. The digits correspond to the system numbers from Table 5.

TABLE 1. The wettability of various oxides by liquid metals and the difference of free formation energies of the solid phase oxide and of the oxide formed by liquid metals

No	System	t°C	θ	$\Delta Z = \Delta Z'' - \Delta Z'$ $\frac{kcal}{g \cdot atom\ O}$	References
1	2	3	4	5	6
1	Beo-Ni	1500	152	82	[18]
2	BeO-Fe	1550	147	64	[18]
3	BeO-PE	900	132	92	[10]
4	CaO-Fe	1550	132	72	[14]
5	CoO-Sn	900	0	7	[10]
6	CoO-Ni	1500	58	4	[10]
7	Fe_3O_4-Cu	1200	57	22	[10]
8	MgO-Fe	1550	130	49	[18]
9	MgO-Ni	1550	152	65	[18]
10	MgO-Cu	1150	160	89	[18]
11	MgO-Sn	1100	139	70	[10]
12	Nio-Cu	1200	68	12	[10]
13	SiO_2-Ni	1500	125	44	[10]
14	ThO_2-Ni	1500	132	86	[18]
15	ThO_2-Fe	1550	111	71	[18]
16	ZnO-Ag	1000	106	48	[55]
17	ZrO_2-Ni	1500	130	55	[18]
18	ZrO_2-Fe	1550	111	52	[18]
19	BeO-V	1800	35	34	[10]
20	ZrO_2-Si	1450	71	13	[18]
21	UO_2-Al	1100	46	2	[55]
22	Uo-Na	500	40	39	[55]
23	UO_2-Sn	1100	110	64	[55]
24	Al_2O_3-Ni	1500	150	67	[10]
25	Al_2O_3-Sn	1100	174	66	[57]
26	Al_2O_3-Fe	1550	141	53	[64]
27	Al_2O_3-Pb	900	132	77	[10]
28	Al_2O_3-Co	1500	125	62	[48]
29	Al_2O_3-Cu	1100	155	82	[10]
30	Al_2O_3-Si	1450	82	25	[18]
31	Al_2O_3-Al	1250	48	0	[55]
32	Al_2O_3-Cr	1900	65	35	[10]
33	BeO-Si	1450	76	41	[18]
34	BeO-Cr	1900	100	48	[10]
35	Cr_2O_3-Fe	1550	88	16	[10]
36	Fe_3O_4-Sn	1000	52	4	[10]
37	MgO-Si	1450	101	20	[18]
38	NiO-Sn	1000	27	7	[10]
39	TiO_2-Si	1450	107	13	[18]
40	CoO-Co	1500	70	0	[71]

Table 1 (continued)

No	System	t°C	θ	ΔZ=ΔZ"-ΔZ' kcal/g·atom 0	References
1	2	3	4	5	6
41	CaO-Ni	1500	135	92	[71]
42	SiO_2-Cu	1100	134	62	[10]
43	SiO_2-Sn	900	127	40	[10]
44	BeO_2-Pt	1780	125	92	[10]
45	ThO_2-Cr	19000	92	52	[10]
46	SiO_2-Au	1100	140	75	[10]
47	Al_2O_3-Au	1100	138	98	[10]
48	UO_2-Cu	1100	125	80	[79]
49	UO_2-Bi	700	140	82	[69]
50	UO_2-Bi	700	155	85	[69]
51	UO_2-Si	1420	90	26	[55]
52	CdO-Ag	970	112	27	[55]
53	MgO-Ag	1235	136	102	[55]

systems. The energy of this type of interaction can be evaluated. The order of magnitude (usually several kcal/mol) of the energy of these forces can be compared with the observed work of adhesion for systems with a low degree of wettability.

Wettability increases as ΔZ positive values decrease, the trend continuing as ΔZ assumes negative values.

A number of pure elements or additions to inactive metals, having a high affinity for oxygen, effectively wet oxide materials. One of the mėtals which has been extensively studied and used in practice is titanium. The wettability of oxides by titanium-containing melts Cu-Ti, Au-Ti, Ni-Ti, Ga-Ti, Sn-Ti and so on, has been investigated in a number of works [57,79,82,101-103,105,179]. Some data are summarized in Tables 2 and 3 and in Fig. 11. A sharp decrease in the contact angle at low elemental concentrations is a characteristic feature for titanium and its alloys which are interfacially active (Fig. 11). Because of influence of adsorption, the titanium concentration at the liquid phase-solid oxide interface is much higher than that within the

TABLE 2. Contact angles for titanium containing melts on high melting oxides.

System	Temperature °C	contact angle (degree), Ti content, % (at.)							
		0	1	2	3	4	6	8	10
Au-Ti-Al_2O_3	1150	135	100	90	81	76	69	66	64
Au-Ti-Ti_2O_3	1150	113	96	85	77	72	67	65	-
Cu-Ti-Al_2O_3	1150	129	88	50	40	32	21	14	-
Cu-Ti-Al_2O_3	1150	148	108	58	40	32	21	14	-
Cu-Ti-MgO	1150	133	95	61	43	36	26	-	-
Cu-Ti-SiO_2	1150	128	72	45	40	35	27	-	-
Cu-Ti-$TiO_{0,86}$	1150	72	67	61	55	49	40	31	-
Cu-Ti-$Ti_{1,14}$	1150	82	75	68	63	58	48	40	-
Cu-Ti-Ti_2O_3	1150	113	90	62	45	32	21	15	-
CuAg(28:72)-Ti-Al_2O_3	980	138	91	85	-	-	-	-	-
Ni-Ti-Al_2O_3	1500	110	104	99	95	92	87	83	77
NiMo(58:42)-Ti-Al_2O_3	1500	114	97	86	77	71	66	64	62
Sn-Ti-Al_2O_3	900	131	80	79	70	67	60	-	-
Sn-Ti-Al_2O_3	1000	127	71	65	60	56	48	-	-
Sn-Ti-Al_2O_3	1150	127	60	49	44	42	38	35	-
Sn-Ti-SiO_2	900	132	60	50	42	35	27	-	-
Sn-Ti-SiO_2	1000	127	52	36	26	18	10	-	-
Sn-Ti-SiO_2	1150	125	47	30	18	11	4	-	-
Sn-Ti-SiO_2*	1150	152	66	37	19	6	0	-	-

*Note: Surfaces of oxide materials are polished. In systems indicated by *) the surface is rough with a microroughnesses height of about 5 microns.

TABLE 3. The work of adhesion in titanum containing melts on high melting oxides at 1150°C, erg/cm^2.

System	Content T_i, % (at.)							
	0	1	2	3	4	6	8	10
Au-Ti-Al_2O_3	308	909	1100	1272	1366	1494	1547	1582
Au-Ti-Ti_2O_3	688	1000	1291	1480	1491	1540	1565	-
Cu-Ti-MgO	422	1152	1894	3314	2317	2432	-	-
Cu-Ti-Al_2O_3	461	1331	2100	2265	2369	2470	2522	-
Cu-Ti-SiO_2	474	1503	2185	2265	2330	2422	-	-
Cu-Ti-Ti_2O_3	739	1280	1861	2184	2365	2475	2517	-
Cu-Ti-$TiO_{1,14}$	1458	1611	1759	1864	1967	2136	2260	-
Cu-Ti-$TiO_{0,86}$	1651	1792	1894	2020	2125	2265	2381	-
Ni-Ti-Al_2O_3*	955	1100	1220	1320	1400	1525	1625	1775
Sn-Ti-Al_2O_3	185	697	770	799	810	831	-	-
Sn-Ti-SiO_2	198	782	867	903	921	928	-	-

*Research temperature 1500°C.

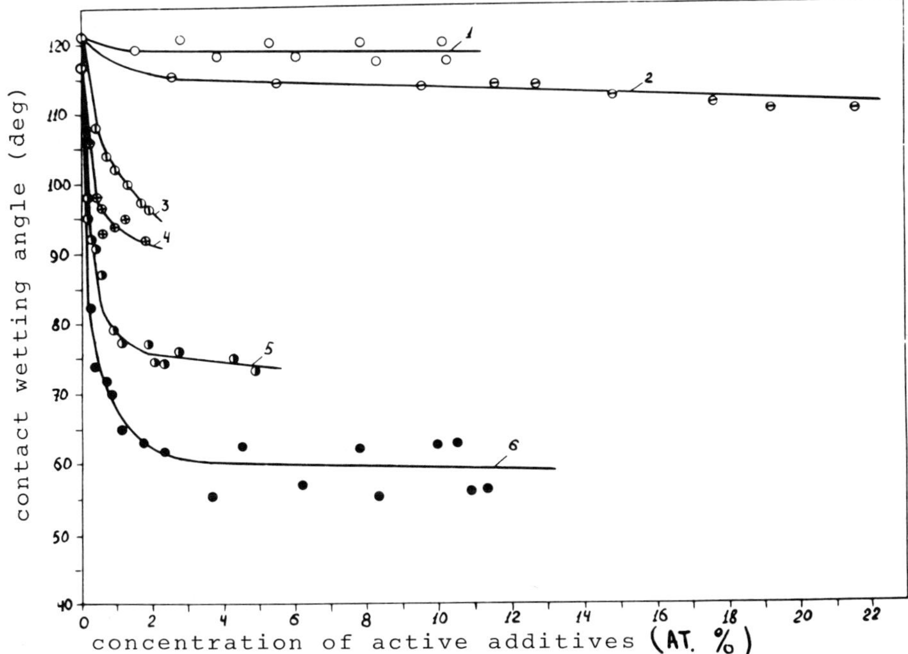

Fig. 11. Wettability of monocrystal aluminum oxide (plane 0001) by gallium-chromium (1,2), gallium-vanadium (3,4) and gallium-titanium (5,6) melts at 900°C (1,3,5) and 1050°C (3,4,6).

liquid phase. This has been established by direct measurements with a local micro-X-ray spectroscopic analysis. X-ray spectroscopic and microscopic studies have found oxide phases of TiO or Ti_2O_3 in the contact zone [102,103].

It is interesting to note that in systems or under conditions when TiO oxide occurs at the boundary having "metal-like" properties (high electrical conductivity, metallic lustre, metallicity of interatomic bond), the wettability is higher than in the systems where the oxide Ti_2O_3 is formed. The latter oxide has interatomic bonds of a substantially ionic nature [104]. The data are presented in Table 4. The wettability of TiO by metals is much greater than that of Ti_2O_3. From Table 2 and 3 it follows that the contact angle of liquid copper is 72° for $TiO_{0,86}$ and 113° for Ti_2O_3.

TABLE 4. The work of adhesion and phase composition of transition layers in titanum containing melts (2%(at.)Ti) or high melting oxides.

Group	System	Temperature °C	Work of adhesion erg/cm^2	Transition layer	Method of transition layer identification
I	$Au-Ti-Al_2O_3$	1150	1100	Ti_2O_3	X-ray
	$Ti-Ti-Al_2O_3$	1500	1220	Ti_2O_3	-"-
	$Li-Mo-Ti-Al_2O_3$	1500	1500	Ti_2O_3	-"-
	$Sn-Ti-SiO_2$	1150	867	Ti_2O_3	Visual*
II	$Cu-Ti-Al_2O_3$	1150	2100	TiO	X-ray
	Cu-Ti-MgO	1150	1894	TiO	Visual
	$Cu-Ti-SiO_2$	1150	2185	TiO	-"-

**Note: Titanium oxides Ti_2O_3 and TiO differ sharply in color. While using transparent solid phase materials (quartz glass, sapphire), the intermediate layer color is observed visually and can serve the indication of preferential formation of this or that titanium oxide.*

Where competitive interaction exists between the liquid metal and the metal which forms the oxide, oxygen can transfer into the liquid phase metal, either forming an intermediate oxide layer or dissolving in the metal. In either case, a contact between the metal-oxygen alloy and the oxide surface is taking place. To understand the mechanism of cohesion between the liquid metal and the solid oxide, it is important to study contact processes in the system solid oxide - liquid metal melt, containing oxygen. Such systems can be considered as a type of model of intermediate states of metal-oxide interactions.

Interesting peculiarities of metal-oxygen alloys are mentioned in the literature. It has been found that on adding oxygen to a liquid-metal melt of iron [64,68,87] or silver [55] the contact angle sharply decreases for the wetting of the oxide surface by the metal. For iron containing 0.04% of oxygen the contact angle on aluminum oxide was 109° against 147° for the pure metal; for silver which wets oxides CdO, ZnO, MgO the angle of contact decreased from

110 to 140° for pure silver to 90° for silver saturated by oxygen from air.

Similar phenomena were observed earlier in steelmaking processes [88,89,90]. Metallurgists noted that "killed" steel wets the bottom worse than "nonkilled" steel. This was confirmed by observations of improvement of the metal-ceramic bond for the Al_2O_3-Cr system [91] as well as the Mn-Al_2O_3 and M_0-Al_2O_3 system [92] after introducing atmospheric oxygen into the metal.

Oxygen effects on surface and contact properties of metals was studied in detail in [93-97,107-109]. In these works, the surface tensions of melts of such systems as Fe-O, Ni-O, Cu-O, Ag-Cu-O have been studied together with their wetting characteristics on aluminium and manganese oxides. The results of some of these studies are presented in Figures 12-15. Oxygen causes an abrupt decrease of the surface tension of the melt and improvement of ability of the melt to wet the oxide surface. The contact angle for copper on aluminium oxide at 1100°C is equal to ∿130° but drops to ∿60° with an oxygen content of 1-2% in copper. Copper-oxygen melts (alloys of copper and copper oxide) intensively wet corundum monocrystal and manganese oxide, as well as some other ceramic oxide phases [108, 109]. As the oxygen content increases, the wettability and adhesion of iron to monocrystal aluminium oxide (sapphire) grows [107]. Surface and interphase activity (relative to oxide phases) of oxy-

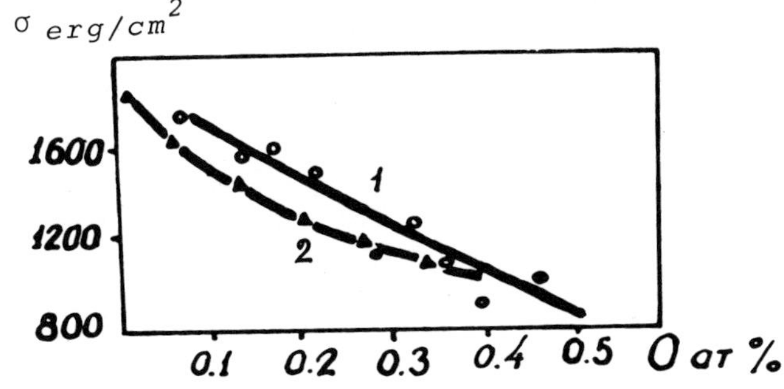

Fig. 12. Surface tension of alloys Ni-O (1500°C, curve 1), Fe-O (1550°C, curve 2).

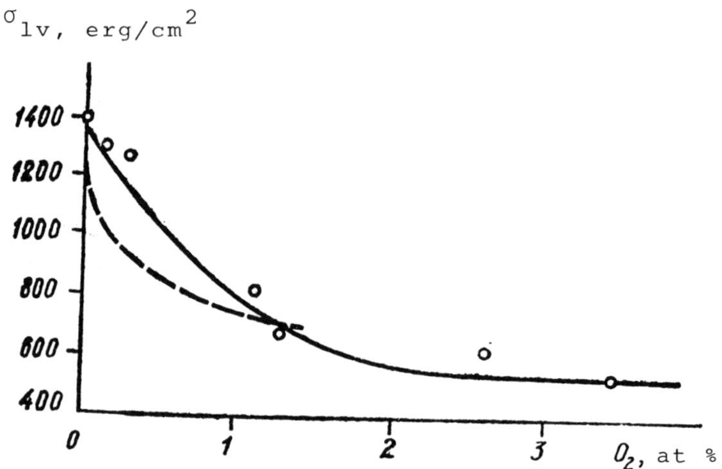

Fig. 13. Dependence of the copper surface tension on oxygen content (1150°C) (dotted curve is the data from [97]).

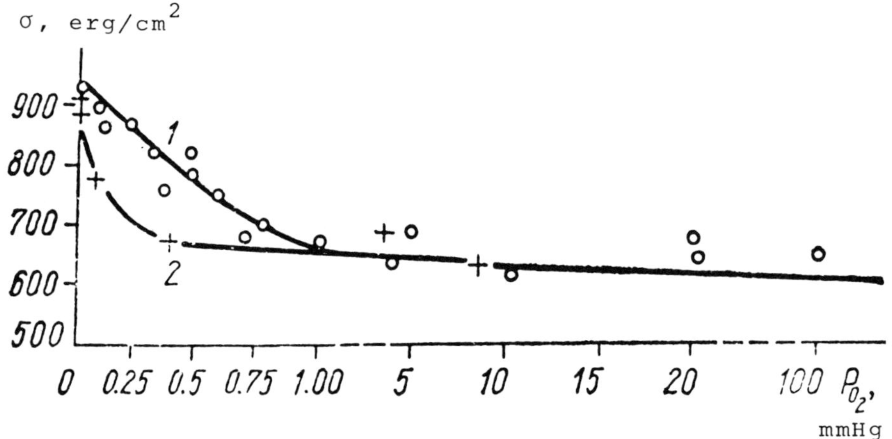

Fig. 14. Oxygen pressure in the gas phase affecting the surface tension of the liquid silver (1) and the alloy Ag-5% Cu (2) (1000°C).

gen becomes apparent for alloys Pb-O [69]; oxygen facilitates better adhesion of gold to a quartz surface [58], as well as that of gallium to both quartz and sapphire surfaces [106]. Absence of oxygen improves adhesion to the glass surface of spray-coated films of

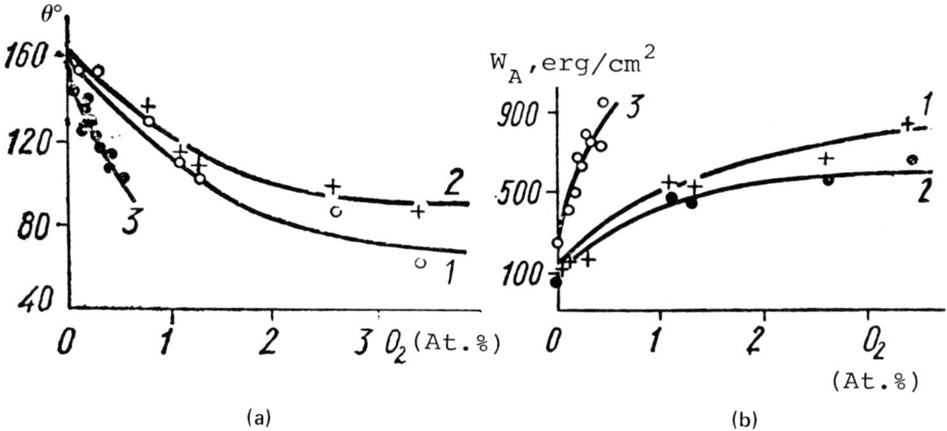

Fig. 15. (a) The effect of oxygen dissolved in the liquid metal on wettability of solid oxides 1-the system copper (oxygen)-aluminium oxide; 2-the system copper (oxygen)-magnesium oxide; 3-the system nickel (oxygen)-aluminium oxide. (b) Adhesion work in the systems: 1-copper (oxygen)-aluminium oxide (1150°C); 2-copper (oxygen)-magnesium oxide (1150°C); 3-nickel (oxygen)-aluminium oxide (1500°C).

lead, nickel, iron, molybdenum, gold, copper, zinc, and other metals [29]. It remains to be shown what the reasons are for the surface metal-gas interface and interphasic metal-oxide interphase activity of oxygen.

In accordance with the statistical molecular theory of adsorption (see [98] for example), the surface properties of binary solutions, the surface activity of the components and even extreme adsorption can be described by means of the value:

$$F = \exp(\Delta/KT)$$

where

$$\Delta = - \bar{E}^{w}_{q1} - \bar{E}_{q1} + \bar{E}^{w}_{q2} - \bar{E}_{q2}$$

$\bar{E}_{q1}$ and $\bar{E}_{q2}$ are the potential energies of molecules of the first and the second components, $\bar{E}^{w}_{q1}$ and $\bar{E}^{w}_{q2}$ the energies of these components in the surface layer and in the bulk solution, i.e. Δ is the difference of the transition work from the bulk volume into the surface of the first and the second component. Δ is included in all the basic formulae and determines the value and the sign of the adsorption of each component. In particular, positive adsorption of the first component can be observed at $F > 1$ and $\Delta < 0$, i.e., the element would be surface-active if its work of transition into the solution surface layer is less than that of another component.

By means of simple calculation for dilute oxygen solutions (component 1) in a metal (component 2), the values of Δ can be expressed approximately through bond energies between the nearest atoms in solution.

For the liquid-gas interface

$$\Delta_{liq.vap.} = \frac{1}{2}\left(Z_v - Z_s\right)\left(U_{22} - U_{12}\right) + \left(Z_s C_1^s - Z_v C_1\right)\left(\frac{U_{11}+U_{22}}{2} - U_{12}\right) \tag{17}$$

For liquid-solid interfaces

$$\Delta_{liq.sol.} = \Delta_{liq.vap.} + Z_s'(U_{1s} - U_{2s}) \tag{18}$$

Here U_{12}, U_{11}, U_{22} are bond energies between adjacent atoms 1-2, 1-1, 2-2; U_{1s} and U_{2s} are the respective bond energies of metal and oxygen with surface atoms of the solid phase; Z_v and Z_s are the coordination numbers inside and at the liquid-gas surface; Z_s' is the number of bonds each metal atom forms with its neighbors at the solid oxide surface; C_1^s, C_1 are the first component concentrations in the surface layer and in the bulk solution.

In considering the freshly formed surface of the diluted solution (the initial stage of the adsorption process), i.e., $C_1^s \approx C_1 \gg 1$, we neglect the second term in formula (17). With a positive value of Δ, the surface activity of component 1 (its tendency to adsorb at the surface) would be observed at $U_{22} > U_{12}$, i.e. when interactions between solvent atoms would be stronger than those between solvent and solute atoms. Subsequently, as the surface is saturated by the component 1 with C_1^s growth, the second term in formula (17) begins to play a role, i.e. in addition to the condition $U_{22} > U_{12}$, a sufficiently high bond energy between component 1 atoms is essential for a high adsorption value. Component adsorption at the liquid-solid interface is determined by combined action of the two factors: displacement of the component from the bulk melt (the tendency of the component 1 atoms to be surrounded by the least number of neighboring atoms of component 2) and the affinity of component 1 atoms for solid body atoms.

One can assume that oxygen in the metal melt (Fe, Ni, Cu and so on) exists essentially in the form of O^{-2} ions or to be more exact in the form of an Me^{+2}-O^{-2} complex, spatial separation of the oxygen and the metal ion being impossible. The oxygen-metal bond in the solution is partially ionic and partially covalent (metalic) with the ionic character prevailing. The ionic bond character increases with growing difference of the electronegativities of oxygen and the metal in which it is dissolved.

There arises a question about the strength of such a complex bond with the matrix metal. This bond obviously involves a positive metal ion. Localization (partial or complete) of this ion's external electrons on the oxygen ion should weaken the intensity of the metallic bond with the remainder of the metal atoms, i.e. a bond between the metallic ion of the metal-oxygen complex and another metal atom should be weaker than that of metallic atoms with each other. Thus the bond energy $U_{compl.-Me} < U_{Me-Me}$ which is in accordance with (17) and is the condition of metal-oxygen complex surface activity, i.e. determines adsorption of the oxygen dissolved in the liquid metal. It is also necessary to find out in which way the oxygen surface activity is affected by the strength of its binding with the metal in which oxygen is dissolved, i.e. the strength of its binding in the complex. Usually as the difference of metal and oxygen electronegativities increases, the ionic character of the bond in the oxide increases, the metallic (covalent) contribution to the bond decreases and the bond strength grows. In the same way the metallic properties of the oxide diminish (in particular this refers to intrinsic electrical conductivity).

The ionicity growth of the ionic character of the metal-oxygen bond in the complex will result in a greater localization of metal electrons near the oxygen ion, i.e. to their not taking part in the metallic bonding of the complex metal ion with the rest of the metal atoms, i.e. to a greater shift of oxygen surface activity $U_{Me-Me}-U_{compl.-Me}$ to higher values. With a very strong bond between the metal and oxygen, the bond energy $U_{compl.-Me}$ can drop to such an extent that the solubility of such complexes (and eventually that of oxygen in the liquid metal would drop sharply; second phase (metal oxide) formation will take place.

Thus as the bond between metal and oxygen becomes stronger the surface activity of the latter should grow when it is dissolved in metal. With very high values of the metal-oxygen bond energy, the surface activity of oxygen can be insignificant due to very low solubility of oxygen in the liquid metal. The value of the heat of oxide formation can be the criterion of the strength of the metal-oxygen bond in the complex dissolved in the metal.

For metal-oxygen systems, the bond energy of the metallic positive ion of the complex $Me^{+2}-O^{-2}$ and the oxide surface formed by negatively charged oxygen ions would be larger than that of the neutral metal ion and oxygen ion of the surface oxide, i.e. $U_{1s} > U_{2s}$ due to the substantial coulombic energy of interaction in the first case. Consequently in systems where there is an oxygen solution in the liquid metal at the boundary with the solid oxide, both factors in formula (18) act in the same direction intensifying oxygen adsorption (complex $Me^{+2}-O^{-2}$ adsorption, to be more exact) at the interface.

The theoretical considerations developed above are designed to meet the following requirements: (1) oxygen dissolved in the liquid metal should be surface active at the liquid-gas interface. Oxygen surface activity at the interface should grow with the strength of metal-oxygen bond (thermal effect of oxide formation); (2) oxygen should be active at the metal melt-oxide interface; it is to be noted that interfacial activity and adsorption of oxygen at the interface should be higher than the corresponding value at the metal-vacuum boundary.

The requirement of oxygen surface activity is confirmed experimentally as has been demonstrated above. Oxygen dissolved in liquid, copper, nickel and iron sharply lowers the surface tension of these metals. A comparison of oxygen surface activity in these systems is presented in Table 5.

The surface activity of oxygen in the silver-oxygen system as a function of concentration is not known; the only known value is that of $\frac{d\sigma}{dp}/p \to 0$ where p is the oxygen partial pressure [95]. Nevertheless from Fig. 14, where $\sigma - p$ isotherms for the system silver-oxygen and silver-copper alloy-oxygen are shown it may be seen that the oxygen surface activity in the latter is much higher. Hence one can state that

$$\frac{d\sigma_{\mathrm{Ag+O}}}{dc}/c \to 0 < \frac{d\sigma_{\mathrm{Cu+O}}}{dc}/c \to 0$$

With allowance for this, it follows from Table 5 that with increase of heat of oxygen formation and of the strength of the metal-oxygen bond, oxygen surface activity grows.

These data can be explained on the basis of electron localization. The degree of electron localization on the oxygen ion in the complex $Me^{+2}-O^{-2}$ depends on the election affinity of the metal

TABLE 5. The surface activity of oxygen in the metal alloy in comparison with heat of metal oxide formation and the affinity of the electron for the bivalent metal ion (the second ionization potential).

System	*ΔH_{298}Me-O kcal/mole*	*I, eV (the second ionization potential of metal)*	*$-\frac{d\sigma_{lv}}{dC}$ $c \to 0$ dyne/cm·at%*
Ag-O	*-7.3*	*21.48*	-
Cu-O	*-41.0*	*20.29*	*720* [94]
Ni-O	*-58.4*	*18.15*	*2400* [93]
Fe-O	*-63.7*	*16.18*	*3000-3500* [87]

ion, in this case on the second ionization potential of metal. A low ionization potential means a greater degree of electron localization on oxygen ion (i.e. an increased ionic character for the complex bond. There will also be a decreased participation of electrons in the metallic bonds of the metal ion with the rest of metal atoms. Such a complex is less strongly bonded with other metal atoms and according to (17) is more surface active. A high metallic ionization potential indicates a decreased localization of electrons on the oxygen ion, a greater degree of covalent bonding in the Me^{+2}-O^{-2} complex (approaching that of the metallic bond), a stronger bond between the complex with other metal atoms and reduced surface activity of oxygen. The values of the second ionization potentials of iron, nickel, copper and silver are given in Table 5. In accordance with the above statements, the surface activity of oxygen in the metal grows as the ionization potential decrease.

Experiments have demonstrated considerable interfacial activity of oxygen at the metal-oxide interface and increased oxide surface wettability by oxygen-containing metal melts (Fig. 15). Note: The surface and interfacial tension decrease was calculated for C = 3 At.% oxygen for the system copper-oxygen and C = 0.5At.% oxygen for the system nickel-oxygen; Γ is the value of adsorption; $G = \left.\frac{\partial\sigma}{\partial c}\right|_{c \to 0}$.

As it is seen from Table 6 interfacial tension decrease at the metal-oxide interface under the influence of oxygen addition, is much stronger than that at the metal-gas interface. The calculated values of adsorption at the solid body-liquid interface are higher under all oxygen concentrations than that at the liquid-gas interface. This agrees with the theoretical concepts being considered.

Great amounts of oxygen localized near the solid-liquid interface (as compared to that at the liquid-gas interface) are indicative of the fact that oxygen ions are bound by additional adhesion bonding to the oxide surface. However, taking into account the

TABLE 6. A comparison of the metal-gas and metal-oxide interfacial activity of oxygen

System	The liquid-gas interface			The liquid-solid body interface		
	G_{lv} dyne/cm·At%	$\Delta\sigma_{lv} = \sigma_{lv_0} - \sigma_{lv_C}$ dyne/cm	Γ^{max}_{lv} mole/cm^2 $\times 10^{10}$	G_{sl} dyne/cm·At%	$\Delta\sigma_{sl} = \sigma_{sl_0} - \sigma_{sl_C}$ dyne/cm	Γ^{max}_{sl} mole/cm^2 $\times 10^{10}$
Cu-O-Al_2O_3	720	800	34	1200	1400	57
Cu-O-MgO	720	800	34	930	1260	52
Ni-O-Al_2O_3	2400	1100	43	6800	1800	56

fact that the oxide surface consists of negatively charged oxygen ions which shield the metal ions, it is difficult to assume that direct adsorption of oxygen anions takes place. It should be considered that at the oxide surface positively charged ions of the liquid metal are adsorbed (one can speak about adsorption of complexes, $Me^{+2}-O^{-2}$ oriented and located in such a way that the metal ion faces the oxide surface and is situated near the oxygen ion of the oxide). Metal ions are bound by strong ionic bonding with the solid oxide anions, thus completing its lattice; oxygen ions from the liquid metal are attached in the positively charged nodes, i.e., a layer of liquid phase metal-oxide is formed at the interface. Such an interfacial structure is confirmed by metal diffusion in the liquid phase observed in some systems by means of microscopic study. Such diffusion takes place in the lattice along the boundaries of the solid oxide grains (copper [96] and iron [97] in aluminium oxide).

The energy of interaction between the oxidized metal and the oxide surface on the basis stated above can be estimated in the following way. Since metal ions of the liquid phase are adsorbed on oxygen ions of the oxide and oxygen of the liquid phase evidently occupies positions between metal ions or close to them, the bond energy of an oxygen containing liquid metal-solid oxide systems may be presented as the energy of interaction between the oxide layer of the liquid metal and a semi-infinite lattice of the solid oxide. This energy is determined by summation of terms of the type $\frac{Zi \cdot Zk \cdot e^2}{r_{ik}}$ - coulomb potentials of interaction between the i-th ion of the liquid phase oxide and the k-th ion of the solid oxide, i.e. this is the analog of Madelung bond energy in ionic crystals. It is possible to calculate the exact sum and to allow for the bond components (repulsive energy, energy of van der Waals forces, etc.) as is done in the theory of ionic crystals. However taking into account the approximate character of computation one can make use of the procedure given below in this particular case.

Let us substitute for liquid phase metal oxide layer, which has a finite thickness (two-three planes of atoms), an infinite oxide layer. Bearing in mind that the main contribution to the bond energy is that of the first layers of atoms of the liquid metal oxide, we shall consider that the bond energy of the oxidized liquid metal with that of the solid oxide to be approximately equal to the bond energy of a semi-infinite crystal of the solid oxide with a semi-infinite crystal of the liquid phase metal oxide, i.e., to be equal to the work of separating those halves of the crystal along the plane of contact. In the case when the lattice parameters of both oxides and the ionic valencies of both metals (metal of the solid phase oxide and the liquid metal) are similar, the given energy can be taken as double the surface energy of the corresponding ionic crystal. The latter has been correctly calculated in many works [85,99]. For lattices of the NaCl-type (oxidized liquid copper or nickel on magnesium oxide), the surface energy of the face (100) is $0.116\ (Ze)^2/d^3$, according to [85] where d is the cube root length of the elementary cell. Assuming $a = 4{\cdot}26A^\circ$ for copper oxide (close to the value of a for magnesium oxide equal to $4{\cdot}21A^\circ$) and $Z = 2$ for the adhesion work we obtain $W_a = 2.\ \ \sigma = 2.0,\ 0.116\ (Ze)^2/d^3 = 2.5{\cdot}10^3$ erg/cm^2, i.e. the order of magnitude of this energy is 10^3 erg/cm^2.

Experimental values of the adhesion work for the system Ni(O)-Al_2O_3 at an oxygen content in nickel of 0.475At.%, are approximately 950 erg/cm^2. As is seen from the plot (Fig. 15) the value of the adhesion work should increase slightly with increasing oxygen content. For the system Cu(O)-MgO, the adhesion work was 630, while for Cu(O)-Al_2O_3 it was 785 erg/cm^2.

Though it was not our aim to obtain numerical values of the adhesium work, the calculations carried out allow us to conclude that the coulombic energy of interaction of the dissolved liquid phase metal and oxygen ions with the solid oxide surface is high enough to explain the adhesion work of the oxidized metal and the solid oxide experimentally observed.

Summarizing the mechanism of action of oxygen dissolved in the liquid metal on the metal adhesion properties the following is to be pointed out:

(a) Neutral atoms at the liquid metal surface cannot be bound by strong chemical bonds with the oxide surface formed by negatively charged ions of oxygen. If it is taken into account that the metal surface is also electrically negative (a layer of electron gas penetrating beyond the metal lattice) then one should suppose that there are repulsive forces acting between the metal and the oxide.

(b) A metal-oxide bond is formed when the atoms of the metal get rid of their valence electrons and give them up to atoms of oxide oxygen. Positive metal ions are then bound by strong ionic bonds to the oxygen ions at the oxide surface.

Thus, ions of oxygen dissolved in the liquid metal are to be treated as the sites of electron extraction (in the liquid metal). The role of oxygen in this case is evidently not specific; it can be substituted by any other metalloid ion (sulphur, halogens, complex anions of the SO_4^{-2} type and many others) soluble in the liquid metal and having sufficient affinity for electrons. Thus, sulphur acts like oxygen in the system Fe-Al_2O_3 [64,100]. Selenium and tellurium in Pb-UO_2 [69] and chlorine in Bi-UO_2 [70] behave in a similar way. The intensity of the metalloid effect on the interfacial tension decrease is determined not only by the factors listed above but also by the relationship of the anionic dimensions (or, to be more exact, of the magnitude of the lattice parameters of the intermediate ionic structure being formed) and those of the solid phase oxide. It should be emphasized that it is necessary for metal atoms to get rid of their valence electrons for bond formation between the metal and the oxide surface to take place. From this point of view, the mechanism of bond (adhesion) between the liquid metal and the solid oxide can be described in the following way:

At absolute zero temperature an ideal metaloxide lattice would have the ionic component of the bond between metal and oxygen ions having the structure of inert gases atoms (closed electron shells). As has been mentioned above, under these conditions the neutral metal atom from the liquid phase cannot form any strong chemical bond with the oxide.

As temperature is increased, thermal motion disturbances in crystal structure are taking place, in particular electrons can pass from the oxygen ion (O^{-2}) to the metal ion. An oxygen ion O^{-} is formed with a local dissociation of the oxide taking place between an atom of oxygen and an atom of the metal. In some oxides electron holes are formed in the electron levels of metal ions ($Ni^{+2} \rightarrow Ni^{+3}$ for instance). At each temperature, a certain equilibrium number of O^{-} ions and oxygen atoms is established. Equilibrium is also established between the number of dissociated ions and of oxygen atoms inside and at the surface of the crystal.

O^{-} ions or oxygen atoms formed at the crystal surface have a positive valency. The metal atom in contact with the oxide on filling the vacant site by its valence electrons (getting rid its valence electrons) can be bound by strong ionic bonds with the oxide surface thus completing its lattice. For monovalent solid oxide metal and liquid metal the reaction which is taking place can be written as follows:

$$Me'_2O + Me'' \rightleftarrows Me' \cdot Me''O + Me' ,$$

where Me' is the solid oxide metal; Me'' is the metal reacting with the oxide (in this case with the site where metal valence electrons are localized is the oxygen atom (ion) of the oxide lattice).

The subsequent development of the bond of the oxygen ion with the liquid phase metal atom is somewhat different. If there is a partial dissociation of the bond Me-O in the solid oxide, for example

$$Me^{+2} - O^{-2} \rightarrow Me^{+} - O^{-}$$

then the oxygen ion remaining at the surface is saturated by the valence electron of the liquid metal atom ($O^- \rightarrow O^{-2}$) and will form the binding bridge between the liquid metal and the oxide.

With complete bond dissociation in the solid oxide the oxygen atom is released, and can pass into the liquid metal, localizing valence electrons of the metal and reforming an O^{-2} ion. A positively charged metal ion is adsorbed at the neighboring oxygen ion of the oxide while the released oxygen ion already adsorbed on the liquid phase side) is fixed on the metal ion. If in the liquid phase metal oxygen is sufficiently soluble, then some of oxygen ions will pass into the metal melt with some locating at the metal-gas interface. Here the role of oxygen is the same as in the case of metal-oxygen systems considered above. If oxygen solubility in the liquid metal is low, then oxygen released due to the metal-oxide reaction and saturated again with metal electrons to reform the O^{+2} state remains localized near the interface on the liquid metal side.

In all of the above cases, a layer of the corresponding metal oxide, completing the lattice of the solid phase oxide, is formed on the liquid metal phase side of the interphase. However when metal ions M^{+2} are available beforehand (e.g. when metal-oxygen solutions readily form) the metal-oxide bond is easily formed and the oxidized metal wets the surface of the ionic compound. In the case of pure metals and oxides, additional energy is needed for oxide dissociation. That is why a liquid metal does not usually wet an oxide with strong interatomic bonding.

A numerical estimation of the bond energy between metal and oxide (the adhesion work) can be carried out using expressions (8)-(11). For this purpose it is necessary to calculate the dispersive energy interaction between the metal and the oxide and the energy of the chemical reaction taking place at the interface, when the reaction takes place according to the equation:

$$Me'' + M_e'O \rightleftarrows Me''O + Me' \qquad (19)$$

If the initial amounts of substances Me'' and $Me'O$ (moles/unit interface area) are equal to n' and n'' respectively, then the concentration of each of the four substances can be expressed through the degree of reaction transformation α:

$$[\mathrm{Me}''] = \frac{n''-\alpha}{n'+n''} ; \quad [\mathrm{Me}'\mathrm{O}] = \frac{n'-\alpha}{n'+n''} ; \quad [\mathrm{Me}'] = [\mathrm{Me}''\mathrm{O}] = \frac{\alpha}{n'+n''}$$

The equation of the reaction isotherm can be written as

$$\frac{dZ}{d\alpha} = RT \ln \frac{\alpha}{(n'-\alpha)(n''-\alpha)} - RT \ln \frac{\alpha_0}{(n'-\alpha_0)(n''-\alpha_0)} \tag{20}$$

where α_0 is the equilibrium degree of the reaction transformation. The adhesion work is calculated by integrating this expression

$$W_a = \int_0^{\alpha_0} \frac{dZ}{d\alpha}\, d\alpha = -RT\left[n'\ln\left(1 - \frac{\alpha_0}{n'}\right) + n''\ln\left(1 - \frac{\alpha_0}{n''}\right)\right] \tag{21}$$

α_0 can be found from the expression for the equilibrium constant and from the isobaric potential change in the course of this reaction

$$\Delta Z = \Delta Z'' - \Delta Z' = -RT \ln \frac{\alpha_0^2}{(n'-\alpha_0)(n''-\alpha_0)} \tag{22}$$

$\Delta Z''$ and $\Delta Z'$ are changes of isobaric potentials on oxidizing the Me" and Me' metals, respectively.

The dispersion energy of the bond between the metal and the oxide W_{AB} can be estimated from relations given in chapter 1 by allowing for the interactions of the nearest atoms disposed in the first contact atomic planes of the metal and the oxide.

The distance R in formula (11) for the bond energy was assumed equal to the metal atom radius (a half of the least distance between the metal atoms in its closest packing structure) added to the radius of oxygen ion $R_{O^{-2}} = 1.4\text{Å}$. According to Pauling, polarizability of the oxygen ion is 3.88Å; in agreement with reference [29], the ionization potential of the oxygen ion (O^{-2}) is

$88 \cdot 10^{-12}$ erg. For metal-oxide systems one can take into account only bonds of metal atoms with oxygen ions. Any additional energy of the dispersion bond of the metal-metallic oxide ion junction due to the low polarizability of the latter and its high ionization potential constitutes only a small correction (about several per cent). It is, however, necessary to take values of the third and the fourth ionization potential for bi-and trivalent metallic ions respectively.

The contact angle was calculated by the formula $\cos\theta = \frac{W_a}{\sigma_{lv}} - 1$. For sufficiently intensive interaction of the contact pair components, the liquid metal is saturated by the products of reaction with the oxide (oxygen in particular). For the system Ni-NiO, the surface tension of nickel is assumed equal to 1000 dyne/cm; the surface tension of titanium and zirconium in contact with oxides is unknown and the angle of contact for those systems has not been determined. The calculation data is summarized in Table 7 from which it is seen, at sufficiently large negative values of ΔZ, that the calculated chemical part of the adhesion work is comparable with the experimental one (the same order of magnitude). At positive ΔZ values the calculated chemical part of the bond energy is small compared to the value of the energy due to van der Waals interaction.

Up to now the discussion has concerned the wettability of metal oxides by metals. It seems possible to now generalize features of contact interaction between the metal and the oxide and to extend them to the case of any ionic compound of metal with non-metal (or complex anion) whose crystals consist of cations and anions (sulphides, various high-melting salts--sulphates, phosphates, halogens, and so on).

The large radius of the anion and its high polarizability are characteristics for such compounds, i.e. their surface is formed by negatively charged ions due to the circumstances considered above. The most important interaction of the liquid metal and the surface is that of the liquid metal and the crystal anion.

TABLE 7. The work of adhesion and wettability in metal oxide systems

System	$T°K$	ΔZ_T kcal/g·atom of oxygen	Radius of metal atom Å	Polarizability (α_{Me}) $cm^3 \cdot 10^{-24}$	W_a(chem) erg/cm^2	W_a(VDW) erg/cm^2	W_a erg/cm^2		θ°	
							calc	exper	calc	exper
$Au\text{-}Al_2O_3$	1370	98	1.44	1.27	0	240	240	270	142	138
$Ag\text{-}Al_2O_3$	1270	100	1.44	1.86	0	380	380	380	125	130
Fe-BeO	1820	64	1.28	1.79	.15	490	490	300	137	147
Si-MgO	1450	20	1.34	1.63	40	320	360	560	120	101
Ni-NiO	1650	0	1.24	1.88	500	480	980	1470	90	70
Sn-NiO	1275	-7	1.58	2.02	715	190	905	955	29	27
Ti-MgO	2000	-6	1.45	2.32	920	350	1270	2000-2500	-	0
Zr-MgO	2100	-13	1.58	2.27	1320	220	1540	2500	-	0

The ionic crystal wettability by the liquid metal will be greater the greater the affinity of the liquid metal for the non-metal (complex anion) and the weaker the bond of non-metal with the solid compound metal, i.e., wettability will be directly connected with the value

$$\Delta Z = \Delta Z' - \Delta Z''$$

where $\Delta Z'$ and $\Delta Z''$ are the changes of thermodynamic potential on forming compounds with the metalloid of the liquid metal phase and with the metal of the ionic crystal of the solid phase. In this case the addition of a metalloid to the liquid metal having sufficient electron affinity to pull electrons away from metal atoms and transform them into positive ions will promote ionic crystal formation and wettability by the metal melt.

There is practically no experimental data on the wettability of various ionic compounds by liquid metals. The author has studied adhesion properties of liquid copper on the high melting salt CaF_2. The tests were carried out at 1320°C in helium. At this temperature, the thermodynamic potential of CaF_2 formation is 115, of CuF_2-45 kcal/g. atom of fluorine. Thus

$$\Delta Z = \Delta Z_{CuF_2} - \Delta Z_{CaF_2} = +70 \text{ kcal/2·atom F}$$

i.e. ΔZ has a large positive value. Based on the curve in Fig. 11 for the system Cu-CaF_2 one should expect nonwettability and a small work of adhesion. Introduction of the "strong" anion (that of oxygen, for example) into copper should increase the degree of wettability.

The experimental study results for the system Cu-CaF_2 are presented in the table:

Solid phase	*Liquid phase*	$\theta°$	W_a *erg/cm*^{2}
CaF_2	*Cu*	*140*	*300*
	Cu+1, 5% O_2	*80*	*620*

Pure copper forms a large angle of contact at the surface of

CaF_2. Addition of oxygen sharply increases the wettability of the solid surface and the adhesion of the alloy. These data are in agreement with the generalizations previously stated.

B. The Wettability of Covalent Solids by Metals

Covalent high-melting solids (elementary substances and compounds) among which such technically important compounds as diamond, graphite, silicon carbide and nitride, boron nitride (hexagonal and cubic), and boron carbide (B_4C) will be considered here. They are characterized like ionic compounds by closed stable electron configurations of atoms with high strength interatomic bonds. The equilibrium part of the adhesion work $W_{a(eq.)}$ arising through mutual saturation of free valences is negligible for the interaction of such bodies with metals. Interaction and high adhesion energy (and wettability) are possible only at the expense of bond dissociation in the solid phase and chemical reaction between the metal and the solid body (contribution of $W_{a(non\text{-}eq.)}$. The metal should possess high chemical affinity to any kind of solid phase atoms.

The data and characteristics of these covalent solid bodies wetting by metals are considered below.

1. Adhesion to, and wettability of, diamond and graphite by metals. From the above discussion it can be assumed that those elements having a high wettability will enter into substantially intensive chemical interaction with carbon, i.e. they will form carbides, dissolve carbon and diffuse inside the solid phase.

The interaction of carbon with metals will depend significantly on the properties of the latter and different types of bonds between atoms of metal and carbon can be established. The electronic structure of carbon atoms ($1s^2 2s^2 2p^2$, there are four valence electrons) is such that carbon can be either a donor or an acceptor of electrons depending on the conditions. Metalloid properties of carbon atoms usually manifest themselves in its interpretation with strong electronegative elements, i.e. with the elements on the left-hand side of the periodic table: alkali and alkaline-earth metals having low ionization potential. These elements form salt like compounds with carbon (carbides) with a substantial degree of ionic bonding between the atoms of Me-C. Thus alkali metals intruded into the graphite lattice in the space between the layers are positively ionized forming polycarbides of the type MeC_8, MeC_{16}, for example [114,140,145,147].

Alkaline-earth metals form also compounds with carbon of the salt like carbide type (MeC_2) where the bonding between the metal and carbon atoms is mainly ionic.

In the interaction of carbon with transition metals (metals with an unfilled d-electron shell) carbon displays its metallic donor properties giving up a part of its outer electrons to the d-band of metal. This situation, formulated first by Ubbelode [148] for hydrides of transition metals, was developed in works by Umansky, Samsonov [141,142] and Kissling [150] for the interaction of carbon and transition metal in carbide metals. Several factors favor the transfer of valence electrons from the carbon atoms to the d-band of the transition metal alloy with positive ionization of the carbon on binding. These factors are the metallic character of carbide phases (phases of intrusion): the small size of carbon ions and their high mobility of diffusion in carbides (much higher than that of oxygen or sulphur anions in oxides and sulphides); the relationship between the heat of formation of the carbide and the occupancy of d-bands of the metal; the substantial difference of lattice parameters of titanium and vanadium carbide experimentally observed and that assumed from calculation of ionic bonding in Ti^+-C^- and V^+-C^-.

According to Bruer [151] the fact that carbon is the donor of electrons, supplying them to d-band of the metal, can be confirmed by a shift of the stability maximum (heat automatization) of solid phases which for transition metals is localized in groups V-VI of the periodic system (formation of stable electron structures d^5), and for carbides in the group IV.

In [152,153] using electron transfer data, the charge of carbon dissolved in transition metals has been determined. The data obtained indicate that carbon is the cation whose value of charge in different alloys is:

Metal	*Ti*	*Ta*	*W*	*Fe*	*Co*	*Ni*
Value of carbon charge	*+4*	*+2.8*	*+0.6*	*+3.8*	*+2.6*	*+1.8*

The degree of carbon ionization is related to the degree of filling of transition metal levels.

The author's investigations allowed him to conclude that carbon should be positively ionized during its dissolution in nickel melts [10]. On the other hand, while some authors consider the ionic component to be the substantial part of the bond between carbide and transition metal in carbides, they ascribe the negative charge to the carbon. According to [143] this is confirmed by the decreasing lattice parameter in an isoelectron row of compounds (structures of the NaCl type): $K^{+}F^{-} \rightarrow Ca^{+2}O^{-2} \rightarrow Sc^{+3}N^{-3} \rightarrow Ti^{+}C^{-}$, namely: 5.33, 4.80, 4.44, 4.31 Å with the growth of both the metal and the anion valency (fourfold negative ionization of carbon is assumed in TiC). Also, in [154] on the basis of an analysis of the electron density distribution in titanium carbide, the author made a conclusion that the carbon ion was negatively charged (about -1).

It should be noted that from this point of view many characteristic properties of carbide phases (their high electric conductivity, the mobility of carbon atoms and so on) prove difficult to treat. Particularly the role of nonoccupancy of the d-electron shell and its relationship with the interatomic adhesion strength for the carbon-metal junction is not clear.

Thus, preference should be given to the first interpretation and it is to be assumed that in the interaction of carbon with transition metals, bonds of metallic character are established and that the carbon atom is positively ionized.

Covalent bonds are usually formed with carbon when bonded to metallic elements with outer 2p or 3p electrons. Boron and silicon are typical representatives of this group. These elements can still form stable carbides. However, when the main quantum number increases to $n \to 4$ then the strength of covalent bond p metal-carbon decreases so sharply that the given elements are already practically inert to carbon.

With regard to their interaction with carbon, metals with complete f-electron shells (lanthanoids and actinoids) should be separated into an individual group. Those elements form "saltlike covalent-metallic" carbides. Small values of the first ionization potentials of lanthanoids result in a substantial ionic contribution to the bonding of the metal with carbon. On the other hand, these metals have nonoccupied shells. As in a "d-transion" this should impart a certain metallic character to the bonding which takes place in such carbides. The latter carbides possess the most characteristic feature of metals, high electrical conductivity of the metallic type. Actinoids give still more "metallic" bonds with carbon.

Finally, all elements interact with carbon through physical forces, i.e., there is always a contribution from dispersion interactions. The bond energy produced by these forces is low, fractions or units of a kcal/mole. That is why the role of these forces is only appreciable in those systems where the level of the carbon-metal chemical interaction is low enough. The metals of the secondary B-subgroups of the periodic system of the fourth, fifth and sixth periods:

I gr.	II gr.	III gr.	IV gr.	V gr.	VI gr.
Cu	Zn	Ga	Ge	As	Se
Ag	Cd	In	Sn	Sb	Te
Au	Hg	Tl	Pb	Bi	Po

fall into this class.

For a high degree of the solid body wettability by liquid metals, intensive interfacial interaction is required. Such interaction favors the formation of bonds of metallic character. Therefore formation at the interface of a new phase with metallic properties is preferable.

Thus the highest degree of graphite and diamond wetting by metals is to be expected for transition metals with d-electron band vacancies.

The intensity of the phase interaction (or the chemical reaction intensity) at high temperatures depends more on the nature of reacting atoms and of bonding forces between them than on the structure they form. Bonding forces between carbon atoms in the layers of the graphite lattice are weaker than in the diamond lattice. This is indicated by the low transition heat (diamond to graphite) of about 0.5 kcal/mole; however, the sublimation heat is about 17 kcal/mole.

The two forms of carbon differ little from one another energetically. This means that the energetic effect of chemical transformation in which graphite takes place should be close to that for those involving diamond. Thus heats of carbon combustion (reaction of carbon with oxygen) for diamond and graphite are close to one another (94.05 kcal/mole for graphite and 94.50 kcal/mole for diamond).

The reaction intensities of diamond and graphite with liquid metals are also close, i.e., we should expect similar metal wettability for both forms. This statement is confirmed experimentally [10] and permits the substitution of inexpensive graphite for diamond in trial studies.

Experimental data and characteristics concerning diamond and graphite wettability by metals are presented below.

While analyzing experimental data the metals being studied are divided into several subgroups in accordance with the types of bonds and the intensity and character of their interaction with carbon.

Metals of Secondary B-sub-groups of the Periodic System IV and VI-th Periods

These elements are practically inert to carbon. They do not form stable carbides [139,144]. In the liquid state up to the boiling point, they usually dissolve only small quantities of carbon (10^{-3}-10^{-1}At.%) [125,135,138,155] and do not corrode graphite materials on long-term contact [136,137].

The wettability of graphite and diamond by copper, silver, gold, gallium, germanium, tin, antimony, and bismuth has been studied in this group [27,118,119,126,132]. These metals do not wet diamond and graphite. The bond energy (the work of adhesion between the liquid metal and the carbon surface) is low: 70-300 erg/cm^2 (0.5-1.9 kcal/mole; Table 8, Fig. 17). The work of ad-

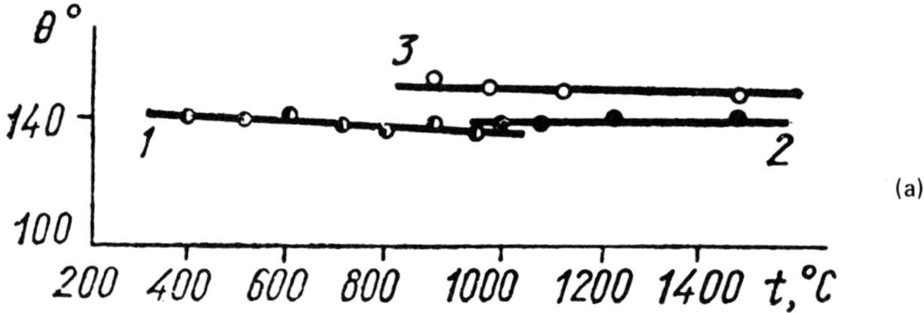

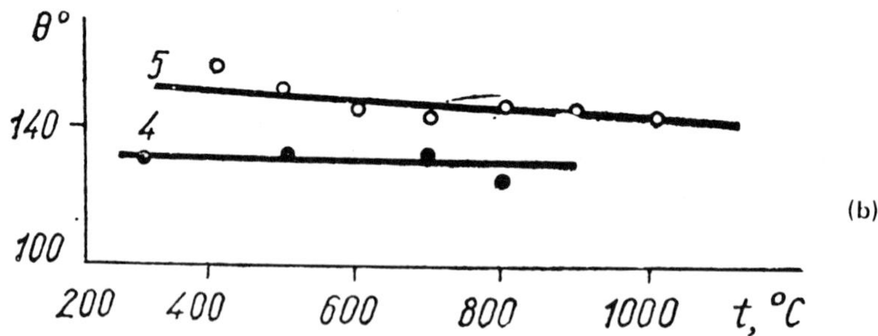

Fig. 16. a,b, Temperature dependences of wetting angles in the systems: 1-indium-diamond; 2-copper-graphite; 3-tin-graphite; 4-tin-diamond; 5-gallium-graphite.

TABLE 8. Wetting angles θ and the work of adhesion W_a of metals inert to carbon at the surface of diamond and graphite [27]

Liquid metal	*t°C*		θ	W_a *erg/cm*^{2}
Copper	*1100*	*Diamond*	*145*	*235*
	1150		*146*	*235*
Silver	*1000*		*120*	*450*
Gold	*1100*		*151*	*120*
	1150		*150*	*120*
Indium	*400*		*156*	*45*
	500		*152*	*60*
	1000		*142*	*100*
Germanium	*1000*		*136*	*115*
	1100		*131*	*145*
	1200		*113*	*390*
Tin	*900*		*125*	*195*
	1000		*125*	*190*
	1100		*125*	*150*
	1150		*124*	*185*
Lead	*1000*		*110*	*265*
Antimony	*900*		*120*	*180*
		Graphite		
Copper	*1100*		*140*	*315*
	1500		*142*	*20*
Silver	*980*		*136*	*255*
Indium	*800*		*141*	*105*
Gallium	*100*		*139*	*180*
	1000		*137*	*170*
Germanium	*1000*		*139*	*100*
Tin	*1000*		*149*	*65*
	1100		*150*	*70*
Lead	*800*		*138*	*75*
Antimony	*900*		*140*	*85*
Bismuth	*800*		*136*	*95*

hesion is independent of temperature for the systems gallium-graphite, indium-diamond and tin-diamond but shows a slight decrease with increasing temperature for the system copper-graphite (Fig. 17).

If it is assumed that this interaction is of a chemical nature then, due to obvious endothermicity of the process, one should expect the increased interaction (an increased work of adhesion) intensity with temperature, which is not the case. This fact can be explained if one assumes that physical interactions are decisive

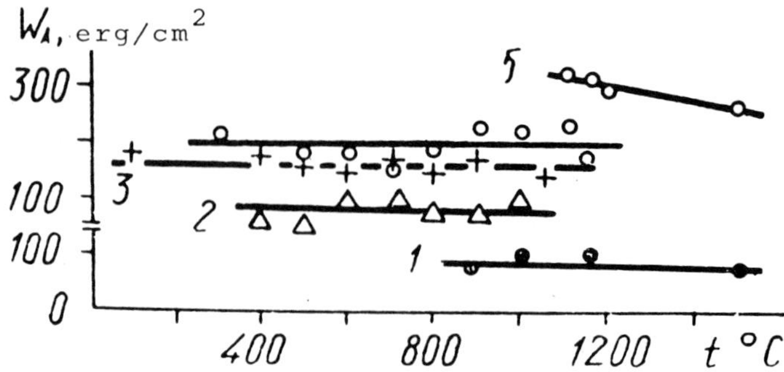

Fig. 17. Temperature dependences of adhesion work in the systems: 1-tin-diamond; 2-indium-diamond; 3-gallium-graphite; 4-tin-graphite; 5-copper-graphite.

for such systems (graphite, diamond, metals of the secondary B-sub-groups).

For the weakly interacting systems being considered here, a thermodynamic equilibrium contact of practically pure components (metal and carbon) exists and is characteristic in the temperature range under study. This means that the value of $W_{a(non\text{-}equil)}$ in expression (4) is small.

With the help of the expression for the dispersion force potential, it is possible to quantitatively estimate the work of adhesion. The value of R in expression (II) is assumed approximately equal to $R_{Me} + R_C$ where R_{Me} and R_C are half the minimum interatomic distances in the metal and graphite respectively. Polarizability was approximately determined by the relation

$$\alpha \cong \frac{e^2 h^2}{4\pi^2 m I^2}$$

where h is the Plank constant, m is the electron mass and I is the first ionization potential.

The energy of interaction $W_{a(equil)}$ per 1 sq.cm of the interfase can be expressed in the following way.

$$W_{a(equil)} \cong nE$$

(n is a number of bonds metal-carbon per 1 sq.cm).

Since 1 sq.cm contains $0.345 \cdot 10^{16}$ carbon atoms in the diamond structure and $0.260 \cdot 10^{16}$ in the graphite structure and since for the metals studied this value fluctuates within the range 0.030-$0.170 \cdot 10^{16}$ At/cm^2 then to a first approximation the number of bonds established at the liquid metal-carbon surface contact will be determined by the number of metal atoms. For this reason the value of n was calculated for each metal respectively.

As can be seen from the data of Table 9, a certain correspondence is observed between the calculated solid-liquid interfacial energies of interaction solid-liquid and the work of adhesion, W_a. At any rate dispersion interactions appears to be sufficient to explain the adhesion energy experimentally observed.

Transition Metals

The elements being studied (Ti, Ta, Cr, V, Mn, Nb, W, Mo, Fe, Co, Ni, Pd, Pt [27,105,126-134,159], Zr [119], U [156], Ce [116], Zr, Ti, Nb, Mo [120-123], Ti, Zr [175,176,179,184,248], La, Ce, Pr, Nd [157], La [158], either in a pure form or in the form of additives to inert elements, show considerable adhesion activity on contact with graphite and diamond (Figs. 18-20, Table 10).

The work of adhesion of such metals and alloys at the graphite or diamond surface is large and reaches 2000-3000 erg/cm^2 or 20-25 kcal/mole when expressed on a unit volume basis.

Increasing temperature enhances the degree of diamond and graphite wettability by such metals (Fig. 21).

Wetting in such systems is determined by chemical interaction, carbide formation or dissolution of the solid in the liquid at the interface. Disperson forces (units of kcal/mole) while clearly present in these systems, contribute insignificantly to the bond energy being observed (about 5-10%).

The results of the calculations of the work of adhesion and wettability in metal-graphite systems carried out by the formulae

TABLE 9. A comparison of calculated and experimental values of dispersion bond energies for the metal-carbon surface

Metal	$R_{me} \cdot 10^8$ cm	I, eV	$\alpha \cdot 10^{24}$ cm^3	$e \cdot 10^{13}$ erg	E· erg/cm^2	W_a at 1000°C erg/cm^2	
						on diamond	on graphite
Cu	1.28	7.7	1.81	2.73	480	230	315
Ge	1.39	6.5	2.54	2.14	320	390	100
Ga	1.39	6.0	2.98	2.80	420	-	160
Ag	1.44	7.6	1.86	1.65	230	450	250
In	1.57	5.8	3.19	1.79	210	100	105
Au	1.44	9.2	1.27	1.33	185	120	-
Sn	1.58	7.3	2.02	1.21	140	190	70
Sb	1.61	8.5	1.49	0.95	105	180	80
Pb	1.75	7.4	1.96	0.81	75	263	75
Bi	1.82	7.2	2.07	0.72	65	-	95

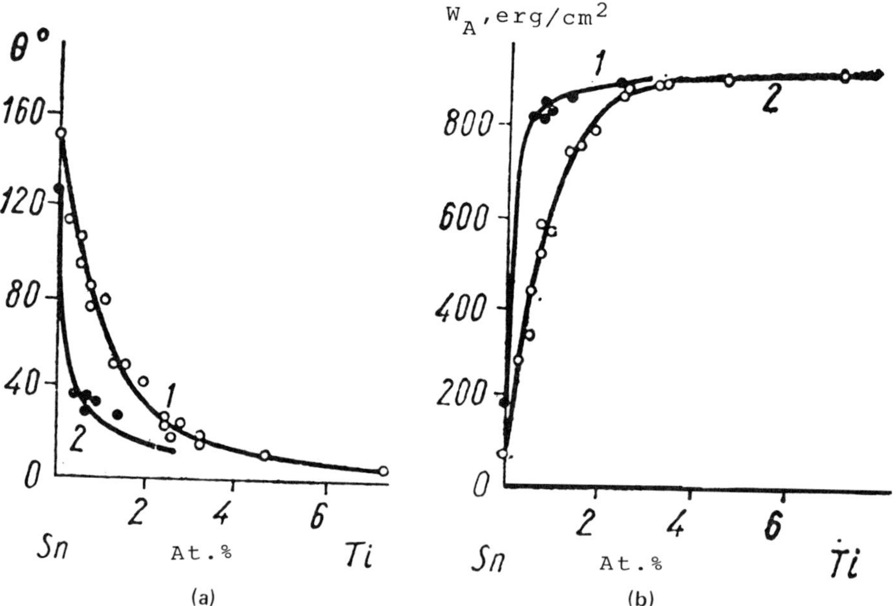

Fig. 18. a,b Concentration dependences of the wetting angle (a) and adhesion work (b) of tin-titanium alloys on graphite (1) and diamond (2) (1150°C.

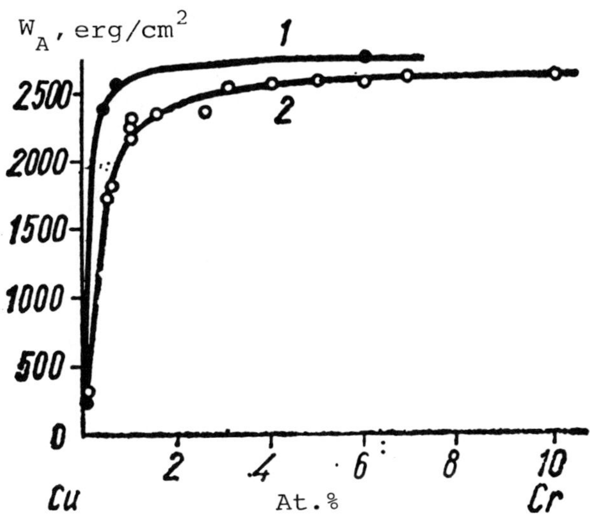

Fig. 19. Concentration dependences of adhesion work of copper-chromium alloys to diamond at (1) 1150°C and graphite (2) at 1250°C.

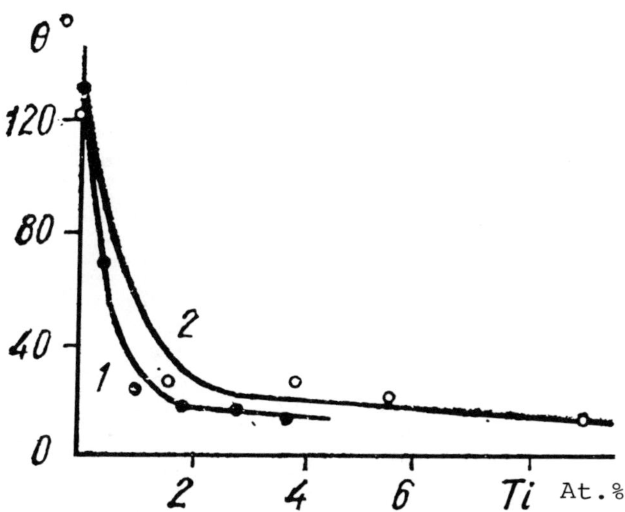

Fig. 20. Angle of contact of wetting diamond (1) and graphite (2) by eutectic alloy 40At.% Cu-60At.% Ag with titanium addition.

TABLE 10. Diamond and graphite wetting by alloys containing transition elements and with pure transition metals

Liquid metal, At%	t°C	θ	W_a erg/cm^2	Reference
		Solid surface-diamond		
Cu+12.8 Ti	1150	0	2660	[10]
Sn+0.5 Ti	1150	38	855	"
Sn+2.5 Ti	1150	22	865	"
Ag+0.1 Ti	1000	45	1555	"
Ag+0.45 Ti	1000	5	1815	"
Cu+0.37 Cr	1150	37	2390	"
Cu+6 Cr	1150	0	2660	"
Au+5 Ta	1150	30	2200	[126]
		Solid surface-graphite		
Cu+6.3 Ti	1150	30	2450	[10]
Cu+10 Ti	1150	0	2660	"
Sn+0.9 Ti	1150	76	560	"
Sn+7 Ti	1150	5	900	"
Ag+0.1 Ti	1000	85	990	"
Ag+1.0 Ti	1000	7	1810	"
Cu+0.6 Cr	1150	84	1380	"
Cu+6 Cr	1150	40	2330	"
Cu+10.2 Cr	1250	0	2620	"
Cu+24 Mn	1200	70	1780	"
Cu+5 Co	1300	138	335	"
Cu+10 Ni	1500	139	310	"
Cu+20 No	1500	134	375	[10]
Cu+6.2 V	1200	60	660	"
(Cu+9.3Ni)+6.2V	1200	50	2490	"
(Cu+9.3Ni)+7Nb	1200	40	2390	"
(79Cu+21Ni)+Mo*	1400	100	950	"
(51Cu+39Ni+105Si)+W*	1420	61	1800	"
La	900	45	1220	[157]
Ce	1090	13	1530	"
Pr	920	20	1310	"
Nd	1000	10	1185	"
Ti	1800	0	2920	[122]
Fe	1550	37	3340	[10]
Co	1500	48	3200	"
Ni	1500	45	2985	"
Pd	1560	44	2500	"
Pt	1800	87	1830	"
Fe+15,7 C	1550	107		[10]
Co+15.5 C	1550	120		"
Ni+13 C	1550	115		"
Pd+C**	1560	116		"
Fe+32 V	1550	35		[254]
Fe+53 V	1550	0		[254]

*The concentration of tungsten and molybdenum in the melt is unknown.

**The concentration of carbon in the melt is close to saturation.

Table 10 (continued)

Diamond and graphite wettability by aluminium, silicium and boron

	$T^{\circ}C$	θ° (on graphite)	θ° (on diamond)	Reference
Aluminium	800	157	150	[10]
	900		145	"
	1000	151	75	"
	1100	142		"
	1200; 13 min	39	-	"
Silicium	1450	15÷0	-	"
Copper+10%Si	1200		70	[253]
Copper+40%Si	1200		24	[253]
Nickel+40%Si	1200		47	[253]
Copper+5%B	1150		36	[10]

cited in Chapter 1 are presented in Table 11. Within an order of magnitude, there is a good correspondence between calculated and experimental values of the work of adhesion.

The adhesion activity of transition metals varies considerably. It can be related with the extent to which the metallic d-band is empty. In alloys formed with copper and gallium [101,105], the wetting activity of alloys with the IVth period elements (titanium, vanadium, chromium, manganese, cobalt and nickel) has been studied.

The adhesion activity as a whole decreases from titanium to nickel (Figs. 22,23) and corresponds to an increase in the filling of the elemental d-band. The work of adhesion of nickel, cobalt and iron in the pure form decreases from iron to nickel: 3340, 3200 and 2985 erg/cm^2, respectively.

For the Vth period the elements studied can be arranged in the order: Zr, Nb, Mo, Rb, Pd, with their decreasing adhesion activity coinciding with the filling of the d-band [10].

On comparing the wetting activity of elements, as with any additive to any solvent, the thermodynamic activity of the element in an alloy is to be taken into account. This is considered in reference [252] where the sharply differing wetting activity of manganese dissolved in various solvents is observed. In copper,

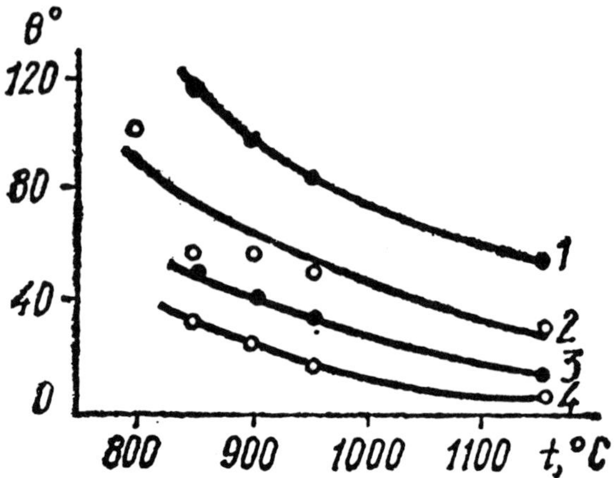

Fig. 21. Temperature dependences of the angle of contact of wetting graphite by tin-titanium alloys. 1-1.3; 2-2.46; 3-4.6; 4-7At.% Ti.

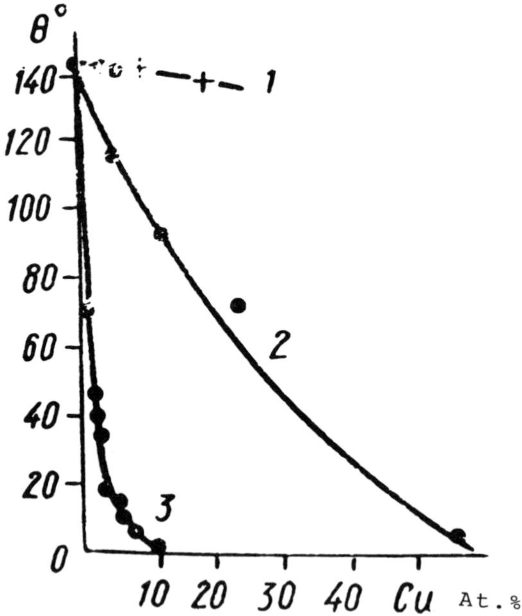

Fig. 22. Wettability of graphite by alloys of copper with nickel and cobalt at 1300-1500°C (1) with manganese at 1200°C (2) and chromium at 1250°C (3).

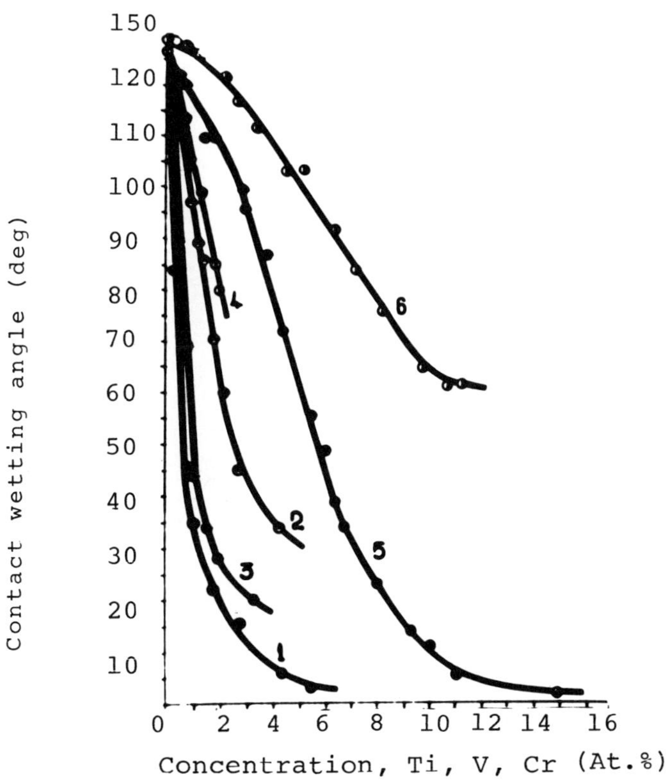

Fig. 23. Wettability of graphite by gallium-titanium (1,2), gallium-vanadium (3,4) and gallium-chromium (5-6) melts at 1050°C (1,3,5) and 900°C (2,4,6).

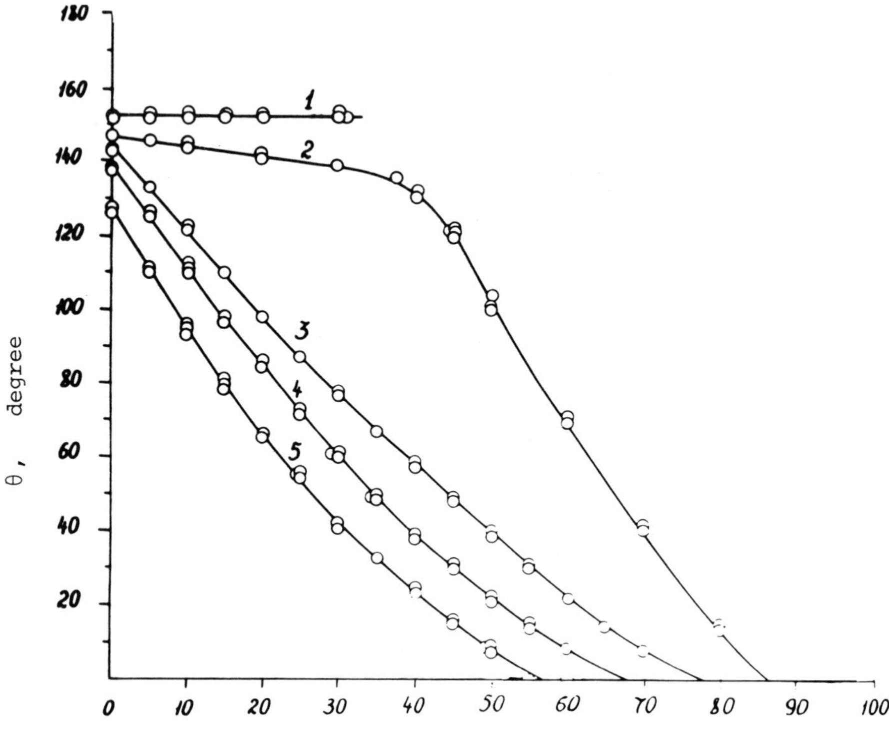

Fig. 24. Adhesion activity of manganese in different alloys at graphite wetting.

TABLE 11. The theoretical calculation of the work of adhesion and the wetting angle in the system metal-graphite (ionization potential of carbon I - 11.2 eV, number of carbon moles per sq. cm n_C - 0.39 $\cdot 10^{-8}$ mole/cm^2)

System	t°,K	Number of metal moles pwer sq. sm n_{Me}, $\frac{mole}{cm^2} \times 10^8$	Ionization potential of metal I_{Me}, eV	Distance between metal and carbon R, cm × 10^8	Change of isobaric potential ΔZ $\frac{kcal}{g\ atom\ C}$	$W_{a(VDW)}$ erg/cm^2	$W_{a(chem)}$ erg/cm^2	$W_a = W_{a(VDW)} + W_{a(chem)}$ erg/cm^2	σ_{lv} $\frac{dyne}{cm}$	θ° calc.	θ° exper.
Ti-C	2000	0.231	6.8	2.15	-37	230	3270	3500	1460	0 positive coeffic. of the spread-ing	0 positive coeffic. of the spread-ing
Zr-C	2200	0.19	6.9	2.34	-38	120	2850	2970	1390	the same	the same
Si-C	1725	0.23	8.1	1.89	-10	400	830	1270	720	45	20-30*
Al-C**	1500	0.223	6.0	2.14	-10.5	280	1100	1380	800	44	39

*Initial value of the wetting angle; at exposure $\theta \to 0$.

**For the purposes of calculation the equation of reaction 4/3 Al + C $\rightleftarrows$ $Al_{4/3}C$ is taken to be Al+C $\rightleftarrows$ AlC.

silver and tin where the thermodynamic activity of manganese is close to its concentration, the addition of manganese decreases the wetting angle. However, on adding different quantities of manganese to gold and germanium the alloys have very low graphite wettability and this does not improve until concentrations of up to 40-50At.% of manganese are added (Fig. 24).

For rare-earth elements, considerable adhesion affinity for diamond and graphite surfaces is observed. In the pure form La, Ce, Pr, Nd [157], Y, La [158] form small wetting angles at graphite surface (Table 10).

The role of solid-liquid phase interactions in wettability phenomena is well illustrated by the systems graphite-metal of the VIIIth group of the periodical system.

In non-equilibrium systems when graphite is in contact with a pure metal (iron, nickel, cobalt, palladium, platinum) and carbon dissolution is taking place at the interface, the wetting angle is small (20-30°) and the work of adhesion is large (∿3000 erg/cm^2). For the same metals saturated under conditions close to those at equilibrium, the wetting angle increases (120°) and the work of adhesion decreases by a factor of three (to 1000 erg/cm^2). Under such conditions (those of equilibrium) the contribution to the adhesion work $W_{a\ \mathrm{chem.non\text{-}equil}}$ disappears.

Metals Forming Covalent Bonds with Carbon.

In this group silicon, germanium, aluminum and boron [116, 118,119,27] have been studied. In the pure form these metals wet graphite (diamond) and the work of adhesion reaches approximately 1000-1200 erg/cm^2 (about 10 kcal/mole). These values, however, are still less than the maximum values of the various bond energies with graphite or diamond (about 2000-3000 erg/cm^2). This means a lower heat of reaction for carbide formation for the given metals compared to those obtained with the transition metals (for example, 12.4 kcal/mole for SiC and 45.5 kcal/mole for TiC). In copper alloys the adhesive activity of boron [27] and silicon [253] on graphite has been studied. Wettability occurs ($\theta \sim 40^\circ$) at a concentration of 5 At.% for boron and 40 At.% for Si.

Metals Forming Ionic Bonds with Carbon

Quantitative data on graphite wettability by alkali and alkaline-earth metals are available for lithium [115] (at 300°C, $\theta = 80^\circ$). It is also known that at temperatures higher than 450°C sodium forms a wetting angle of less than 90° [136] on graphite. Alkali metals corrode graphite surfaces on prolonged contact [113,136,137].

On contact of graphite with alkali metals the formation of metal polycarbides of MeC_8, MeC_{16} type takes place with negative heats of reaction (though with low absolute values).

For instance, for reaction of graphite with liquid potassium

$$C_{sol} + K_{liq} \rightarrow C_4K$$

ΔH = -1500 kcal/g - atom of carbon [114].

Taking into account the low values of alkali metal surface tensions (100-400 dyne/sm) and negative values of the reaction heat it may be assumed that in these systems wettability should be generally observed (wetting angle not less than 90°). This is found to be true for lithium and sodium.

There is no data available in the literature about the wetting properties of alkaline-earth metals on graphite and diamond. High negative values of the heats of formation of carbides with the appropriate metals (-28.7 kcal/mole for BeC_2; -7 kcal/mole for CaC_2; -6 kcal/mole for BaC_2) should result in small wetting angles for these systems. The relationship between the adhesion of a solidified liquid metal melt to a solid phase and the mechanical strength of the contact between the solidified metal and the solid body surface is of practical importance and has been studied for metal-diamond systems in a number of works [10,248].

2. The wettability of covalent high-melting carbides, borides and nitrides by liquid metals (SiC, B_4C, BN, Si_3N_4).

The analysis of wettability in the systems formed by metal and binary compounds can be, to a first approximation, carried out by considering the interactions between the liquid metal and each kind of components atoms of the solid phase (in cases considered here, with silicon and carbon for SiC, with carbon and boron for B_4C, and with nitrogen and silicon for Si_3N_4, all as if each substance was in a pure state). The actual intensity of the interaction of the metal and the compound will, of course, be weakened by the bonding between the components in the compound lattice [10].

Systems: inactive metal - SiC. From the metals of secondary subgroups of the elements of the periodic table (periods IV-VI) the following elements have been studied in contact with silicon carbide: Cu, Au, Ag, In, Ge, Ga, Sn. These metals are inert to carbon. In this subclass of elements only copper, arsenic and selenium form chemical compounds with silicon which are stable [125,181]. The rest of the metals form either simple eutectic systems with silicon in the absence of solubility in the solid state or form immiscible systems. Germanium and silicon form a system with unlimited mutual near-ideal solubility [181] (Table 15). In such systems positive or close to zero heat of alloy formation is observed, i.e. definitely less than the heat of SiC formation (equal to 15.2 kcal/mole according to the data of [195] and to 12.4 kcal/mole according to [18].

Thus the level of chemical interaction of such metals as Au, Ag, In, Sn, Ga, Ge with SiC is insignificant. This results in a low adhesion energy, high wetting angles and a virtual independence of these values on temperature. All of this is confirmed by experimental data [10,57,190,191] (Fig. 23, Table 12).

There are data cited in [57] for the wettability of silicon carbide and silicon nitride by liquid alloys; at 1100°C wetting angles are 165 and 164°, respectively. Obtuse wetting angles on

TABLE 12. The wetting angles and work of adhesium in the systems metal - silicon carbide

Metal	*t,°C*	*θ°*	*W_a, erg/cm^2*	*Reference*
	Elements non-interacting with carbon and weakly interacting with silicon			
Ag	*1100*	*128*	*350*	[*190*]
Au	*1150*	*138*	*260*	"
Ga	*800*	*118*	*350*	"
In	*800*	*130*	*175*	"
Ge	*1050*	*113*	*365*	"
Sn	*1050*	*135*	*145*	"
Sn	*1100*	*165*	*80*	[57]
	Elements forming carbides and silicides			
Cu	*1100-1250*	*Interaction at the interface,* [*10*] *Formation of intermediate phases* [*191*]; *wetting*		
Al	*1100*	*34*	*1450*	[*190*]
Ni	*1460, 1 sec.*	*65*	*2500*	[*190*]
Si	*1480*	*36*	*1320*	[*190*]

aluminium nitride and silicon nitride form with metals of the iron group: copper, aluminium, tin and silver. Silicon nitride is wetted ($\theta < 90°$) by silicon and calcium [185,186,188].

Systems: inactive metal-boron carbide. Most of the metals of the secondary group of the IV-VIth periods show low reactivity relative to boron (Table 15). A certain amount of boron dissolves copper [40,192]. It is known [40,192] that germanium forms an unstable compound GeB_4. Such elements as Cu, Ag, Ge, Sn studied in contact with boron carbide have rather low chemical affinity for boron. This affinity is much lower than that of boron and carbon in the solid phase substance, B_4C: 12.2 kcal/mole.

As has been already mentioned carbon does not contribute significantly to the chemical portion of the adhesion energy with metals from the subclass listed here. This allows us to understand the low values of the work of adhesion observed during investigations (150-380 erg/cm^2) and the high wetting angles [10] (Table 13, Fig. 25).

TABLE 13. Wettability of boron carbide B_4C by liquid metals

Metal	t°C	θ°	W_a, erg/cm^2	Reference
	Elements inactive to carbon and weakly interacting with boron			
Cu	1100	136	380	[10]
Ag	1300	137	240	"
Ge	1100	130	255	"
Sn	1100	135	140	"
	Elements forming carbides and borides			
Al	1150 (5 min.)	33	1500	[10]
Si	1430	0	1600	"
Ni	1500	87	1845	"
Co	1500	46	3150	"
Fe	1500	36	3310	"
Cu+5% Cr	1150	30	-	"
Sn+5% Ti	1000	0	-	"
Cr	1900	0	-	[177]

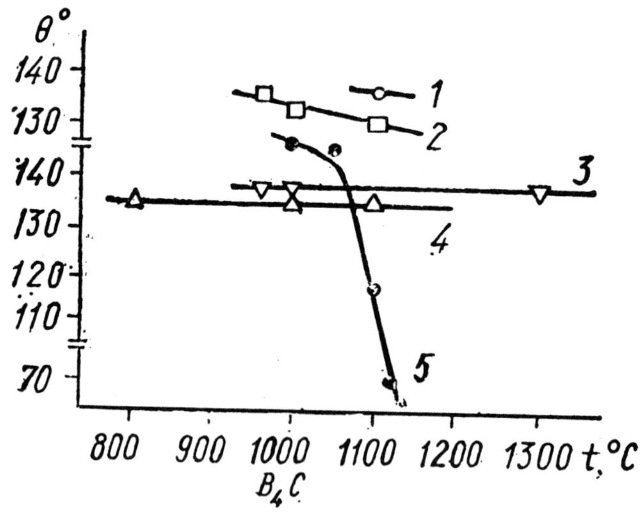

Fig. 25. Wettability of boron carbide by liquid metals 1-Cu; 2-Ge; 3-Ag; 4-Sn; 5-Al.

Systems: inactive metal-boron nitride. Compounds with nitrogen and such metals as copper, silver, gold, indium and tin are unstable (their heat of formation has either a small positive or small negative value) they decompose or even explode with gentle heating. Stable nitrides are formed by either silicon or gallium. However their heat of formation (45 and 25 kcal/mole respectively) is much less than that of boron nitride (60 kcal/mole). Only aluminium forms a compound which has a greater heat of formation than boron nitride (see Table 15).

Such elements as silver, gold, gallium, indium and tin essentially do not fuse with boron. For this reason it may be assumed that at the interface between copper (silver, gold, germanium, gallium, indium and tin) and boron nitride there is little chemical activity. Because of this, high values of wetting angles and low adhesion of these metals to boron nitride [10,196,197] can be observed (Fig. 26, Table 14). The systems lead-aluminium nitride (wetting angle is 143° at 540°C) and silver-aluminium nitride (wetting angle is 133° at 1000°C [250,251]) belong to such systems.

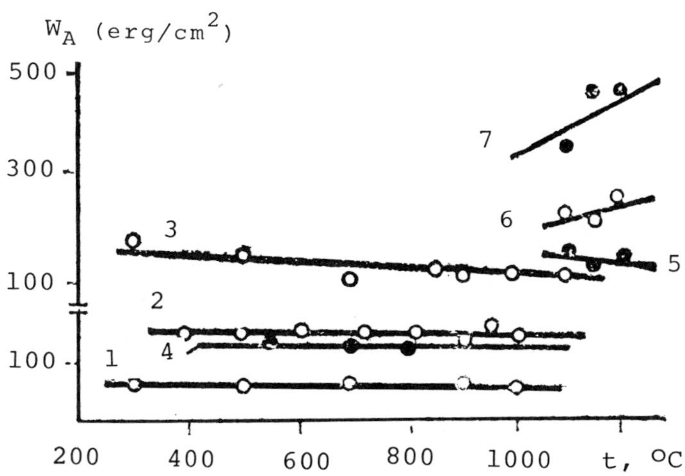

Fig. 26. Adhesion of hexagonal (1-Sn,6-Cu) and cubic boron nitride (2-In, 3-Sn, 4-Ga, 5-Ge, 7-Cu) to liquid metals.

TABLE 14. The wetting angles and work of adhesion of liquid metals on the surface of cubic and hexagonal boron nitride [10]

Liquid metal	Cubic boron nitride			Hexagonal boron nitride		
	t, °C	θ°	W_a, erg/cm2	t, °C	θ°	W_a, erg/cm2
Elements weakly interacting with boron and nitrogen						
Copper	1100	137	360	1100	146	225
Silver	1000	146	160	1000	140	200
Gold	1100	145	175	-	-	-
Boron	-	-	-	2200	133	350
Gallium	1100	130	225	-	-	-
Indium	1000	110	300	1000	136	115
Germanium	1100	138	175	1100	139	165
Tin	1100	137	115	1000	150	60
Elements actively interacting with boron and nitrogen						
Silicon	1500	95	855	1500	110	565
Aluminium	1100	60	1220	1100	<90	>810
Cobalt	-	-	-	1500	35	3100
Nickel	-	-	-	1500	75	2150

From the systems studied one can single out those in which the liquid metal has a low chemical affinity for the component atoms of the solid phase (to be more exact those having a lower affinity than the mutual affinity of the components and interacting rather weakly with the crystal surfaces of such compounds).

Small values of the work of adhesion are observed (50-400 erg/cm^2 or 0.1-1.5 kcal/mole) and are almost unaffected by either temperature or time of contact and evidently have "non-chemical" origin.

Physical, van der Waals, forces (dispersion interactions) can be responsible since this interaction is small and of the correct magnitude (fractions or units of kcal/mole). Such forces become apparent despite their universality, only for non-reacting systems.

It is rather difficult to obtain an exact value of the bond energy as has been done previously for the metal-graphite system, since effective ionization potentials and the atomic (ionic) size of the atoms/ions size in crystals of carbides and nitrides are

TABLE 15. The Character of Metal Interaction with Boron, Nitrogen, Carbon, and Silicon

Metal	Boron [192]	Nitrogen	Silicon [181]	Carbon
Cu	Liquid copper dissolves 10At.%B at 1100°C	Compound Cu_3N (ΔH_{298} = 17.8 kcal/mole) decomposes at heating with explosion [125,78].	Formation of silicides CuSi, Cu_3Si and the other.	
Ag	No alloys are formed	Compound Ag_3N (ΔH_{298} = decomposes at heating with explosion [125, 78].	Simple eutectic system with no solubility in the solid state.	Formation of unstable carbides (of CuC_2, AgC_2 type) which decompose at heating with explosion. Insignificant solubility of carbon in metal at high temperatures. Unwettability of diamond and graphite by metals.
Au	The same	There is no solubility [189]	The same	
Ga	"	Compound GaN (ΔH_{298} = -25 kcal/mole) is formed only at reaction with ammonia [195,78]	Eutectic diagram without solubility in the solid state. Eutectic point is strongly shifted to Ga.	
In	"	Compound InN (ΔH_{298} = -4.6 kcal/mole) is formed only at reaction with ammonia [125,78].		
Ge	Compound GeB_4 is formed	Compound Ge_3N_4 (ΔH_{298} = -39 kcal/mole) [189,194].	Complete mutual solubility in the solid and the liquid state.	
Sn	No alloys are formed		Eutectics without solubility in the solid state. Eutectic point is strongly shifted towards tin.	
Si	Liquid silicon dissolves 30At.%B at 1500 C. There is a compound SiB_3.	Compound Si_3N_4 (ΔH_{298} = -245 kcal/mole) [125,78].		
Al	Liquid aluminium dissolves 8At.%B at 1500°C. There are compounds AlB_2 and AlB_{12}.	AlN (ΔH_{298} = -76.5 kcal/mole [125,78].	Eutectics; low solubility of silicon in aluminium in the solid state.	Strong carbides Al_4C_3 and SiC are formed.

not known. These dimensions however can be given an approximate value (as close as possible to the true value) such that the character and direction of bond energy changes from one metal to another for the same type of solid phase.

For such calculations the following expression has been used:

$$E = N_{Me} \cdot \frac{3}{2} \frac{\alpha_1 \cdot \alpha_2 \cdot I_1 \cdot I_2}{R^6 (I_1 + I_2)} ,$$

where N_{Me} is the number of metal atoms per unit surface area. It was assumed that $R = R_{Me} + K_1$, where R_{Me} is the metallic atomic radius equal to one half of the least interatomic distance in the structure of metal. K_1 is a constant characterizing the distance from the atom to the high-melting phase surface

$$N_{Me} \sim \frac{1}{R_{Me}^2} ; \quad \alpha = \frac{e^2 R^2}{4\pi^2 m I^2} ; \quad d \sim \frac{1}{I^2}$$

(e and m are the electron change and mass, h is Plank constant). Then

$$E = A \frac{1}{I_{Me}(I_{Me}+K_2)(R_{Me}+K_1)^6 R_{Me}^2} = A\emptyset \qquad (23)$$

An analysis of this dependence allows us to arrange the metals in the order of their changing potentials due to dispersive interaction with the solid surface. In particular, within a group of the periodic table, for metals of the secondary group of the IV-VIth periods, the following sequence of decreasing dispersion bond energies can be listed: Cu > Ag > Au, Ga > In > Tl, Ge > Sn > Pb. The values of $\emptyset$ are proportional to the dispersion potential of the metallic element with respect to the same solid substance (SiC) are given in Table 16. Calculations have been carried out using $K_1 = 1.34 \cdot 10^8$ cm and $K_2 = 8.1$ eV (for the radius and the ionization potential of the silicon atom).

Thus, it should be expected that the degree of wettability (or

TABLE 16. Values of interaction dispersion potentials between the metal and the solid surface (exact except for the constant)

I group		III group		IV group	
Element	$\phi \cdot 10^{-57}$ $eV^{-2} \cdot cm^{-64}$	Element	$\phi \cdot 10^{-57}$ $eV^{-2} \cdot cm^{-64}$	Element	$\phi \cdot 10^{-57}$ $eV^{-2} \cdot cm^{-64}$
Cu	15.3	Ga	14.6	Ge	9.4
Ag	8.8	In	8.4	Sn	5.7
Au	6.6	Tl	5.0	Pb	3.3

the work of adhesion would decrease as we proceed within a group from the top to the bottom of the periodic table. A similar dependence can be observed for both diamond and graphite and in general for nonmetallic crystals such as oxides and sulphides.

The adhesion energies of liquid metals to the surface of boron carbide, silicon carbide, boron nitride (hexagonal and cubic) as well as to that of diamond and graphite are compared in Fig. 27 and indicate a dependence on the site of the element in a given group in the periodic table. As is seen from the figure, there is decrease in the work of adhesion of the metals of groups I, III and IV as we proceed from the top to the bottom of the table. The opposite direction of change of W_a is observed only in three pairs of fourteen sequences (fifty pairs in all): Sn-Pb for graphite, Cu-Ag for diamond and Ag-Au for boron nitride (dashed line in Fig. 27). Evidently this should be attributed to experimental error. The latter may be large for such nonreactive systems and small values of adhesion energies are subject to the influence of trace amounts of additives and impurities both in the metal and solid phases and at the contact surface.

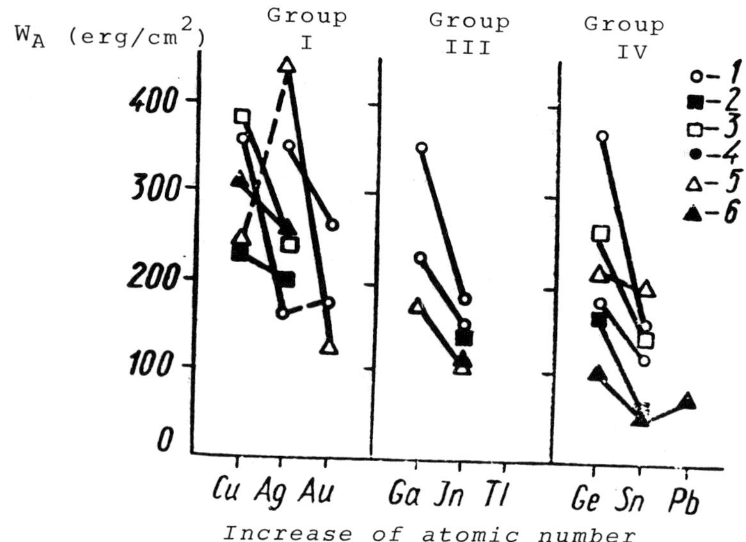

Fig. 27. Adhesion in the system covalent crystal-metal of side B-subgroups of the periodic system depending on the metal location in the group (T=1000-1100°C):1-BN(cubic), 2-BN(hexag.), 3-B_4C, 4-SiC, 5-diamond, 6-graphite.

Systems with Chemically Active Metals

From Tables 12, 15 and 15 it follows that metals with substantial chemical affinity to the elements constituting the solid phase (boron, nitrogen, carbon, silicon) will readily wet the solid compounds under study. Such elements are aluminium and silicon, metals of the iron group, chromium and titanium.

With the formation of a reaction zone, metals of the iron group in the solid and liquid state exhibit moderate interaction with silicon carbide surface. Wetting angles of these metals range from 45-90° on silicon carbide surfaces at 1500-1600°C; that for melted chromium is 5° [189]. Nickel, cobalt and iron at 1600°C interact vigorously with B_4C [10], liquid chromium spreads completely over the carbide B_4C surface [177].

The degree of wettability of solid carbides, borides or nitrides depends on the relative values of the heats (or better free energies) of formation of metal-solid compounds and of the solid compound itself. Thus the contact angles of boron nitride wetted by aluminium and silicon are 60 and 90°, respectively according to reference [196] (Figs. 28, 29), i.e. aluminium wets boron nitride better, in accordance with the higher heat of formation of aluminium nitride AlN (-76.5 kcal%g-atom of nitrogen) [78]. At the same time the wettability of silicon carbide by aluminium and silicon differ very little ($\theta_{Al-SiC} = 34°$, $\theta_{Si-SiC} = 36°$). This would be expected from the approximately same heat of formation of silicon carbide and aluminium carbide (-12 and -12.4 kcal/g-atom of carbon, respectively). Contact angles on wetting boron with silicon and aluminium are also close. At its melting point silicon spreads over B_4C with a wetting angle close to zero. Aluminium at 1150° and a 5 minute exposure exhibits an angle of 33° and a tendency to spread further [10]. In line with the above reasoning, it is assumed that the interaction of silicon and aluminium with nitrogen and carbon is determining as compared to these elements interactions with boron and silicon. Really, aluminium and silicon nitrides and carbides are more stable than the corresponding borides. This is even more true when they are compared to the silicides. By forming reasonably stable silicides, copper interacts with silicon carbide and wets it after a long-term

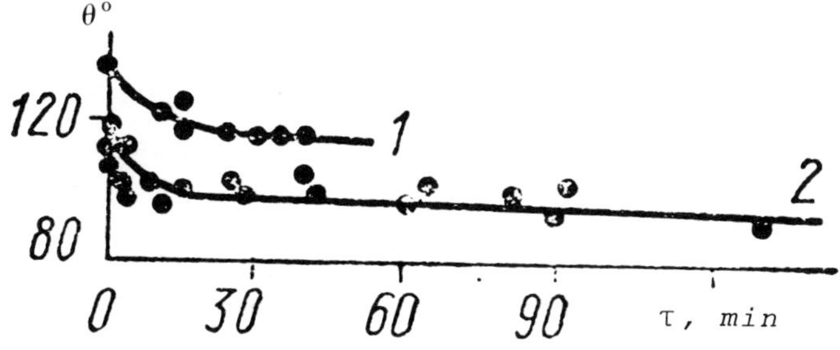

Fig. 28. Wetting angle of silicon at the surface of hexagonal (1) and cubic (2) boron nitride depending on time.

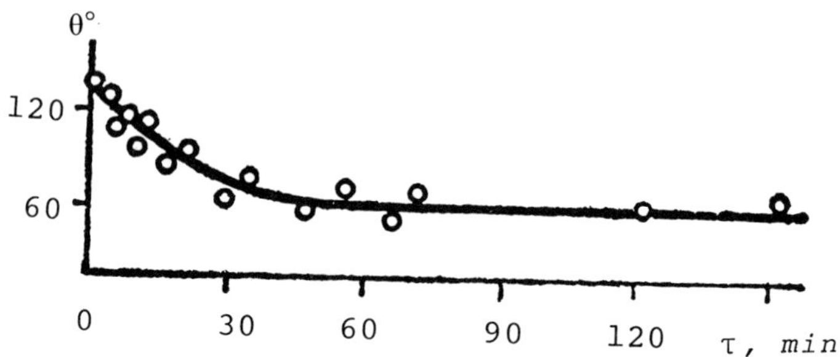

Fig. 29. Aluminium spreading over boron nitride.

contact (it is difficult to measure the wetting angle because of a two phase metal melt and the distorted form of the liquid surface). Liberation of free carbon, the product of the reaction of copper with silicon carbide [10,191], is observed at the surface of the liquid phase.

The system Si-SiC should be considered separately. Unlike the systems Ni-SiC and Cu-SiC, for instance, where interaction is rather intensive, the system Si-SiC is stable and in spite of high temperature (up to 1580°C) long-term exposure (up to one and a half hours) the wetting angle tends to increase with increasing temperature and exposure. Evidently this is brought about by the fact that at temperatures close to the silicon melting point the latter dissolves small amounts of carbon, i.e. practically pure silicon and silicon carbide form the equilibrium system.

Evidently the system B-BN where boron forms a large wetting angle (-134° at T = 2200°) [10] (Table 14) is similar to the system Si-SiC. The phase diagram of this system has not yet been developed; one can only assume that liquid boron in contact with BN forms a near-equilibrium system.

While comparing wetting properties of such compounds as BN, B_4C and SiC one can notice the regular increase of wetting by metals of these compounds in the order given. This was observed for all of the metals under study. Table 17 shows a comparison of the metal wettability of these compounds. This order of wetting

TABLE 17. *Contact angles of wetting of covalent carbide and boride surfaces by metals*

Covalent crystal	ΔH_{298}, kcal/mole	θ°									
		Cu	Ag	Ga	Ge	Au	Sn	In	Ni	Si	Al
SiC	-12.4	wetting, corrosion of the surface	129	116	110	138	138	130	65	36	34
B_4C	-12.2	136	137	-	130	-	135	-	87	0	33
BN_{cub}	-60.0	137	146	130	138	145	137	-	-	95	60
$BN_{hecsag.}$	-60.4	146	140	-	139	-	150	136	75	110	-

TABLE 18. The wettability of metal-like compounds by metal melts

Compound	Metal	t°C	Atmosphere	θ°	Reference
Cr_3C_2	Ni	1500	Argon	0 rapid spreading	[16]
	Ni	1500	Vacuum	0	[165]
	Cu	1100	"	47	[165]
	Cu	1150	"	44	[10]
Cr_7C_3		150	Argon	instantaneous spreading, corrosion, of carbide surface	[16]
HfC	Cu	1100-1200	Vacuum	132-132	[165]
	Ni	1380	"	23	[165]
	Ni	1450	Argon	37	[178]
	Mn	1300	Vacuum	86	[178]
	Co	1420	"	40	[165]
	Co	1500	Argon	40	[178]
	Fe	1490	Vacuum	45	[165]
$NbC_{0.97}$	Cu	1100-1200	Vacuum	70-48	"
	Ni	1380	"	18	"
	Co	1420	"	14	"
	Fe	1490	"	15	"
	Mn	1300	"	90	[178]
Mo_2C	Cu	1100-1200	Vacuum	18-0	[167]
	Ge	1000	"	76	[178]
	Ni	1380	"	0	[165]
	Co	1420	"	0	"
	Fe	1490	"	0	"
	Mn	1300	"	90	[178]
	Cu	1100-1300	Vacuum	108-70	[55]

TiC	Cu	1100-1200	"	112-109	[165]
	Pb	400-1000	Argon	152-90	[55]
	Pb	660	Vacuum	120	[55]
	Pb	400	"	113	[250]
	Pb+0.5%Ni	660	"	98	[55]
	Bi	300-600		138-122	[55]
	Ag	980	"	108	[55]
	Al	700	Argon	118	[163]
	Al	1150	Vacuum	20	[10]
	Zn	550	Argon	120	[163]
	Ni	1450	Hydrogen	17	[162]
	Ni	1450	Helium	32	[162]
	Ni	1450	Vacuum	30	[162]
	Ni	1380	"	23	[165]
	Ni	1500	Argon	0	[10]
	Ni	1450	"	25	[178]
	Co	1500	Hydrogen	36	[162]
	Co	1500	Helium	39	[162]
	Co	1500	Vacuum	5	[162]
	Co	1420	Argon	6	[178]
	Fe	1550	Hydrogen	49	[162]
	Fe	1550	Helium	36	[162]
	Fe	1490	Vacuum	28	[165]
	Mn	1300	"	15	[178]
	Si	1450	Vacuum	0	[10]
	Ni+10%Ti	melting point	"	25	[163]
	Ni+10%Cr	"	"	23	[162]
	Ni+10%Mn	"	"	22	[162]

Table 18 (continued).

Compound	Metal	t°C	Atmosphere	θ°	Reference
	Ni+10%Nb	"	"	22	[162]
	Ni+10%V	"	"	21	"
	Ni+10%Ta	"	"	15	"
	Ni+10%W	"	"	14	"
	Ni+10Mo	"	"	0	"
TaC	Ag	1140	Hydrogen	113	[249]
	Cu	1100-1250	Vacuum	75-36	[55]
	Cu	1100-1200	2	78-60	[165]
	Cu	1140	Hydrogen	80	[249]
	Mn	1300	Vacuum	90	[178]
	Ni	1380	"	16	[165]
	Co	1420	"	13	[165]
	Fe	1490	"	23	[165]
UC	Cu	1100-1240	Vacuum	113-69	[55]
	Cu+6%Ni	1100-1330	"	93-45	"
	BO	300-700	"	140-93	"
	Bi+0.3%Ni	350-650	"	141-52	"
	Na	200-400	Argon	165-141	"
	Na	200-400	Argon	158-90	"
	Cu	1090-1200	Vacuum	54-39	"
	Mn	1300	"	90	[178]
$VC_{0.88}$	Cu	1100-1200	Vacuum	50-40	[165]
	MnNi	1380	"	17	"
	Co	1420	Vacuum	13	[165]
	Fe	1490	"	20	"
WC	Cu	1100	Vacuum	30	"
	Cu	1200	"	20	[161]
	Cu	1100-1200	"	20-7	[165]
	Ge	1000	"	63	[178]
	Sn	500-1300	Argon	120-30	[55]

	Bi	700-1100	"	140-52	[55]
	Ni	1380	Vacuum	0	[165]
	Co	1500	Hydrogen	0	[48]
	Co	1500	"	0	[161]
	Co	1420	Vacuum	0	[165]
		1300	"	90	[178]
ZrC	Cu	1100	Vacuum	135	[55]
	Cu	1100-1500	Argon	140-118	[55]
	Cu	1100-1200	Vacuum	127-125	[165]
	Si	1500	"	22	[178]
	Ni	1380	"	24	[165]
	Ni	1450	"	32	[178]
	Co	1420	"	36	[165]
	Co	1500	Vacuum	15	[178]
	Fe	1500	"	45	[165]
	Mn	1300	"	75	[178]
TiB_2	Al	800	"	67	[251]
	Cu	1100-1500	Argon	158-154	[166]
	Cu	1120	Vacuum	142	[186]
	Ni	1480 at the moment of melting	Helium	100	[166]
	Ni	1480 20 min	Helium	38.5	[166]
	Ni	1500	Vacuum	0( )	[186]
VB_2	Cu	1100-1400	Argon	150-114	[166]
NbB_2	Cu	1100-1400	Argon	131-122	[166]

Table 18 (continued).

Compound	Metal	$T°C$	Atmosphere	$\theta°$	Reference
ZrB_2	Cu	1100-1400	"	42	"
	Ni	1500	"	42	[186]
	Co	1500	"	39	[186]
ZrB_2	Fe	1550	"	55	[186]
	Sn	280	"	104	[186]
TaB_2	Cu	1100-1400	Argon	77-47	[166]
	Ag	1300	"	118	[166]
	Ni	1500	"	43	[166]
CrB_2	Cu	1430	Helium	50	[166]
	Ni	1480 1 min	"	11	[166]
MoB_2	Ni	1480 1 min	"	8	[166]
$MoSi_2$	Sn	1100-1400	Hydrogen	154-28	[157]
TiN	Cu	1500	Ammonia	148	[171]
	Cu	1300	Vacuum	155	[186]
	Al	900	"	135	[186]
	Al	800	"	53	[251]
	Fe	1550	Ammonia	132	[171]
ZrN	Cu	1130	Vacuum	148	[171]
	Cu	1500	Ammonia	146	[186]
	Al	900	Vacuum	167	[186]
	Fe	1550	Ammonia	140	[171]
	Fe	1590	Vacuum	49	[186]
	Co	1540	Vacuum	7	[186]
NbN	Cu	1500	Ammonia	150	[171]
	Cu	1130	Vacuum	130	[186]
	Al	900	"	156	[186]
VN	Cu	1500	Ammonia	150	[171]
Cr_2N	Cu	1130	Vacuum	36	[186]

can be attributed to a lower thermodynamic strength of SiC and B_4C compared to that of BN (lower formation heat).

C. The Wettability of High-Melting Metal-Like Compounds by Metals

The wettability of metallike carbide, boride and nitride phases by metals has been studied in a number of works [10,16,55,160-174, 178,180,249-251]. The data are summarized in Table 18. In general metallike compounds are wetted better than covalent ones. Thus the wetting angles of nickel at the surface of SiC, B_4C and BN are approximately 65, 37, and 75° while at the surface of metallike phases they range from 0 to 30°. This indicates the extent of the metallic character of the solid phase and a rather high value of W_a(equil). Nontransition metals wet metallike compounds much worse than transition metals. Nevertheless in some systems these metals form quite small wetting angles (less than 90°). Copper for example will wet zirconium and tantelum borides [16], vanadium carbide and chromium carbide [165] on heating up to 1200-1300°. Tin also forms a wetting angle of less than 90° at temperatures higher than 1200°C at the surfaces of WC (it should be noted that acute wetting angles with pure copper and tin were never observed for ionic compounds wetting where the compound had high strength ionic bonds: oxides of magnesium, berillium, aluminium and covalent crystals boron nitride, diamond).

While interacting with, dissolving in or forming intermediate compounds with carbides, borides, nitrides and so on [182,183], the transition metals form non-equilibrium contact systems with the solid phase substance. This means that the value of $W_{a(\text{non-equil})}$ (the energy of the chemical boundary interaction accompanied by bond rupture in the solid phase) is added to the value of $W_{a(\text{equil})}$. The latter can be rather high for metallike phases at the interface with metals.

From the above general statements, one can assume that compound having high-strength interatomic bonds should interact with, and be wetted by, the liquid metal to a lesser degree. This statement is confirmed by the behavior of the metal borides [10,166]. Less stable diborides (MoB_2, CrB_2) are better wetted by metals (copper, nickel) than more stable compounds (TiB_2, ZrB_2, VB_2).

According to the data from [178], a regular increase of the contact angle of wetting by liquid nickel, manganese, cobalt in the order TiC, ZrC, HfC is observed the greater the compounds stability.

Remquist [165] has found that wettability of carbides by metals (nickel, cobalt, ferrum, copper) decreases with the increasing heat of formation of the carbide from its elements. The same observation has been made for a number of carbides (WC, NbC, VC, TiC) [174]. A decrease of the compound heat of formation can explain the data of reference [165] which shows a decreasing contact angle of wetting of titanium and tantalum carbides by liquid iron, cobalt and nickel on lowering the amount of carbon in the carbide below that of its stoichiometric content in the metal-carbide compound.

G. V. Samsonov [142,178] considered the intensity of the contact interaction and the degree of wettability from the atomic point of view through electron exchange to form stable electron states of atoms, which for transition d-metals are the states with d^0, d^5 and d^{10} d-level structures [187]. In particular according to [178], the greater degree of wettability and the lower stability of the VIth group metal carbides to molten metals is attributed to the idea that the transitive metals of this group disturb the sp^3 electron states of carbon in the carbides favoring a greater percentage of nonlocalized electrons in the solid phase substance which can then bond with the liquid phase metals.

Carbides, borides and nitrides should be wetted by liquid metals to a different degree due to certain differences in the nature of the interatomic bonds in these compounds. Of the three types of compounds the carbides of the transitive metals [142] are characterized by a uniform distribution of electron density and by the greatest metallic character of bonds and properties. Due to the presence of localized p-electrons there is a tendency to form covalent bonds between boron atoms (lattices and frameworks are formed by boron atoms) and perhaps between the atoms of boron and the metal. That is why these compounds are characterized by a reduced metallic character. In the nitrides of transitive metals, great asymmetry and non-uniformity of electron density distribution exist in the compound lattice. Thus the bonding should include a significant ionic component. This arises because of the higher ionization potential of nitrogen (14.47 eV) as compared to that of carbon (11.2 eV) and of boron (8.5 eV). This is why it can be expected that a higher value of the equilibrium part of the adhesion work - $W_{a(equil)}$ is characteristic for metal carbides while a low value is more characteristic of borides and nitrides.

The thermodynamic stability of the compounds being considered increases from carbides to borides to nitrides. Evidently W_a will also decrease in the same order. Therefore one can suppose that wettability by liquid metals will decrease in the order carbide-boride-nitride. This is what is usually found experimentally (Table 19).

TABLE 19. A comparison of the wettability of metal carbides and borides by nickel at 1500°C [10] (The contact angles are given in degrees.)

Carbide	Contact angle	Boride	Contact angle
TiC	0-30°	TiB_2	40°
ZrC	15-30°	ZrB_2	42°
TaC	16°	TaB_2	43-59°
Cr_3C_2	0° (positive coefficient of spreading	CrB_2	11
Mo_2C	0°	MoB_2	8°

D. Wettability in the Solid Metal-Liquid Metal Melt System

Among the contact systems formed by metal melts, there are numerous examples which are of importance where the metal is the solid phase. Such combinations are used in various technological and physical processes, among which the following ones are of significance: (1) The manufacture of products by impregnating a solid skeleton with a liquid metal or by sintering in the presence of a liquid phase; (2) soldering, welding, and coating; (3) heat transfer in steam power plants and nuclear reactors [113,199,200]; (4) melt crystallization, modification, etc.; (5) the alteration of mechanical properties of solid metals under the influence of a liquid-metal medium [200-208]. In all of these processes, the phenomena occurring at the liquid metal-solid metallic interface, the value of the interfacial energy and the degree of wettability are of great importance.

Among numerous examples studied of contact between the liquid and the solid metallic phases [199,209-226,228-235] relatively few are devoted to quantitative measurement of wetting angles and interfacial energies [10,209,216,218,219,228-230,236]. The following methods of estimating the liquid metal wettability were applied in different works: (1) Visual observation of the liquid metal drop behavior at the solid surface and the establishment of the qualitative characteristics of the contact system ("spreading", "non-spreading") [232]; (2) determination of the speading areas of standard weighed quantities of metal [210,217]; (3) determination of the possibility of obtaining stable films of liquid on the solid surface on immersing the latter in the melt or of the liquid complete running-off from it [210]; (4) direct measurement of the contact angle of the solid metal wetting by the liquid one; (5) measurement of the interfacial energy of the solid metal-liquid metal interface.

Works of this latter type are relatively few in number; measurements were carried out by the somewhat disputable method of dihedral angle or neutral drop [215,218,219,223,236]. In [222] the method of zero creep is used and the interfacial energy of zincum coated by gallium is determined.

The most general observation concerning wetting in the systems under study is thc relation between the wettability and the degree of bonding between the solid and the liquid metal, i.e. formation of chemical compounds, solid and liquid soluitons and even the diffusion of the liquid metal into the solid one.

Hildebrand [193,212] has pointed to this relation. He notes that absence of wetting of the solid metal by the liquid metal is to be expected in cases when there are no chemical compounds formed and no solubility between the metals.*

Just as formation of solutions is observed in metals with similar properties (in particular, specific bond energies, i.e. bond energy between atoms on a unit volume (c.c.) basis), non-wetting according to Hildebrand is to be observed in metals with sharply differing bond energies (if these metals do not form chemical compounds).

Tamman and Ruhenbach [211], Tamman and Arnstz [213], Beiley and Watkins [210] underline the wetting angle dependence on the degree of diffusion of liquid into the solid metal. These scientists have carried out successive comparisons of wettability with the constitution diagram between the components of the contacting pair and interaction between them. Cases of non-wettability are found essentially in systems where there is neither interaction nor solubility [221].

The first studies of wettability in intermetallic systems were carried out using mercury. Tamman and Arntz [213], Alty and Clark 214 have visually observed the spreading of mercury over the surfaces of iron, copper, silver and tin. They have found that almost always a type of halo forms around the liquid drop which can be explained by surface diffusion of the liquid over the solid surface.

**The interrelation between the solubility of solid bodies in liquids and wettability in such systems is first mentioned in works by P. A. Rebinder [227].*

Since the existence of the halo could be also attributed to evaporation and condensation of the liquid atoms, Alty and Clark have also investigated the rise of mercury along a tin capillary in an oil medium when there is no evaporation. In both cases a band of diffusible mercury atoms is formed over the front of the liquid.

Formation of a "matt spot" (halation) at the polished zinc surface near the edge of mercury drop (surface diffusion of mercury atoms) and difference between this process and the spreading proper of the compact layer of the liquid metal are shown in [202]. Regularities of spreading the halation front are generalized in [226].

Tamman and Ruhenback [211] describe their observations of liquid drops at solid surfaces for many systems within a substantial temperature range. The following systems have been studied: tin, lead, bismuth, silver - on iron; tin, lead, bismuth - on nickel; tungsten, molybdenum, tantalum, mercury - on copper, silver, gold.

According to the authors' definition, "non-wetting" means that, $\theta > 90^\circ$, "wetting" that $\theta < 90^\circ$. It has been found that mercury wets the surface of copper, gold and silver, forming a halo, at room temperature. At 350°C, tin, lead and bismuth do not wet the surface of iron. At 600-700°C wetting by these metals (transition of the wetting angle through 90°) takes place in 30 minutes. In 60 minutes the drop has spread ever more forming a substantial halo.

Tin begins to wet nickel at temperatures about 400°C. Tin and lead behave similarly on nickel, but in the case of tin a great "crown" is formed in accordance with the fact that the solubility of tin in nickel is higher than that of lead.

In the Ni-Bi system when the temperature is increased up to 650°C, spreading is slowed down rather than accelerated. The authors attribute this phenomenon to decomposition of unstable compounds with temperature growth; at 470°C decomposition of the compound $NiBi_3 \to NiBi$ can take place, at 640°C $NiBi \to Ni+Bi$.

A similar temperature effect on the wetting characteristics is observed in the system (tin-lead)-copper [210]. A stable coating of tin-lead alloy is formed at temperatures below 380°C (the temperature of decomposition of Cu_6Sn_5 phase). This coating is destroyed as the temperature increases.

Usually a temperature increase facilitates wetting and the wettability increase is irreversible. In some cases spreading occurs on reaching a certain "critical" temperature.

The authors [211] have found that the wetting of molybdenum, tungsten and tantalum by lead and bismuth increases sharply on reaching a temperature equal to one third of the melting temperature of the solid phase. Perhaps this wetting is conditioned by an enhanced diffusion mobility of the solid phase atoms providing improved interaction in the contact zone.

The existence of a "wetting temperature" has been found for a eutectic solder of Sn-Pb on copper under a flux [56] as well as for the wetting of solid metals by liquid sodium [209].

A detailed study of the wetting of solid metals by liquid metals of spreading velocity and of coating formation has been carried out by Bailey and Watkinson [210]. Numerous data were obtained (though of qualitative nature) which relate the degree of wettability with the extent of interaction between the components of the system. During tests, the solid surface was coated with the liquid metal. After a certain time delay, the liquid metal was allowed to flow down the solid surface. The following states of the solid surface were observed (using the authors' terminology): (1) Wetting, formation of a thin continuous film of the liquid metal (in this case the wetting angle can be considered to be $\theta \cong 0$); (2) poor wetting, the liquid metal flowed down the surface but did not form a continuous film though adhering to the solid surface as individual drops (the wetting angle is finite but less than $90°$); (3) non-wetting, the liquid does not adhere to the solid surface neither as a film not as drops (evidently in this case $\theta > 90°$).

These authors' data [210] about contact properties in these systems, organized in accordance with the observed degree of wetting are given below:

1. Stable coatings are formed in the system ($\theta = 0$):

Liquid	- Solid	t°C	Liquid	-Solid	t°C
Chemical compounds are formed between metals					
Al	- Fe	700	Zn	- Fe	500
Sb	- Fe	700	Zn	- Ni	500
Sb	- Ni	700	Zn	- Cu	500
Sb	- Cu	600	Zn	- Au	450
Sb	- Ag	550	Zn	- Ag	500
Fe	- Fe	500	Pb	- Au	400
Fe	- Ni	500	Bi	- Ni	400
Fe	- Cu	400			
Fe	- Ag	500			
Cd	- Ni	400	Sn	- Fe	400
Cd	- Cu	350	Sn	- Ni	400
Cd	- Au	350	Sn	- Cu	400
Cd	- Ag	400	Sn	- Ag	275
The solid metal is dissolved in the liquid metal					
Ag	- Ni	1000	Pb	- Ni	
Ag	- Cu	850	Pb	- Ag	400
Ag	- Au	1000	Bi	- Ag	300

2. No stable coatings are formed. Liquid metal adheres to the solid surface in the form of individual drops ($\theta \neq 0$, but less than 90°). This state is realized in the system:

Liquid	- Solid	t°C
Ag	- Fe	1000
Pb	- Cu	400
Bi	- Cu	400

3. Liquid metal flows completely off the solid metal ($\theta > 90°$)

Liquid	- Solid	t°C
Pb	- Fe	400
Cd	- Fe	400
Bi	- Fe	400

It is seen that a high degree of wettability is observed in the presence of substantial interaction, the formation of a chemical compound or a solid solution; this is confirmed by the data of [226] where wetting angles are compared with a kind of phase diagram. When the phase diagram changes from the simple one corresponding to complete immiscibility of the liquid and solid states (systems Al-Pb, Fe-Bi, Al-Cd) to a more complex one with eutectics, solubility and chemical compounds (Zn-Pb, Zn-Bi, Zn-Cd, Zn-Fe, Sn-Cu, Al-Zn) the wetting angle decreases from 120-140 to 10-30°.

Bondi [56] has treated the data of a number of works determining the time of spreading of metal drops over solid metals and established the time needed to reach a given value of the wetting angle.

It has been stated that the time of spreading depends logarithmically on temperature and that the time decreases with increasing temperature. Hence it is possible to estimate the "activation energy" of the process. Values of this "activation energy of spreading" (V_A) for different systems are presented below:

System	$t°C$	V_A, *kcal/mole*
Bi-Cu	*400-700*	*5*
Pb-Ni	*400-450*	*29*
Sn-Cu	*300-350*	*31*

Taking into account the uncertainty of the exact value of "activation energy of spreading,: large errors of measurement, the use of different parameters by which the degree of wettability can be judged (in some cases the area of spreading, in other cases the wetting angle or the time of formation of a stable coating), the values obtained can be considered to be approximate. As to the order of magnitude (tens of kcal/mole) these values correspond to an activation energy of diffusion rather than to viscous flow of liquid metals.

On prolonged contact, the properties of the contacting phases and the degree of wettability can change. The following cases are possible:

(1) The solid phase tension can change due to adsorption of the liquid vapor. While being adsorbed on solid copper, lead can reduce the surface energy of copper from 1200-1300 to 700 erg/cm^2 [56]. At the moment of contact the energetic conditions are such that lead should spread over copper ($\theta \sim 0$), but adsorption is

taking place rapidly and soon the wetting angle becomes greater than zero (before the drop has time to spread completely), and finally the wetting angle of lead on copper reaches a value of between 40 and 60° (at T = 400 to 600°C).

At the same time, according to Lewis's data [217], lead under colohony or halloid flux at 350°C spreads rapidly over copper (the wetting angle value is not given). This is evidently attributed to the fact that lead cannot be adsorbed on copper through the flux and that the surface tension of copper remains high.*

(2) The degree of wetting is found to decrease with time following solid-liquid contact in systems forming chemical compounds. In this case a layer of an intermetalloid is formed at the interface which changes the interfacial energy. The final degree of wetting is determined by the relationship of the liquid metal properties and the intermetalloid being formed. As Milner states [220], the intermetalloid always resembles an ionic compound more than does the pure metal. Since ionic compounds are poorly wetted by metals, intermetalloid formation can adversely affect wetting. Examples of such systems are Ag-Sn, Ni-Sn, Fe-Sn [210] and Cu-Sn [210-218]. While immersing the solid metal in the liquid metal, the latter forms a coating on the former which reassembles into separate drops (de-wetting).

Studies of wetting on alloys has revealed the effect of intermetallic affinity and interaction. Thus lead being added to tin sharply increases the degree of wetting of steel and iron [221] (tin and iron form a stable intermetalloid); sodium at a concentration of 0.3% in lead decreases the interfacial surface energy at the boundary of liquid lead and zinc from 128 to 82 dyne/cm [224] (zinc and sodium also form an intermetalloid $Zn_{13}Na$); nickel added to lead (at a concentration of 0.1%) stabilizes the lead coating

In general, if a σ_{SV} change takes place together with a substitution of the solid body then the σ_{SV} increase does not necessarily lead to an improvement in wetting.

on iron and copper [210] (nickel forms solid solutions with copper and iron).

Addition of indium to lead decreases the wetting angle of germanium [216]. Solubility of germanium in the liquid indium is higher than in lead; pure indium wets the solid surface of germanium more intensively than pure lead [216,225].

Frolov [221] has noted that wettability in the system metal-metal is the better the smaller the temperature difference of the fusion temperatures of the two metals, i.e. the more the metals resemble each other. This is confirmed by the data of Bailey, Fox and Watkins cited by Taylor [218]about interfacial energies and wetting angles of iron, nickel and copper for wetting by low melting metals (the data is obtained by the sessile drop and dihedral angle methods in vacuum):

Solid metal	*Fe*	*Ni*			*Cu*		
Liquid metal	*Sn*	*Sn*	*Pb*	*Cd*	*Sn*	*Pb*	*Cd*
Wetting angle, °	*72*	*48*	*40*	*9*	*31*	*27*	*13*
Interface energy, erg/cm^2	*640*	*470*	*540*	*280*	*200*	*180*	*140*

It can be assumed that differences in the properties of low melting elements and those of iron, nickel and copper, decrease from iron to copper (melting temperatures, electronic structure). The wetting angle and interfacial energy behave in the same way.

Large differences in the values of specific bond energies (heat of evaporation expressed on a unit volume basis and called by Hildebrandt the parameters of solubility) produce immiscibility or areas of stratification in liquid metal melts. Stratification occurs when

$$\frac{V_1 + V_2}{2}(\delta_1 - \delta_2) \geq 2RT \qquad (24)$$

where $\delta = \frac{Q}{V}$, Q is the sublimation energy of the metal and V the atomic volume.

Thus when the properties of the contacting metals approach each other, there is a greater tendency for mutual solubility and diffusion, the interfacial energy is reduced, and the wettability of the system increases. With large differences between the properties of the components the opposite can occur; if metals differ in their specific volumes and energies but still have similar electrochemical properties then the wettability will decrease and interfacial energy will increase. In addition, when differences in their electronegativities increase there will be an increasing tendency to form intermetalloids which, at least initially, will intensify the degree of wettability.

From the data cited above, the relation between the wettability in metallic systems and the intensity and character of intermetallic interaction, as well as the components properties in the contacting pair is clear. On another hand, studies [209, 220, 228, 229, 232, 235,234] and experience in soldering and welding show that the adhesion properties of a liquid metal on a solid metal depend substantially on the presence of any oxide film. Such films are practically always present on contacting surfaces and hinder the true contact of metals. In [20] there is an example given which indicates considerable stability of such surface compounds. The oxide film at the surface proved to be stable at a temperature of 930°C and an oxygen pressure of about 1 atm, while the equilibrium constant for oxide dissociation is $1.7 \cdot 10^{11}$ atm at this temperature. In [10] the results are given of studying tungsten and molybdenum wettability by liquid tin, copper and silver, and the wettability of iron by lead. Some data are given in Figs. 30-32. Increasing the temperature decreases the contact angle sharply and irreversibly.

With continuous heating of molybdenum in contact with tin in vacuum or hydrogen and of iron with lead in hydrogen, the contact angle decreases and wetting occurs at temperature above 750 and 650° respectively. At temperatures below these "critical" values the contact angles are large (∿120-130°).

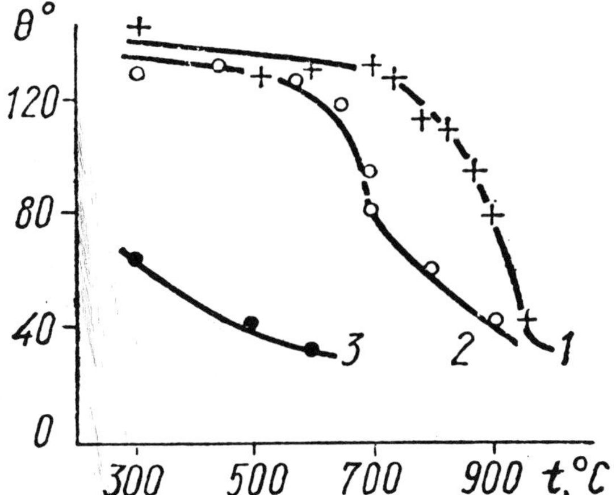

Fig. 30. Temperature dependence of the angle of contact of wetting of tungsten (1) and molybdenum (2) by liquid stannum; wettability in the system molybdenum-stannum after thermovacuum treatment of contact surfaces (3).

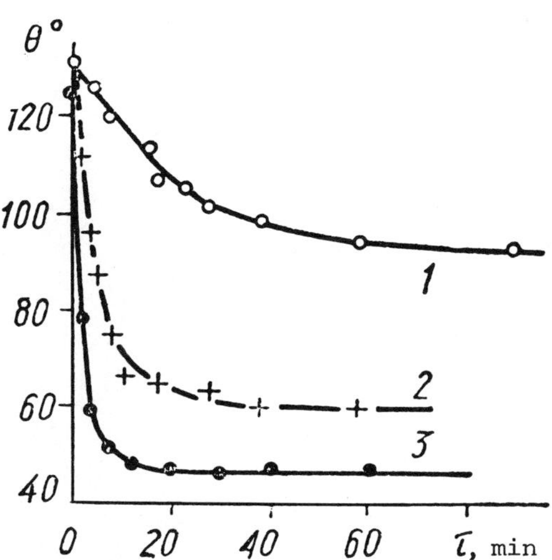

Fig. 31. Time dependence of molybdenum wettability by tin (vacuum): 1-700°C, 2-800°C, 3-850°C.

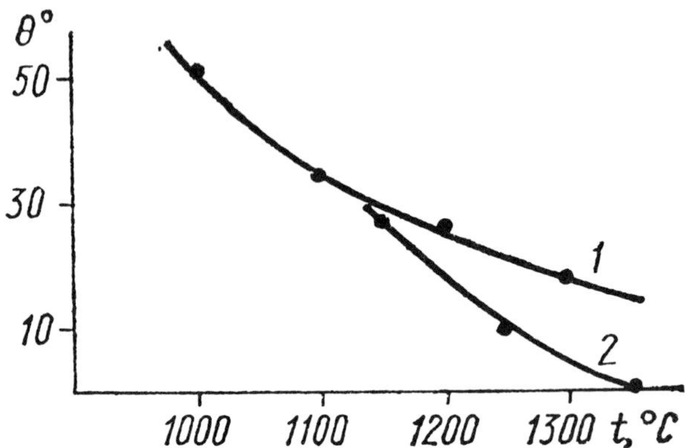

Fig. 32. Wettability of tungsten by liquid silver (1) and copper (2).

With a thermovacuum treatment of the surfaces of molybdenum iron, tin and lead carried out by separately heating the solid and liquid metals to temperatures above the "critical" values, wetting can be achieved. The contact angle being 15-40° (at low temperatures (300-400°C) and high values of wetting angles were observed before thermovacuum treatment (Fig. 30, curve 3). Iron is wetted by lead at T - 400°C with a wetting angle of about 20° [10]. Evidently heating molybdenum and iron surfaces in a vacuum and a hydrogen atmosphere eliminates absorbed films from these metallic surfaces. These films are supposed to be oxides.

Molybdenum oxides are volatile in vacuum and are readily reduced by hydrogen. The oxide nature of these films is also shown by the following experiment with the system iron-lead. Short-term action of atmospheric air at room temperature on thermally clean iron surfaces reduces wetting and to reproduce wetting behavior, repeated thermal treating is needed.

Non-wetting in the systems tin-molybdenum and lead-iron, as well as abrupt changes of adhesion and wetting angle with temperature and time are, to a great extent, brought about by the presence of oxide films inhibiting direct contact of the liquid and solid metal. Pure surfaces of molybdenum and iron are intensively wetted by tin and lead at the melting temperatures of the latter.

This conclusion can evidently be extended to other metallic systems, i.e. interaction between the solid and liquid metal in the contact layer, even for nonreacting systems, is sufficient to provide a high degree of wettability.

Both from the point of view of research methods and from comparison with the systems described above, the results of studying solid copper wetting by liquid aluminium are of great interest [239,240]. The tests were carried out by placing molten aluminium onto the solid copper surface at the given temperature. A device with a "spoon" was used, the latter being made from beryllium oxide. The drop shape was recorded with the help of a camera.

Taking into account the tendency of aluminium to oxidize, appropriate measures were taken as to improve the experimental component purity. While the furnace was heating and initial equipment decontamination was proceeding, the vacuum in the system was still too low. During this time the spoon containing the aluminium (weighed quantity 0.6 g) was isolated from the furnace and remained cold. Only after the equipment was thoroughly heated and the high vacuum was at least $1\text{-}2 \cdot 10^{-5}$ Torr was the spoon introduced into the furnace. It was possible in this way to obtain a pure shining aluminium surface which was very active in wetting solid bodies. If the spoon with aluminium had stayed in the furnace during the heating of the furnace and equipment decontamination, then even under high vacuum conditions (10^{-6} Torr) the aluminium surface would be covered with oxides and the tests would not have succeeded.

It should be noted that even ultrahigh vacuum conditions (10^{-10} -10^{-11} Torr) though being favorable from the point of view of the experimental cleanliness, do not themselves lead to any substantial improvement of the contact surfaces (solid and liquid metal) in these studies. It is stated in [243] that liquid indium at $T = 160°C$ forms an angle of contact of $170°$ at an aluminium surface at a pressure of 10^{-10} Torr. This unusually high value of the angle of contact corresponds to the system In-Al_2O_3 and not to In-Al. The angle of contact of melted silver on niobium and tantalum was > 90° at 10^{-10} Torr [244].

In the system Al-Cu, due to special measures taken against contamination of the contacting surfaces and the active interaction of the solid and the liquid metal, good wetting was already observed at a temperature of 700°C and a vacuum of $1 \cdot 10^{-5}$ Torr. (It should be noted that in the system Al-Mo [245] the "wetting temperature" is more than 900°C. This is attributed to the less intensive interaction of aluminium and molybdenum).

Experimental data for the system Al-Cu are given in Fig. 33 [238]. Aluminium spreads over copper. Increasing the temperature accelerates the process but its influence is much weaker than for the systems lacking interaction between the components (Mo-Sn). Practically, the activation energy of the wetting process is estimated by measuring the time to achieve a certain value of the wetting angle (for the given system this value was chosen within the interval 30-50°) and plotting $\ln t\theta_o$ Vs $\frac{1}{T}$ (see Fig. 34). This indicates that the activation energy for the system copper-aluminium (17±4 kcal/mole) is much lower than that for the system Mo-Sn: 45±5 kcal/mole.

While treating the phenomena of wetting in accordance with the above procedures, the contact systems solid metal-metallic liquid should be divided into two classes: (1) non-equilibrium systems where

$$\mu_i^s \neq \mu_i^l$$

(2) equilibrium systems where

$$\mu_i^s = \mu_i^l$$

Systems where the components are completely immiscible in the liquid and solid state below to type (2). For instance, pure iron (solid phase) is in equilibrium with pure lead (liquid). Diffusion, dissolution of one metal in another, is absent in these systems as is any directed flow of particles from one metal to another.

Previously for such systems non-wettability of the solid metal by the liquid one ($\Theta > 90^\circ$) was often found. But as experiments have shown [10], non-wettability in such systems is conditioned by the presence of contaminating films which hinder the true contact of the metals. With proper cleaning of the surface in such systems it is possible to obtain a high degree of wettability. This means that the contact of two metals due to metallic properties of the both is sufficient to slightly reduce the interfacial energy ($W_{a(equil)}$ is close to cohesion work), i.e., the presence and influence of the field formed by the system of ions and electrons of a mobile metal and by the second metal charge density at the interface are such that they smooth sharp changes of electron density taking place at the metal-vacuum interface. In this respect, the theoretical computations of the surface interfacial energy with a simple (but perfect) metal contact taking no account of diffusion, dissolution and formation of intermetalloids [241], i.e. of the equilibrium part of the interfacial energy are of great importance. Using these results one can estimate the wetting angle value in a system if it is assumed that one of the metals is liquid while the other is solid. To estimate the surface tension of the latter, Zadumkin's relation may be used:

$$\sigma_{sv} \cong 1.15\ \sigma_{lv}.$$

Knowing this value and the value of interfacial tension from [241] it is possible to find the wetting angle from Young's equation. The results of such estimations for various pairs of metals are presented in Table 20. As is seen in most cases the interfacial tension values are so small that complete wetting (usually with a large positive coefficient of spreading) or at least wetting angles less than 90° will occur. Zero wetting angles with small positive or even negative values of the coefficient of spreading are typical of systems which incorporate metals with a low bond-

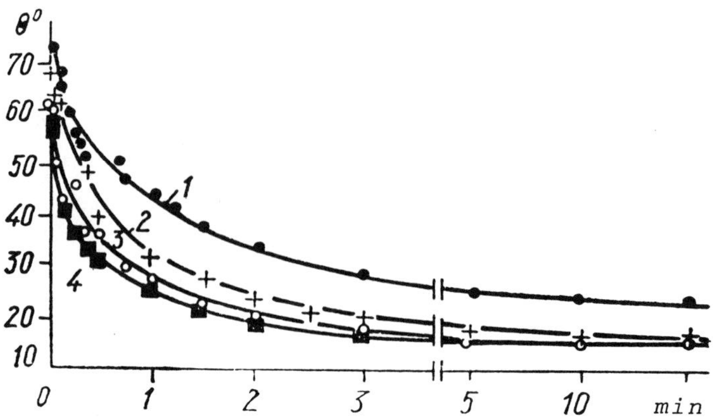

Fig. 33. Kinetics of liquid aluminium spreading over copper: 1-700°C, 2-800°C, 3-900°C, 4-1000°C.

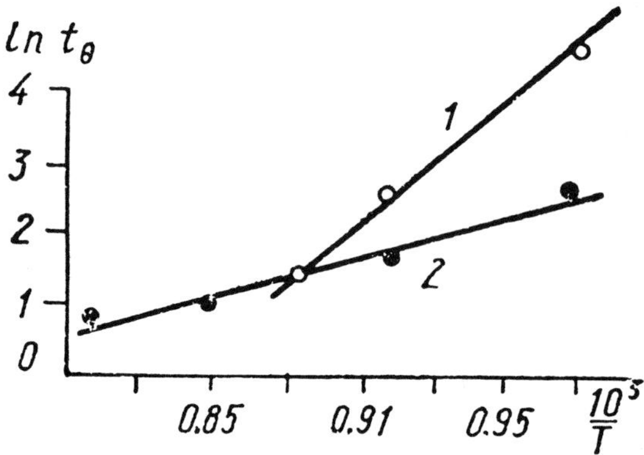

Fig. 34. Dependence of the period of "half-spreading" on temperature for systems tin-molybdenum (1) and aluminium-copper (2).

energy density (alkalai metals like rubidium and cesium). Only few of the pairs formed by Ca, Sr and Ba have wetting angles close to 90° as can be observed where the solid phase metal tension is below that of the metal forming the liquid phase. It should be noted that the wetting angle values deter-

TABLE 20. Interfacial tension calculated theoretically [241] and estimated wetting angles in metal-metal systems.

Metal	σ_{Zn-Me} dyne/cm / θ°_{Zn-Me}	σ_{Cd-Me} dyne/cm / θ°_{Cd-Me}	σ_{Sn-Me} dyne/cm / θ°_{Sn-Me}
Zi	200/0	50/0	180/0
Na	550/0	310/0	380/0
K	710/0	360/0	630/60
Rb	750/0	560/0	650/65
Cs	790/0	600/0	730/90
Cu	300/0	780/0	830/0
Ag	25/0	180/0	270/0
Au	20/0	250/0	390/0
Al	150/0	350/0	360/0
Ga	30/0	130/0	120/40
In	20/0	0/0	20/0
Be	300/0	520/0	520/0
Mg	70/35	5/0	5/0
Ca	140/70	50/55	60/50
Sr	340/90	160/72	230/80
Ba	370/90	20/85	260/90
Zn	-	50/0	20/0
Cd	50/0	-	350/50
Hg	140/0	30/0	30/0
Sn	20/0	0/0	-
Pb	100/0	10/0	20/0

mined by Young's equation as the difference of two approximately calculated values (σ_{sl} and σ_{sv}) depends strongly on errors of calculation and assumptions. Moreover it should be emphasized that in calculating interfacial tension [241] the authors did not take into account ionic, covalent or van der Waals component of the bond. Even under such conditions, zero wetting angle values are obtained in most cases. This confirms the conclusion that metal contacts form metallic bonds between the boundary layer atoms. This produces high values of $W_{a(equil)}$ and a high degree of wettability.

The estimation of the interfacial energy in intermetallic systems carried out in [242] relative to the equilibrium part of the interfacial tension of equilibrium systems also predicts high wettability in the systems studied. At the same time, in systems where a simple contact is the single cause of an interfacial energy decrease (i.e. in systems where there is no mutual diffusion or formation of intermetalloids) the establishment of intermetallic bonds at the interface can be easily destroyed by the presence of contaminating films which inhibit perfect contact. An oxide film at the interface of these two metals increases the distance between the contacting surfaces and decreases the superimposition of electron orbits of boundary atoms while at the same time lowering drastically the intermetallic interaction. It is important that the system metal-film-metal remains stable since there is no tendency to movement and exchange of atoms between the phases: $\mu_i^s = \mu_i^l$. The solid metal is in thermodynamic equilibrium with the liquid one.

Quite another picture can be observed with the interactions of nonequilibrium systems. For such systems beside the purely contact process, interfacial energy can be decreased due to interfacial interaction (diffusion of one metal into another, formation of intermetalloids at the interface or dissolution of the solid metal in the liquid metal, i.e. entrainment into interaction with deeper metal layers). Even in the presence of an oxide film between the contacting surfaces such a system is unstable. One or the both of metals tend to penetrate through the film as $\mu_i^s \neq \mu_i^l$. Such penetration, either spontaneous or initiated, was observed in published experiments [232]. If one of the pair of metals in the liquid state was placed on the oxidized surface of another metal and the film was not very thick then mutual penetration of atoms of both metals (or one of them) through the film can occur and further reaction (diffusion, dissolution) of metals may be observed.

The mechanism of penetration and formation of a good contact is as follows. According to [232] one of the metals, the liquid one, for example, reaches the solid metal surface (through the gas phase, for instance or through the faults in the oxide film). If the liquid metal dissolves the solid one, further reaction will take place preferably by dissolving the solid metal layers adjacent to the oxide film (these layers are energetically rich, and dissolve quicker), i.e. the liquid spreads over the solid metal-oxide layer interface separating the interlayer from the solid metal. Subsequently this process results in wetting and spreading of the liquid metal over the oxidized surface of the solid metal. If the film is too thick and the process of penetration does not proceed spontaneously it may be initiated by the local break of the oxide film under the liquid metal layer (for example by punching it with the needle). This creates a point contact between liquid and solid after which the process takes place spontaneously.

According to [209], a peculiar tunneltype mechanism of penetration through a faultless oxide layer of sufficiently small thickness is possible for the solid metal atom. Capillary flow of the liquid metal through cracks and pores of the oxide film, if the wettability of the oxide by the liquid metal is sufficiently high, and simple dissolution of the oxide in one of the metals are also possible. Such mutual penetration of metals towards each other is accompanied by a decrease in the system energy and by a reduction of the difference between the chemical potentials of the components. These processes cannot take place in noninteracting equilibrium systems. Thus, in the system Cd-Al the system is as close to complete immiscibility of the components as possible according to calculations (see Table 20) and has a near zero value of the wetting angle unwettability of aluminium by liquid cadmium is found experimentally; the wetting angle is 145° [226]. This is attributed to the presence of an oxide film on the aluminium surface. Scratching the oxide layer at the interface, i.e. the creation of contact between the liquid and the solid metal initiates spreading in the interacting system. Similar scratching proves useless for the system Cd-Al where spreading does not occur [232].

Rather large wetting angles in noninteracting systems (W-Sn, Mo-Sn, Fe-Pb) and their change with temperature and time are conditioned by the presence of intermediate oxide layers at the interface. These layers however are not always stable. Oxides of the

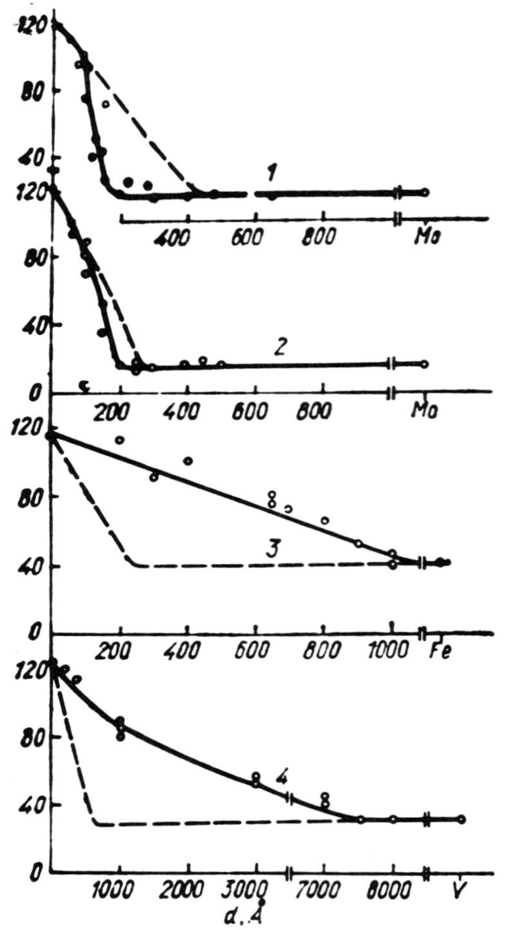
θ°
120
80
40
0
1
2
3
4
5
6
7
Mo
V
Fe
200
400
600
800
1000
d, Å

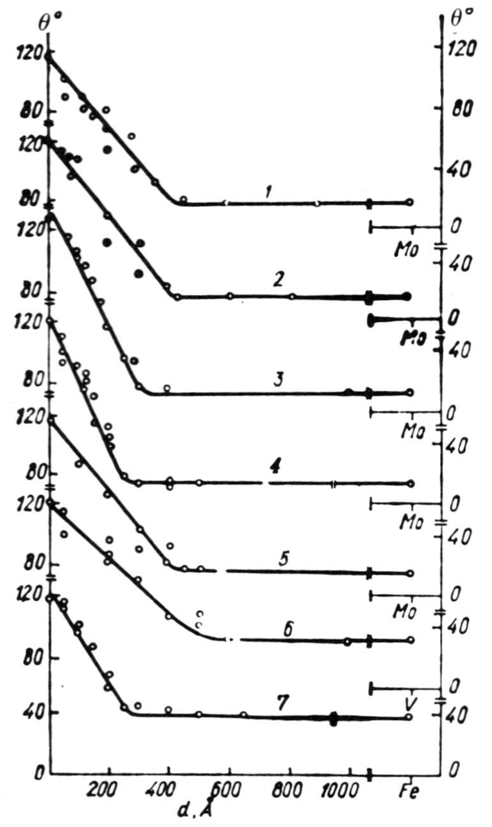
120
80
40
0
1
2
3
4
Mo
Fe
V
200
400
600
800
1000
2000
3000
7000
8000
d, Å

corresponding solid metals are readily reduced in hydrogen and sublime at high temperatures, a factor which can explain a wettability increase with increasing temperature. A high "activation energy" of the wetting process (45 kcal/mole for Mo-Sn) is of the same order of magnitude as the dissociation energy of the corresponding oxide. $\Delta H^{298}_{MoO_3} = -60$ kcal/g-atom of O_2 and is evidently conditioned by the destruction of the oxide layers.

In the copper-aluminium system a partly reducible film is formed at the liquid metal surface. Due to mobility of the latter, one should expect numerous faults in the oxide film and good mutual penetration of metals through the film (especially if it is sufficiently thin; in a "poor" vacuum of 10^{-4}-10^{-3} Torr, when the film becomes thicker, there is no wettability or spreading in this system). Thus, destruction of the oxide film is not the limiting stage of the process (evidently it is partially dispersed in the liquid metal and in the surface). Really the activation energy of the spreading process for the system under study is not high and is about 17±4 kcal/mole, a value which correlates well with the activation energy value of copper dissolution in liquid aluminium (16 kcal/mole) [237].

It may be assumed that the process of copper dissolution in liquid aluminium defines wetting and spreading in the system.

From the above discussion one can see the relation between the degree of wettability and the character and intensity of interaction between metals on one hand and oxide and other layers at the contact surface on the other. Both factors are acting jointly.

So the conclusion is that the metallic character of the contacting substances and a good intermetallic contact are, as a rule,

Fig. 35. Wettability by metals of oxides (a) and graphite (b) depending on the metal coating thickness: (a) 1-Al_2O_3-Mo-Cu (1150°C); 2-SiO_2-Mo-Cu (1150°C); 3-Al_2O_3-Mo-Ag (1000°C); 4-SiO_2-Mo-Sn (900°C); 5-SiO_2-Mo-Sn (900°C); 6-SiO_2-V-Sn (900°C); 7-SiO_2-Fe-Pl (700°C); (b) 1-C-Mo-Cu (1150°C); 2-C-Mo-Sn (900°C); 3-C-Fe-Pl (700°C); 4-C-V-Sn (900°C).*

**Preliminary annealing of the film at 1150°C (dotted lines present wetting in the systems similar to SiO_2)*

sufficient to provide a low interfacial energy and a high wettability in the system. Cases of non-wettability in intermetallic systems are not observed on contact of pure surfaces. Nonwettability ($\theta > 90^\circ$) observed experimentally on contact of the liquid and a solid metal is due to the presence of intermediate oxide and other layers at the interface.

The Wettability of Metallized Surfaces of Solid Bodies

When it is necessary to increase wettability of solid metal bodies by metal melts, the former are subject to metallizing-coating with thin metal films. In such complex systems, wettability and adhesion depend on many factors (film thickness, the character and intensity of its interaction with the substrate and the liquid phase, etc.). Some questions of this kind are studied in [246, 247].

Studies have been made of the wettability of oxides (monocrystal Al_2O_3-sapphire, quartz, glass) and graphite coated by evaporation of molybdenum, vanadium and iron films of different thickness and by melts of copper, silver, tin and lead. The dependency of the wetting angle on the coating thickness is shown in Fig. 35. The character of the dependence for most systems is the same; the wetting angle shows a neat linear drop from its value for the substrate pure surface to the value for the surface of the compact material film achieved at a certain coating thickness. The threshold value is different for various systems (from 150 Å to 7000 Å).

These results can be treated in the following way [246]. During the condensation of metals and film growth the film is not continuous while still thin, but has an insular or network structure. As the amount precipitated grows, depending on the annealing temperature; the islands are also growing (at the expense of coagulation) and the film becomes continuous only at a certain critical thickness, which depends on several factors, in particular on adhesion of the deposited metal to the substrate. Thus metallic condensates in the form of continuous films on metallic substrates can be obtained at much smaller total film thickness than on nonmetallic substrates. These substrates are "unwetted" by solid crystallites or condensate nuclei, and their relative area of contact with the substrate is smaller.

For heterogeneous surface in contact with the liquid phase, the total adhesion work can be put down as follows:

$$W_a = W_{a1} \cdot S_1 + W_{a2} \cdot S_2 \ldots \tag{25}$$

where W_{a1} and W_{a2} are specific adhesion works of the melted metal on different sections of the solid phase surface (either pure or coated by deposited metal), S_1 and S_2 are, respectively, the portion of the substrate surface still uncoated and that covered with condensate ($S_1+S_2=1$).

S_2 growth alone will cause an increase of the degree of wettability of the substrate by the liquid metal as the film thickness on the substrate increases. The critical thickness is smaller at a lower annealing temperature, with greater adhesion of the film to the substrate and good wettability of the substance reaction products with the substrate. From Fig. 35a it follows that for a molybdenum film the critical thickness on oxides SiO_2 and Al_2O_3 with wetting by copper is 250-450 Å, depending on the film annealing temperature. For graphite coated by molybdenum (Fig. 37b) this thickness is smaller and is about 150 Å (there is a high adhesive interaction between Mo and graphite). Because of the poorer wettability of vanadium carbides and iron by metals as compared to pure vanadium and iron, the critical thickness of films in the graphite-vanadium system and the graphite-iron system (liquid metallic phase tin and lead) shifts to about 10^3 Å.

An improved formula for the wetting angle of a solid body surface coated by an insular structure film, taking into account the contribution of the nuclear form of the film to its adhesion to the substrate is given in [246].

V. CONCLUSION

The study of adhesion and wettability of solid bodies by metal melts has recently made substantial progress. A number of adequately improved methods of measurement have been developed. A great amount of experimental data on wettability in a wide variety of systems is now available. A general analysis of the data obtained has permitted the derivation of theoretical statements which have allowed us to understand and explain some common features of the process.

A chemical interpretation of the phenomenon of solid body wettability by liquid metals and an interpretation of the process as a chemical reaction in the interphase area makes it possible to calculate quantitative values of the strength of adhesion and of the wetting angle value in specific systems.

Recently some studies have been devoted to elucidating the connection between the degree of wettability, the character and type of interphase interaction)dissolution, diffusion or chemical compound formation) and the nature and properties of newly-formed interface layers.

The technique of measuring interfacial energies at the boundaries of solid phases is being improved though these data are not yet in complete agreement with the values of interfacial energies obtained from the wetting angle equation.

The results of these basic studies of the wettability of solid bodies by metals are used now to solve many technical problems.

The author is grateful to Prof. D. A. Cadenhead for preparing the English text of this article.

References

1. Thomas Young (1805). *Trans. R. Soc., London 94,* 65.
2. Gromeka, I. S. (1879). Ocherk teorii kapillarnih javlenij/ Teoria poverhnostnogo czeplenia gidkostei/, Moscow.

3. Gibbs, J. (1950). Termodinamicheskie raboti. IL., Moscow-Leningrad, p. 400-414; Scientific papers of J. W. Gibbs, in "Thermodynamics," Vol. 1. Dover, New York, 1961.
4. Bakker, G. (1928). Kapillarität and Oberflächenspännung. Akadem. Verlagsges. Leipzig.
5. Rebinder, P. A. (1937). Fizicheskij slovar̆, 3. ONTI, Moscow, 287, 640.
6. Laplace, P. (1805). Traite de mechanique celest, 1-er Suppl. to 10th book in Vol. IV. Paris, p. 45.
7. Davidov, A. (1851). Teoria kapillarnih javlenij, Moscow.
8. Frenkel, Ja.I. (1948). *J. Exp. Theor. Phys. 18,* 7, 659.
9. Jonson, R. E. (1959). *J. Phys. Chem. 63,* 10, 1655.
10. Naidich, Ju. V. (1972). Kontaktnie javlenia v metallicheskih rasplavah, Naukova Dumka, Kiev.
11. Olsen, D. A., Soyner, P. A., and Olsen, M. D. (1962). *J. Phys. Chem. 66,* 5, 883.
12. Bikerman, J. J. (1950). *J. Colloid. Sci. 5,* 349.
13. Scherbakov, L. M., and Razanzev, N. P. (1963). Poverhnostnie javlenia v rasplavah i prozesah poroshkovoi metallurgii. Izd-vo AN Ukranian, Kiev, p. 152.
14. Levin, A. M. (1952). Sbornik trudov Dnepropetrovskogo metallurgicheskogo instituta, Dnepropetrovsk, vip. 28, p. 89.
15. Naidich, Ju. V. (1968). *J. Phys. Chem. (USSR) 42,* 8, 1946.
16. Eremenko, V. N., and Naidich, Ju. V. (1958). Smachivanie redkimi metallami tverdih poverhnostei tygoplavkih soedinenij. Izd-vo AN Ukranian, Kiev, p. 20.
17. Kingery, W. D. (1955). *J. Amer. Cer. Soc. 36,* 11, 362.
18. Humenik, M., and Kingery, W. D. (1954). *J. Amer. Cer. Soc. 37,* 1, 18.
19. Eremenko, V. N., and Naidich, Ju. V. (1959). *J. Phys. Chem. 33,* 6, 1239.
20. Jordane, D. O., and Lane, I. E. J. (1962). *Austral. Inst. Metals 7,* 1, 27.
21. Ono, S., and Kondo, S. (1963). Molekularnaja teoria poverhnostnogo natjajenia v jidkostjah. IL, Moscow, p. 73. (Molecu-

lar theory of surface tension in liquids, in Handbuch der physik, Herausgegeben von S. Flügge, Band X (1960). Springer-Verlag, Berlin-Göttingen-Heidelberg).

22. Juhovizki, A. A., Grigorjan, V. A., and Mihalik, E. (1964). DAN USSR, *155*, 2, 392.

23. Naidich, Ju. V., and Perevertajlo, V. M. (1971). *J. Phys. Chem. 45*, 7, 1806.

24. Naidich, Ju. V., and Perevertajlo, V. M. (1977). V sb. Smachivaemoct e poverhnostnie svoystva rasplavov e tverdih tel. Naukova Dumka, Kiev.

25. Aksay, J. A., Hoge, C. E., and Pask, J. A. (1974). *J. Phys. Chem. 78*, 12, 1178

26. Naidich, Ju. V., Perevertajlo, V. M., and Nevodnik, G. M. (1972). Poroshk. metallurg. 7, 51.

27. Naidich, Ju. V., and Kolesnichenko, G. A. (1968). Vzaimodeystvie metallicheskih rasplavov s poverhnostju almaza e graphita. Naukova Dumka, Kiev.

28. Guggenheim, T. (1940). *Trans. Faraday Soc.* 397.

29. Benjamin, P., and Weaver, C. (1959). *Proc. R. Soc. (London) 252A*, 418.

30. MacDonald, I. E., and Eberhart, I. G. (1965). *Trans. Metal Soc. AJME 233*, 3, 512.

31. Kunin, L. L. (1955). Poverhnostnie javlenia v metallah. Metallurg/izdat, Moscow.

32. Adam, N. K. (1947). Physika i chimia poverhnostei. Gostehizdat, Moscow (Physics and Chemistry of Surfaces, 1941, London).

33. Ferguson, A. (1941). *Proc. Phys. Soc. (London) 53*, 554.

34. Bashforth, F., and Adams, J. (1883). An attempt to test the theories of capillary action, Cambridge.

35. Fowler, R., and Guggenheim, E. (1949). Statisticheskaja termodinamika JL, Moscow, (Statistical Thermodynamics, 1949, 2nd ed., London).

36. Popel, S. I., Ivanov, S. M., and Nikitin, Ju. P. (1961). Izv. Vuzov, Chernaja metallurgia, 11.

37. Dorsey, N. E. (1928). *J. Wach. Acad. Sci. 18*, 505.

38. Guastalla, J. (1970). *Bull. Soc. Chim. (France)*, p. 3211.

39. Volkova, Z. V. (1939). *J. Phys. Chem. 12*, 12, 224.

40. Bikerman, I. I. (1944). *Ind. Eng. Chem. Anal. 13*, 443.

41. Wolfram, E., and Weber, K. (1962). Kolloid Z. u. Z. Polymere 81, 3, 73.

42. Naidich, Ju. V., and Eremenko, V. N. (1961). Phys. metallov i metallovedenie, 11, 6, 883.

43. Wilhelmy, L. (1863). *Ann. Physik 119*, 177.

44. Naidich, Ju. V., and Juravlev, V. S. (1971). Zavodskaja laboratoria, 4, 52.

45. Naidich, Ju. V., Chuvashov, Ju. N. and Juravlev, V. S. (1975). Zavodskaja laboratoria, 7, 825.

46. Naidich, Ju. V., Perevertailo, V. M., and Nevodnik, T. M. (1976). Zavodskaja laboratoria 1, 39.

47. Olsen, H. O., Smith, C., and Grittenden, E. (1945). *J. Appl. Phys. 16*, 405.

48. Baxter, I., and Roberts, P. (1954). *Powder Metal. Symp.* London (Iron and Steel), 63.

49. Brace, P. H. (1948). *J. Electrochem. Soc. 94*, 170.

50. Wooster, W., and Mac Donald, O. (1947). *Nature 160*, 4060, 260.

51. Pulliam, O. R. (1957). *J. Science 31*, 3, 503.

52. Weyl, W. A. (1948). *J. Soc. Glass. Technol. 32*, 148, 247.

53. Kingery, W. D. (1956). *Amer. Cer. Soc. Bull. 35*, 3, 108.

54. Cabrera, N., and Mott, N. (1948-49). *Report on Progress in Physics 1*, 12, 163.

55. Livey, D., Murray, P. (1956). Wärmefeste und Korrosionbeständige Sinterwerkstoffe, Plansee Seminar, Reut-Tyrol.

56. Bondi, A. (1953). *Chem. Rev. 52*, 2, 417.

57. Allen, B., and Kingery, W. (1959). *Trans. Metall. Soc.* AIME, *215*, 30.

58. Thornton, H. R. (1959). *Ceramics 10*, 123.

59. Novikov, A. N. (1956). In "Physiko-chimicheskie osnobi keramiki." Metallurgizdat, Mowcow.

60. Presnov, A. A. (1953). Izv. vuzov, *Physika 1*, 103.

61. Economous, G. E., and Kingery, W. D. (1953). *J. Amer. Cer. Soc. 36,* 12, 403.

62. Kingery, W. (1958). *J. Phys. Chem. 62,* 7, 961.

63. Kirkjian, C., and Kingery, W. D. (1956). *J. Phys. Chem. 60,* 7, 961.

64. Halden, G. A., and Kingery, W. D. (1956). *J. Phys. Chem. 59,* 6, 557.

65. Kingery, W. D. (1954). *J. Amer. Cer. Soc. 37,* 2, chapter 1, 42.

66. Armstrong, W. A., Chaklader, A. C. D., and Clark, I. G. (1962). *J. Amer. Cer. Soc. 45,* 3, 115.

67. Armstrong, W. A., and Chaklader, A. C. D. (1962). *J. Amer. Cer. Soc. 45,* 9, 407.

68. Zarevski, B. V., and Popel, S. I. (1961). In "Poverhnostnie javlenia v metallah i ih rol b prozessah poroshkovoi metallyrgii. Izd-vo AN Ukranian, Kiev.

69. Bradhurst, D. A., and Buchanan, A. S. (1959). *J. Phys. Chem. 63,* 9, 1486.

70. Bradhurst, D. H., and Buchanan, A. S. (1961). *Anstr. J. Chem. 14,* 3, 409.

71. Tikkanen, M. (1963). Jerkontorets Annaler 147, 1, 37.

72. Zarevski, B. V., and Popel, S. I. (1961). In "Physiko-chimicheskie osnovi proizvodstva stali." Izd-vo AN USSR, Moscow.

72. Zarevski, B. V., and Popel, S. I. (1960). Izv. Vuzov, Chernaja metallurgia, 8, 15.

74. Zarevski, B. V., Popel, S. I., and Domojirov, B. Ph. (1965). In "Poverhnostnie javlenia v rasplavah i voznikajushih iz nih tverdih phazah." Kabardino-Balkarskoe knijnoe izdatelstvo, Nalchik, 316.

75. Smirnov, L. A., Popel, S. I., and Zarevski, B. V. (1965). In "Poverhnostnie javlenia v resplavah i voznikajushih iz nih tverdih phazah." Kabardino-Balkarskoe knijnoe izdatelstvo, Nalchik, 320.

76. Popel, S. I., *et al.* (1968). Poverhnostnie javlenia v rasplavah. "Naukova dumka," Kiev, 86.
77. Van Vlach, L. H. (1965). *Metals Eng. Quart. Amer. Soc. Metals, 11*, 7.
78. Zephirov, A. P. (1965). Termodinamicheskie svoistva neorganicheskih veshestv, pod red. Atomizdat, Moscow.
79. Milner, A. (1964). *J. Inst. Metal. 10*, 10, 48.
80. Carnahan, R. D., Johnston, T. L., and Li, C. H. I. (1958). *J. Amer. Cer. Soc. 41*, 9, 343.
81. Volkov, S. E., Mchedlishvili, V. A., and Samarin, A. M. (1963). DAN USSR 149, 5, 1131.
82. Naidich, Ju. V., and Juravlev, V. S. (1974). Ogneypori 1, 50.
83. Naidich, Ju. V. (1965). In "Poverhnostnie javlenia v rasplavah i voznikajushih iz nih tverdih phazah." Kabardino-Balkarskoe knijnoe izd-vo, Nalchik, 30.
84. Madelung, F. (1919). *Phys. Z. 20*, 494.
85. Born, M., and Heppert-Majer, M. (1938). Teoria tverdogo tela. ONTI, M.-L., Moscow-Leningrad, p. 325.
86. Weyl, W. A. (1952). *Ceram. Age 60*, 5, 28.
87. Van-Zin-Tan, Karasev, G. A., and Samarin, A. M. (1960). Izv. AN USSR, OTN, Metallurgia i toplivo, 1.
88. Sapiro, S. I. Stal (1947) 3, 395; (1946) 7-8, 449; (1950) 4, 311.
89. Geld, P. V., and Esin, O. A. (1954). Phyzicheskaja chimia pirometallurgicheskih prozessov, 2. Metallurgizdat, Moscow, Pr. 573.
90. Nehedzi, Ju. A. (1948). Stalnoe lite, Metallurgizdat, Moscow, p. 52.
91. Bleakburn, A., Shevling, T., and Lowers, H. (1949). *J. Amer. Cer. Soc. 32*, 3, 81.
92. Helgesson, C. T. (1966). *Trans. of Chalmers University of Technology*, Gotheuburg, Sweden, 311.
93. Eremenko, V. N., and Naidich, Ju. V. (1960). Izv. AN USSR, OTN, Metallurgia i toplivo 1, 53.

94. Eremenko, V. N., Naidich, Ju. V., and Nosonovich, A. A. (1960). *J. Phys. Chem. 34,* 5, 1018.

95. Eremenko, V. N., and Naidich, Ju. V. (1959). In "Poverhnostnie javlenia v rasplavah i prozessah poroshkovoi metallurgii." Izd-vo AN Ukranian, Kiev.

96. Eremenko, V. N., Naidich, Ju. V., and Nosonovich, A. A. (1960). *J. Phys. Chem. 34,* 6.

97. Momma, K., and Suto, H. (1960). *J. Japan Inst. Metals 24,* 374, 377, 544.

98. Semenchenko, V. K. (1957). Poverhnostnie javlenia v metallah i splavah. Gostehizdat, Moscow, 170.

99. Frenkel, N. I. (1924). Elektricheskaja teoria tverdih tel, gl. 13.

100. Kingery, W. D. (1958). *J. Phys. Chem. 62,* 7, 878.

101. Naidich, Ju. V., Chuvashov, Ju. N., and Juravlev, V. S. (1976). Adgezia rasplavov i paika materialov, vip. 1, Naukova dumka, Kiev, 50.

102. Naidich, Ju. V., Juravlev, V. S., and Chuprina, V. G. (1973). Poroshkovaja metallurgia, 11.

103. Naidich, Ju. V., Juravlev, V. S., and Chuprina, V. G. (1974). Poroshkovaja metallurgia, 3.

104. Naidich, Ju. V. (1972). Smachivaemost i poverhnostnie svoistva rasplavov i tverdih tel. Naukova dumka, Kiev, 7.

105. Chuvashov, Ju. N., Juravlev, V. S., and Naidich, Ju. V. (1976). *J. Phys. Chem. 50,* 4, 993.

106. Naidich, Ju. V., Juravlev, V. S., and Chuvashov, Ju. N. (1974). Poroshkovaja metallurgia 12, 66.

107. Ogino Kazumi, Nodi Kiyoshi, and Koshida Yukio (1973). *J. Iron Steel Inst. (Jap.) 10,* 1380.

108. Kuzmin, L. I. (1972). Sb. Smachivanie i poverhnostnie svoistva rasplåvov i tverdih tel, Kiev, Naukova dumka, 233.

109. Kuzmin, L. I. (1973). Ogneupori 12, 30.

110. Sveshkov, Ju. B., Mogilevski, V. I., Polonski, Ju. A., and Valdman, O. A. (1976). Jurnal prikladnoi chimii 12, 2629.

111. Kalmikov, V. A., Svechkov, Ju. V., and Elezov, S. A. (1976). Phisiko-chimia obrabotki materialov 6, 64.

112. Kurbatov, G. A., Kalmikov, V. A., and Baldman, O. (1973). Jurnal prikladnoi chimii 9, 2101.

113. Kutateladze, S. S. *et al.* (1958). Jidkometallichesrie Teplonositeli. Atomizdat, Moscow, 171-189.

114. Ubbelode, A. P., and Luis Ph.A. (1956). Graphit i ego kristallicheskie soedinenia, "Mir," Moscow.

115. Belaev, A. I., and Jemchujina, E. A. (1960). In "Nauchnie dokladi In-ta svetnoi metallurgii," 33, 132-142.

116. Humenik, M., Hall, H. D. W., and van Alsten, R. L. (1960). *Iron Age 10,* 11, 171-173.

117. Humenik, M., and Kingery, W. D. (1954). *J. Amer. Cer. Soc. 37,* 1, 18.

118. Humenik, M., Hall, H. D., and Van Alsten, R. L. (1962). *Metall. Progress 81,* 4, 10.

119. Ford, A. R., and White, A. E. (1963). *Chem. Eng. 166,* 61-67.

120. Eljutin, V. N., Kostikov, V. I., and Maurah, M. A. (1964). Izv. vuzov Chernaja metallurgia 11, 5.

121. Eljutin, V. N., Kostikov, V. I., and Maurah, M. A. (1965). Izv. vuzov, Chernaja metallurgia 1, 3, 58.

122. Eljutin, V. N., Kostikov, V. I., and Maurah, M. A. (1965). In "Poverhnostnie javlenia v pasplavah i voznikajushih iz nih tverdih phazah." Kabardino-Balkarskoe knijnoe izdatelstvo, Nalchik, 345, 352.

123. Eljutin, V. N. *et al.* (1965). Neorganicheskie materiali 1, 12.

124. Williams, Z., and Murray, P. (1954). *Metallurgia 49,* 836-837.

125. Hansen, M., and Anderko, K. (1958). Constitution of binary alloys. McGraw-Hill, N.Y.

126. Naidich, Ju. V., Kolesnichenko, G. A., and Perevertaillo, V. M. (1974). Sinteticheskie almazi v promishlennosti, Naukova dumka, Kiev, 32.

127. Naidich, Ju. V., and Kolesnichenko, G. A. (1968). Izv. AN USSR, Metalli, 4.

128. Naidich, Ju. V., and Kolesnichenko, G. A. (1961). Poroshkovaja metallurgia, 6.

129. Naidich, Ju. V., and Kolesnichenko, G. A. (1963). Poroshkovaja metallurgia, 1.

130. Naidich, Ju. V., and Kolesnichenko, G. A. (1964). Poroshkovaja metallurgia, 3.

131. Naidich, Ju. V., and Kolesnichenko, G. A. (1966). Poroshkovaja metallurgia, 2.

132. Naidich, Ju. V., and Kolesnichenko, G. A. (1968). Poroshkovaja metallurgia, 2.

133. Naidich, Ju. V., and Kolesnichenko, G. A. (1968). Poroshkovaja metallurgia, 7.

134. Naidich, Ju. V., and Kolesnichenko, G. A. (1965). In "Poverhnostnie javlenia v rasplavahi vosnikajushih iz nih tverdih phazah" Kabardino-Balkarskoe knijnoe izd-vo, Nalchik, p. 564.

135. Bever, M. B., and Floe, C. F. (1946). *Trans AIME 166*, 128.

136. Bowman, F. E., and Cygan, R. (1965). *Liquid Metals Handbook*, 152.

137. Kelman, R., Wilkinson, W. O., and Jaggee, T. L. (1950). *Liquid Metals Handbook*, 77.

138. Rutt, O., and Berg-dal, B. (1919). *Z. anorg. chem. 106*, 91.

139. Meerson, G. A., and Umansky, Ja. S. (1953). Izv. sektora physiko-chimicheskogo analiza IONH AN USSR 22, 104.

140. Samsonov, G. V. (1965). Poroshkovaja metallurgia 1, 98.

141. Umanski, Ja. S. (1949). Karbidi tverdih splavov. Metallurgizdat, Moscow.

142. Samsonov, G. V., and Umansky, Ja. S. (1957). Tverdie soedinenia tugoplavkih metallov. Metallurgizdat, Moscow.

143. Novotny, G. (1963). In "Electronic structure and alloy chemistry of the transition elements," ed. P. A. Beck, Interscience publishers, J. Wiley, New York, p. 166.

144. Meerson, G. A. (1963). Kratkaja chimicheskaja enziklopedia 2, "Sovetskaja enziklopedia," Moscow, p. 423.

145. Setton, R. (1960). *Bull. Soc. Chim. (France) 3,* 521.

146. Setton, R. (1960). *Bull. Soc. Chim. (France) 10,* 1758.

147. Solzano, F. I., and Aronson, S. (1964). *J. Inorg. Chem. 26,* 8, 1456.

148. Ubbelohde, A. R. (1938). Uspehi chimii 7, 1692.

149. Umansky, Ja. S. (1943). Izvestia sektora physiko-chimicheskogo analiza 161, 1, 127.

150. Kiessling, R. (1957). *Met. Rev. 2,* 77.

151. Brjuer, L. (1963). In "Electronic structure and chemistry of the transition elements," ed. P. A. Beck, Interscience Publishers, J. Wiley, New York, p. 211.

152. Franzevich, I. N., and Kovensky, I. I. (1961). DAN Ukranian 11, 1169.

153. Franzevich, I. N., and Kovensky, I. I. (1961). DAN Ukranian 8.

154. Arbuzov, M. P., and Haenko, B. V. (1966). Poroshkovaja metallurgia 4, 74.

155. Meka, M. A., Chaus, I. S., and Mitjureva, T. T. (1963). Gally, Gostehizdat Ukranian, Kiev.

156. Swartz, E. L. (1959). *Trans. Met. Soc. AIME 215,* 4, 522-554.

157. Kononenko, V. I. *et al.* (1977). Izvestia AN USSR, Metalli i, 52.

158. Vershinin, N. P., Barajan, N. I., and Denisov, Ju. D. (1975). Sbornik "Splavi redkoh metallov s osobimi physiko-chimicheskimi svoistvami," Nauka, Moscow, p. 142.

159. Naidich, Ju. V., Perevertailo, V. M., and Nevodnik, G. M. (1972). Izvestia AN USSR, Metalli, 3, 240.

160. Kingery, W. D., and Halden, F. A. (1955). *Amer. Cer. Soc. Bull. 34,* 4, 117.

161. Gurlland, S. J., and Norton, J. (1952). J. Metals 4, 10, 1051.

162. Humenik, M., and Parikh, N. (1956). *J. Amer. Cer. Soc. 39,* 2, 60.

163. Belaev, A. I., and Jemchujina E. A. (1952). Poverhnostnie javlenia v metallurgicheskih prozessah. Metallurgizdat, Moscow.

164. Takachi, Tamura, and Josida (1956). *J. Jap. Inst. Metals 20*, 7, 375.

165. Ramkvist, L. (1965). *Intern. J. Powder Metallurgy 1*, 4, 2.

166. Eremenko, V. N., and Naidich, Ju. V. (1959). *J. Neorg. Chem. 4*, 931.

167. Whalen, T. J., and Humenik, . (1950). *Trans. Met. Soc. AIME 218*, 252.

168. Tinklepaugh, I. R., and Crendall, W. B. (1960). (Cermets, ed.) Reinhold, New York, p. 66.

169. Tangermann, I. (1961). Neue Hütte 6, 12, 767.

170. Eremenko, V. N., and Phesenko, V. V. (1961). Poverhnostnie javlenia v metallah i splavah i ih pole v prozessah poroshkovoi metallurgii. Izd-vo AN USSR, Kiev, p. 178.

171. Kamishov, V. M., Esin, Ju. N., and Chuchmarev, S. K. (1965). *J. Phys. Chem. 40*, 1, 262.

172. Samsonov, G. V. (1963). In "Poverhnostnie javlenia v rasplavah i prozessah poroshkovoi metallurgii." Izd-vo AN USSR, Kiev, p. 90.

173. Tumanov, V. N., Phynke, V. Ph., Belenkaja, L. I. (1963). In "Poverhnostnie javlenia v rasplavah i prozessah poroshkovoi metallurgii." Izd-vo AN USSR, Kiev, p. 167.

174. Phunke, B. Ph., *et al.* (1965). In "Poverhnostnie javlenia v rasplavah i voznikajushih iz nih tverdih phazah." Kabardino-Balkarskoe knijnoe izd-vo, Nalchik, p. 397.

175. Mortimer, D. A., and Nicholas, M. (1973). *J. Mater. Sci. 8*, 640.

176. Milner, A., and Tottle, C. (1964). *J. Inst. Metals 93*, 48.

177. Hamjian, H. J., and Lidman, W. G. (1952). *J. Amer. Cer. Soc. 35*, 2, 44.

178. Samsonov, G. V., Panasjuk, A. D., and Kozina, G. K. (1968). Poroshkovaja metallurgia 11, 42.

179. Standing, R., and Nicholas, M. (1978). *J. Meter. Sci. 13*, 1509.

180. Samsonov, G. V., Yasinskaja, G. A., and Tai-Shou-Vei (1968). Ogneupori 25, 1, 35.

181. Samsonov, G. V. (1957). Silizidi i ih ispolzovanie v tehnike. Izd. AN Ukranian, Kiev.

182. Engel, . (1951). *Metal Progress 59*, 5, 664.

183. Metcalf, O. G. (1957). *Mach. mod. 50,* 672, 49.

184. Scott, P., Nicholas, M., and Devar, B. (1975). *J. Mater. Sci. 10,* 1833.

185. Kotsch, H., and Putrky, G. (1966). Abhaudl. Dtsch. Akad. Wiss. Berlin, Math. Phys. und Techn. 1, 249.

186. Kotsch, Heinz (1967). Neue Hütte 12, 6, 350.

187. Samsonov, G. V. (1965). *Ukr. chim. J. 31,* 1233.

188. Meyer, O. (1930). Ber. Deusch. Keram. Soc. 11, 333.

189. Toda Gedzo and Zunoda Teruo (1966). *Chemei 4,* 10, 322.

190. Naidich, Ju. V., and Nevodnik, G. M. (1969). Heorganicheskie materiali 5, 12, 2066.

191. Gnesin, G. G., and Naidich, Ju. V. (1969). Poroshkovaja metallurgia 2.

192. Samsonov *et al.* (1960). Bor i ego soedinenia i splavi, Izd-vo AN Ukranian, Kiev.

193. Hildebrand, I. H., and Scott, R. (1950). *The Solubility of Non Electrolytes, New York.*

194. Kisljakov, I. P., Beilina, L. V., and Kuzin, A. N. (1963). Izv. Vuzov, Svetnaja metallurgia 1, 117.

195. Davis, S. G., Anthrop, D. F., and Searcy, A. W. (1961). *J. Chem. Phys. 34,* 2, 659.

196. Naidich, Ju. V., Kolesnichenko, G. A., Pheldgun, L. I., and Drui, M. S. (1970). Neorganicheskie Materiali 6, 10, 1758.

197. Naidich, Ju. V., Kolesnichenko, G. A., Drui, M. S., Pheldgun, L. I., and Shpotakovski, D. Ph. (1970). *Abrazivi 3,* 1.

198. Lashko, N. Ph. and Lashko-Avakan, S. V. (1939). Paika metallov, Mashgiz, Moscow.

199. Frost, B. R. T. (1957). *Atomics Nucl. Energy 8,* 10, 97.

200. Gudzov, N. T., and Gavze, M. N. (1956). Vozdeistvie rtuti kak teplonositela na stal v energeticheskih ustanovkah, Izd-vo AN USSR, Moscow.

201. Lihtman, V. I., Rebinder, P. A., and Karpenko, G. V. (1952). Vlianie Poverhnostnoi sredi na prozessi dephormazii metallov. Izd-vo AN USSR, Moscow.

202. Gorjunov, Ju. V. *et al.* (1963). In "Poverhnostnie javlenia v rasplavah i prozessah poroshkovoi metallurgii." Izd-vo AN USSR, Kiev, p. 309.

203. Pebinder, P. A., Lihtman, V. I., and Kochanova, L. A. (1956). DAN USSR, 111, 1984.

204. Shukin, E. D. (1958). DAN USSR, 118, 1105.

205. Kochanova, L. A., Andreeva, I. V., and Shukin, E. D. (1959). DAN USSR, 126, 1304.

206. Kishkin, S. T., and Nikolenko, V. V. (1959). DAN USSR 110, 1018.

207. Rostoker, U., Mak-Kochu, J., and Markus, G. (1962). Hrupkost pod vlianiem jidkih metallov. IL, Moscow.

208. Lihtman, V. I., Shukin, E. D., and Rebinder, P. A. (1962). Physiko-chimicheskaja mehanika metallov. Izd-vo AN USSR, Moscow.

209. Adisson, C. C., *et al.* (1956). *J. Chem. Soc. 6,* 1454.

210. Beiley, G., and Watkins, H. (1951). *J. Inst. Metals 5,* 2, 57.

211. Tamman, G., and Ruhenbach, . (1935). Z. anorg. allgem. Chemie 223, 192.

212. Hildebrandt, I., Hognass, T., and Taylor, H. (1923). *J. Am. Chem. Soc. 45,* 9129, 2828.

213. Tamman, G., and Arnstz, F. (1930). *Z. anorg. Chem. 192,* 43.

214. Alty, F., and Clark, A. (1935). *Trans. Faraday Soc. 31,* 648.

215. Sears, G. (1950). *J. Appl. Phys. 21,* 721.

216. Mironzeva, S., and Rozova, V. V. In "Poverhnostnie javlenia v rasplavah i voznikajushih iz nih tverdih phazah. Kabardino-Balkarskoe knijnoe izd-vo," Nalchik, 1965.

217. Ljuis, V. In "Adhesion," ed. N. Debroyn and R. Huvink (1954). JL, Moscow, p. 488.

218. Taylor, J. W. (1955). *J. Nuclear Energy 2*, 15.

219. Fischer, I. C., and Dunn, C. G. (1952). In "Imperfection in Nearly Perfect Crystals," John Wiley, New York, p. 317.

220. Milner, D. H. (1958). *Brit. Welding Journal*, 90, march.

221. Phrolov, V. V. (1953). In "Zachita metallov ot korrozii i obrazovania nakipi, 24. Mashgiz, Moscow.

222. Lihtman, V. I., Brjuhanova, L. S., and Andreeva, I. (1965). In "Poverhnosnie javlenia v rasplavah i voznikajushih iz nih tverdih phazah." Kabardino-Balkarskoe knijnoe izd-vo, Nalchik, p. 438.

223. Van Vlach, L. H. (1951). *J. Metals 3*, 251.

224. Geld, P. V., and Chuchmarev, S. K. (1952). DAN USSR 83, 6, 877.

225. Presnov, V. A., and Vytkin, A. Pa. (1961). In "Poverhnostnie javlenia v metallah i splavah i ih role v prozessah poroshkovoi metallurgii." Izd-vo AN Ukranian, Kiev, p. 91.

226. Gorjunov, Ju. V. (1964). Uspehi chimii, 33, 9.

227. Rebinder, V. A. (1930). *J. Phys. Chem. 1*, 4-5, 553.

228. Addisson, C. C., Iberson, E., and Manning, I. A. (1962). *J. Chem. Soc.* 2699.

229. Addisson, C. C., and Iberson, E. (1965). *J. Chem. Soc. 116*, 1437.

230. Gilliland, R. (1964). *Welding Journal 43*, 6.

231. Klein, R. I., and Wassink, . (1967). *J. Inst. Metal. 95*, 2, 38.

232. Wall, A. I., and Milner, D. K. (1961-62). *J. Inst. Metal. 90*, 10, 394.

233. Eremenko, V. N., and Lesnik, N. D. (1961). In "Poverhnostnie javlenia v metallah i splavah i ih role v prozessah poroshkovoi metallurgii." Naukova dumka, Kiev, p. 155.

234. Bergh, A. A. (1963). *J. Electrochem. Soc. 110*, 8, 908.

235. Bergh, A. A., and Holschwander, L. H. (1963). *J. Electrochem. Soc. 110*, 8, 904.

236. Rogus, B.I. (1966). *Trans. Quart. Amer. Soc. Metals 59*, 2, 240.

237. Lichko, I. I., Voropai, N. M., and Shantarin, V. D. (1967). Avtomaticheskaia svarka, 3.

238. Naidich, Ju. V. *et al.* (1968). In "Poverhnostnie javlenia v rasplavah." Naukova Dumka, Kiev, p. 354.

240. Rabkin, D. M. *et al.* (1967). Avtomaticheskaia svarka, 11.

241. Karashev, A. A., and Zadumkin, S. N. (1965). In "Poverhnostnie javlenia v rasplavah i voznikajushih iz nih tverdih phazah." Kabardino-Balkarskoe knijnoe izdatelstvo, Nalchik, p. 79.

242. Popel, S. I., *et al.* (1968). In "Poverhnostnie javlenia v rasplavah." Naukova dumka, Kiev, p. 86.

243. Aldrich, D. A., and Keller, D. V. (1965). *J. Appl. Phys. 36*, 11, 3546.

244. Gretz, R. D. (1966). *Surface Science 4*, 494.

245. Shebo, P., Gutka, P., Govalda, A., Ivan, I., and Taborski, L. (1975). Kavale mater. 13, 5, 654.

246. Naidich, Ju. V., Kostjuk, B. D., Kolesnichenko, G. A., and Shaikevich, S. S. (1975). "Phizicheskaia chimia kondensirovannih phaz, sverhtverdih materialov i ih graniz razdela." "Naukova dumka," Kiev, p. 15.

247. Naidich, Ju. V., Kolesnichenko, G. A., and Kostjuk, B. D. (1973). Phisika i chimia obrabotki materialov, 6.

248. Evens, D., Nicholas, M., and Scott, P. M. (1977). *Industrial Diamond Rev.*, September, p. 306.

249. Rhee, S. K. (1972). *J. Amer. Cer. Soc. 55*, 3, 157.

250. Rhee, S. K. (1971). *J. Amer. Cer. Soc. 54*, 7, 332.

251. Rhee, S. K. (1970). *J. Amer. Cer. Soc. 53*, 7, 386, 426.

252. Naidich, Ju. V., Perevertailo, V. M., and Lebovich, E. M. (1976). Adgezia rasplavov i paika materialov 1, 28.

253. Kostikov, V. I., Andropov, Ju. I., Otopkov, P. P., and Nojkina, A. V. (1974). "Adgezia rasplavov," "Naukova dumka," Kiev, p. 129.

254. Kostikov, V. I., Maurah, M. A., and Nojkina, A. V. (1971). Phyzicheskaia chimia poverhnostnih javleni pri visokih temperaturah, Kiev, Naukova dumka, p. 142.

INDEX

N

O

X